Assessing Neurodegenerative Diseases: High-Sensitivity Technologies That Assay Pathological Biomarkers in Blood

Charles Shieh-Yueh Yang • Marwan Noel Sabbagh

Assessing Neurodegenerative Diseases: High-Sensitivity Technologies That Assay Pathological Biomarkers in Blood

Springer

Charles Shieh-Yueh Yang
MagQu Co., Ltd.
New Taipei City, Taiwan

Marwan Noel Sabbagh
Alzheimer's and Memory Disorders Division
Barrow Neurological Institute
PHOENIX, AZ, USA

ISBN 978-3-032-23862-7 ISBN 978-3-032-23863-4 (eBook)
https://doi.org/10.1007/978-3-032-23863-4

This Springer imprint is published by the registered company Springer Nature Switzerland AG
The registered company address is: Gewerbestrasse 11, 6330 Cham, Switzerland

Preface

Diagnosing Alzheimer's disease (AD) and other neurodegenerative diseases such as dementia with Lewy bodies (DLB), Parkinson's disease (PD), frontotemporal dementia (FTD), and amyotrophic lateral sclerosis (ALS) poses significant challenges to healthcare, often resulting in delayed or inadequate patient care. Most conditions have relied heavily on clinical phenotypic descriptions, which are not always accurate.

An added complication is, despite the high prevalence of mild cognitive impairment, dementia, and other neurodegenerative diseases, up to 92% of patients with early cognitive decline remain undiagnosed or misdiagnosed in healthcare settings. Moreover, the scarcity of neurological specialists in the USA exacerbates the situation, with many patients failing to access specialty care, and those who are referred experiencing prolonged wait times. This challenge is even more pronounced in other parts of the world, where access to specialists is very limited. Consequently, because most of the neuroimaging or cerebrospinal fluid (CSF) biomarker testing for neurodegenerative diseases takes place in specialty care settings (i.e., secondary or tertiary care), only a small fraction of patients receive such testing.

Traditional biomarker modalities for confirming neuropathology in conditions such as AD, PD, and DLB include positron emission tomography (PET) or a lumbar puncture to collect CSF to measure concentrations of Aß and tau. PET and CSF biomarkers have a number of shortcomings including high cost, limited accessibility, and perceived invasiveness, making them unsuitable for widespread use across care settings. Recent advancements in blood-based biomarker (BBM) tests for AD and potentially for PD and ALS offer a promising alternative. These tests are less costly, more accessible, more acceptable to patients, and more practical for serial collection to monitor disease progression.

Blood-based biomarkers (BBM) for Alzheimer's disease (AD) and other neurodegenerative diseases are being increasingly used in clinical practice to support a clinical diagnosis. In contrast to traditional diagnostic modalities, such as amyloid positron emission tomography and cerebrospinal fluid biomarkers, BBMs offer a more accessible and lower-cost alternative. Their unique scalability addresses the anticipated surge in demand for biomarker testing with the emergence of disease-modifying treatments that require confirmation of target pathology. The major change in the field is that the detection assays have improved from nanomolar (10^{-9}) to femtomolar (10^{-15}), which allows for detection of proteins previously not considered measurable in the blood spaces.

The clinical integration of blood-based biomarkers (BBMs) for AD and other neurodegenerative diseases holds promise in enabling early detection of pathology and timely intervention. Similar investigations are being considered for other neurodegenerative diseases. However, several critical considerations remain, such as limited understanding of BBM test characteristics, insufficient evidence on BBM performance across diverse populations, and how individual test parameters performed on different assay platforms can be meaningfully compared.

This book summarizes the current assays in development for neurodegenerative diseases. Each chapter describes the technical aspects of the individual assays. Also included are a clinical perspective for each assay and potential challenges.

New Taipei City, Taiwan Charles Shieh-Yueh Yang
PHOENIX, AZ, USA Marwan Noel Sabbagh

Contents

About the Authors

Charles Shieh-Yueh Yang, PhD, finished a PhD program in physics at National Taiwan Normal University in 1999. His major focus is superconducting electronics, especially superconducting quantum interference devices (SQUIDs). From 1999 to 2004, he served as a postdoctoral research fellow at the Institute of Physics, Academia Sinica, Taiwan. The main topic of his postdoctoral work was bio-functionalized magnetic nanoparticles. As a faculty member at the Institute of Electro-optical Science and Technology at National Taiwan Normal University, Dr. Yang et al. invented the so-called immunomagnetic reduction (IMR) technology and demonstrated the impact of IMR on unmet demands in the clinic. In 2008, he quit the faculty and started a private biotech company. He has been the CEO of the biotech company for more than 10 years.

He was the editor of the bimonthly magazine "Physics" of the Taiwan Association of Physics in 2006. He was the secretary of the 2006 Taiwan International Conference on Superconductivity and the 8th Workshop on Low-temperature Physics. He has been the principal investigator of research/technology development projects supported by the Ministry of Science and Technology, the Ministry of Economics, and the Ministry of Health and Welfare, Taiwan, since 2002. He has published more than 200 papers in SCI journals and has issued more than 30 patents in the USA, Taiwan, Europe, Japan, and China.

Due to his excellent academic experiences and outcomes, he was rewarded with a Research Award by the

Taiwan Association of Magnetic Technology in 2007. In 2020, he was rewarded with the Carl H. Rosner Entrepreneurship Award issued by the IEEE Council on Superconductivity because of his outstanding entrepreneurship skills in the field of applied superconductivity, particularly for founding MagQu Co., Ltd., the commercialization of SQUID-IMR alternating current (ac) magnetosusceptometers, and for transforming SQUID-IMR assay technology and ac magnetosusceptometry into clinical uses for quantifying ultralow concentrations of proteins in human blood specific to Alzheimer's disease, Parkinson's disease, and frontotemporal dementia.

Marwan Noel Sabbagh, MD, FAAN, a board-certified neurologist and geriatric neurologist, has dedicated his career to finding a cure for Alzheimer's and other age-related neurodegenerative diseases.

He is a leading investigator for many prominent national Alzheimer's treatment trials. He is on the editorial board for *Journal of Alzheimer's Disease* and *Alzheimer's and Dementia TRCI*. He is editor-in-chief of *Neurology and Therapy*. He has authored and co-authored over 500 medical and scientific articles on Alzheimer's research.

He is the author of *The Alzheimer's Answer: Reduce Your Risk and Keep Your Brain Healthy*, with foreword by Justice Sandra Day O'Connor, and of *The Alzheimer's Prevention Cookbook: 100 Recipes to Boost Brain Health*. He has edited *Palliative Care for Advanced Alzheimer's and Dementia: Guidelines and Standards for Evidence Based Care and Geriatric Neurology*, published in 2014, and *Fighting for My Life: Living in the Shadow of Alzheimer's Disease* in 2019 and *Strong Heart, Sharp Mind* published in 2022.

He earned his undergraduate degree from the University of California, Berkeley, and his medical degree from the University of Arizona in Tucson. He received his residency training in neurology at Baylor College of Medicine, Houston, Texas, and completed his fellowship in geriatric neurology and dementia at the University of California, San Diego School of Medicine, where he served on the faculty as assistant professor.

Before joining the faculty of the Barrow Neurological Institute and Creighton University School of Medicine, he was the Camille and Larry Ruvo Endowed Chair for Brain Health, the Director of the Cleveland Clinic Lou Ruvo Center for Brain Health in Las Vegas, and the founding Principal Investigator of the NIH funded Nevada Exploratory Alzheimer's Disease Research Center. From 2000 to 2015, he was the director of the Banner Sun Health Research Institute in Sun City AZ. From 2015 to 2018, he was the Karsten Solheim Endowed Chair for Alzheimer's research at the Barrow Neurological Institute. Presently, in addition to being vice chair of Clinical Research at Creighton University School of Medicine, he is a professor of Neurology at the Barrow Neurological Institute.

Chapter 1
Ultrasensitive Assay Technologies

Abstract With the demands of precise measurement of blood-based biomarkers at ultralow concentrations, e.g., pg/ml-fg/ml, in the assessment and diagnosis of neurodegenerative diseases, many ultrasensitive technologies have been developed. Some of these assays have already been commercialized. Several relevant assay kits have been approved for clinical use in the USA, Taiwan, Europe, and other regions. However, versatile detecting mechanisms are utilized in different assay technologies. It is difficult to have a complete understanding of the details of these complicated technologies. Furthermore, people are frequently confused by the measured results obtained using different technologies. In order to have a correct understanding of the measured results, we should clearly know how the measurements work and what the technology measures. In this chapter, 17 assay technologies, grouped into immunoassay and mass spectrometry, are introduced. The detecting mechanism, feature of the assay, sample preparation, commercialization, target biomarkers, and key specifications of assay kits for each technology are reviewed. Readers can have a comprehensive understanding of these complicated technologies for assaying blood-based biomarkers associated with neurodegenerative diseases.

Keywords Assay technology · Immunoassay · Mass spectrometry · Blood-based biomarker · Neurodegenerative disease

1.1 Introduction

Pathological hallmarks reveal the associations between several types of peptides or proteins and neurodegenerative diseases [1–4]. For example, amyloid-β peptide and tau protein are pathological biomarkers for Alzheimer's disease [5–7]. Parkinson's disease is characterized by the α-synuclein protein [8, 9]. TAR DNA-binding protein 43 (TDP-43) is highly related to frontotemporal dementia (FTD) [10, 11].

C. S.-Y. Yang, M. N. Sabbagh, *Assessing Neurodegenerative Diseases: High-Sensitivity Technologies That Assay Pathological Biomarkers in Blood*,
https://doi.org/10.1007/978-3-032-23863-4_1

In addition to pathological biomarkers, some proteins, such as neurofilament light chain (NfL), are associated with brain neuron injuries and could be promising indices of disease severity [12, 13]. Glial fibrillary acidic protein (GFAP) is a structural protein of microglia in the brain and is suggested as an early-stage biomarker for neurodegenerative diseases [14, 15]. Other proteins that are promising biomarkers of neurodegenerative diseases have been identified in published papers. Researchers are exploring new biomarkers that are directly related to the pathogenesis of neurodegenerative diseases.

Once the brain degenerates, the levels of relevant biomarkers become abnormal. Most of these biomarkers exist in the brain or cerebrospinal fluid (CSF). Assays of biomarkers in CSF are the main trend for assessing or diagnosing neurodegenerative diseases. For example, in 2018, the Alzheimer's Association announced a research framework for the use of CSF amyloid-β peptide and tau protein to assess Alzheimer's disease in clinical practice [16]. This framework represents a major step toward biological definitions of Alzheimer's disease.

The concentrations of these biomarkers in CSF are sub-ng/ml or several ng/ml [17–20]. Conventional assay technologies, such as enzyme-linked immunosorbent assay (ELISA), are able to detect these biomarkers in CSF precisely. Unfortunately, the drawbacks of lumbar puncture, such as headache, back pain, bleeding, or clear fluid leakage and swelling where the needle travels into one's back, limit its use in clinical practice [21, 22]. Demands to assay biomarkers in blood have motivated researchers. However, the levels of biomarkers in blood are much lower than those in CSF, i.e., pg/ml or fg/ml [23–27]. Conventional assay technologies are not sensitive enough to assay biomarkers in blood. The development of ultrasensitive assay technologies is needed.

Researchers have developed ultrasensitive assay technologies for exploring neurodegenerative disease biomarkers in blood. Many technologies have been proposed and commercialized for research or clinical use. Many studies have reported the clinical significance of blood biomarkers in disease assessment and intervention monitoring. With convincing clinical evidence, the Alzheimer's Association revised the criteria for the diagnosis and staging of Alzheimer's disease in 2024 [28]. In the revised guidelines, plasma amyloid-β peptide and tau protein levels are listed as fluid analytes to assist in the diagnosis of Alzheimer's disease. Plasma biomarkers are officially recognized as diagnostic information for Alzheimer's disease in clinical practice.

Obviously, without progress in the development of ultrasensitive assay technologies, evolving from CSF biomarkers to blood biomarkers for diagnosing neurodegenerative diseases is impossible. Although review articles introducing assay technologies have been published [29–40], new technologies are still being developed. An update of the state-of-the-art assay technologies for identifying blood biomarkers of neurodegenerative diseases is necessary. In this work, commercialized technologies for assaying plasma biomarkers associated with neurodegenerative diseases are reviewed. Notably, this review describes the technology, not clinical performance. It could help readers gain a comprehensive and deep understanding of currently available assay technologies for identifying blood biomarkers of neurodegenerative diseases.

1.2 Assay Technologies

Several ultrasensitive technologies have been developed to measure blood biomarkers associated with neurodegenerative diseases. According to published papers, trade shows at the Alzheimer's Association International Conference or International Conference on Alzheimer's and Parkinson's Diseases and related neurological disorders, websites, etc., seventeen commercially available technologies have been developed, as listed in Table 1.1. Some technologies are applied to research-use-only markers, whereas others are used in clinical practice.

The fundamental mechanisms of these 17 technologies are categorized into 2 groups: immunoassays and mass spectrometry assays. In immunoassays, the sensing of signals that are converted to the concentrations of analytes of interest includes optical detection, magnetic detection, and quantitative real-time polymerase chain reaction. In optical detection, the triggering techniques for emitting optical signals include chemical reactions, electricity, and optical excitation. The trigger technique in magnetic detection involves a magnetic drive. In order to amplify the sensing signals, technologies involving DNA/RNA polymerase chain reactions of

Table 1.1 Reviewed assay technologies that are commercially available in the research-use-only or diagnostic markets

Group	Sensing signals	Trigger	Technology
Immunoassay	Optical detection	Chemical reaction	Chemiluminescent enzyme immunoassay (CLEIA) High-sensitivity chemiluminescence enzymeimmunoassay (HISCL)
		Electricity	Electrochemiluminescence (ECL) Magnetically enhanced electrochemiluminescence (ME-ECL)
		Optical excitation	Fourier transform infrared (FT-IR) Multimer detection system (MDS) Single molecule array (SIMOA) Single molecule counting (SMC) Surface plasmon resonance (SPR) Multi-analyte profiling (xMAP)
	Magnetic detection	Magnetic drive	Immunomagnetic reduction (IMR)
	Quantitative real-time polymerase chain reaction		Proximity ligation assay (PLA) Nucleic acid-linked immuno-sandwich assay (NULISA) Successive proximity extension amplification reaction (SPEAR)
	Seed amplification assay		Real-time quaking-induced conversion (RT-QuIC)
Mass spectrometry	Charge-to-mass spectrum		Liquid chromatography tandem mass spectrometry (LC/MS/MS) Immunoprecipitation mass spectrometry (IP/MS)

specifically designed sequency or seeding amplification of target molecules are adapted. In mass spectrometry assays, the charge-to-mass spectrum of the target analyte is inspected. The working mechanism of the technology, features of the assay, plasma preparation, commercialization, and assay kits for plasma biomarkers of neurodegenerative diseases are individually introduced for these 17 assay technologies in Table 1.1.

1.3 Chemiluminescent Enzyme Immunoassay (CLEIA)

- Mechanism

  The chemiluminescent enzyme immunoassay (CLEIA) is an assay technique used to quantify the antigens to be detected. In CLEIA, the light/luminescence produced from a chemical reaction is measured and converted to the concentrations of antigens to be detected [41, 42]. Figure 1.1 illustrates the flow chart of a typical sandwich CLEIA [43]. Capture antibodies are immobilized at the bottom of a well, followed by the formation of a complex between the capture antibody and the antigen to be detected. After the unbound molecules are washed out, a biotinylated detection antibody is added to specifically label the captured antigen to be detected. The excess biotinylated detection antibody is subsequently removed by washing. Then, streptavidin–horseradish peroxidase (HRP) is added to bind with the biotinylated detection antibody. Unbound streptavidin–HRP is washed out, followed by the addition of a luminol (3-aminophthalhydrazine)

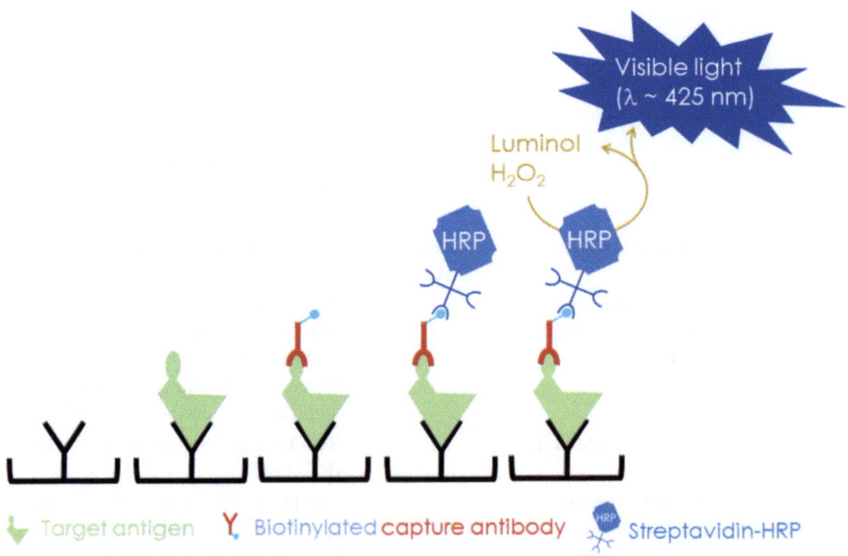

Fig. 1.1 Assay of sandwich CLEIA

substrate and oxidants such as H_2O_2. HRP acts as an enzyme in the chemical reaction between the luminol substrate and the oxidant H_2O_2.

The reaction produces a product or an intermediate that is unstable and releases a photon as it transitions to a lower energy state. Consequently, light/luminescence signals (wavelength $\lambda \sim 425$ nm) are generated. A typical HRP–H_2O_2–luminol substrate reaction is shown in Fig. 1.2 [44].

The intensity of the light/luminescence signals depends on the concentration of HRP, i.e., the concentration of the antigen to be detected. Hence, the antigen to be detected can be quantitatively detected.

Alkaline phosphatase (ALP) is also a common enzyme used in CLEIA. The related oxidant is AMPPD, CDP-Star, or APS-5 [45–47]. Comparisons between HRP and ALP are provided in the *High-Sensitivity Chemiluminescence Enzyme Immunoassay* **(HISCL)**.

Magnetic particle substrates or enhancers have been utilized in CLEIA to increase the light/luminescence signals and sensitivity [48, 49], referred to as magnetic particle-based CLEIA (MP-CLEIA). Instead of using a solid substrate coated with capture antibodies, magnetic particles are utilized as substrates in CLEIA, the so-called magnetic particle-based CLEIA, as illustrated in Fig. 1.3 [50]. With the use of magnetic particles, the assay becomes homogeneous, and the antigens to be detected are concentrated through magnetic separation. Hence, the sensitivity of CLEIA is enhanced.

An alternative method to increase the detection sensitivity of CLEIA is the use of enhancers, such as 4-(1-imidazolyl)phenol (IMP) or modified phenols, which act as mediators in the peroxidase-induced catalysis of the chemiluminescent reaction. The enhancers do not act as more efficient emitters but exert their action earlier in the complex reaction between the peroxidase, oxidant, and luminol [42]. The detailed mechanism of enhanced CLEIA using enhancers was clarified via the "ping-pong" model [51–53]. When enhancers are used, CLEIA benefits from not only increased signals but also a prolonged luminescence time and increased intensity [54]. Compared with that of traditional enzyme immunoassays, the low limit of detection (LLoD) of the enhanced CLEIA is improved by three to five orders of magnitude, i.e., from 10^{-21} to 10^{-18} mol/ml [55, 56].

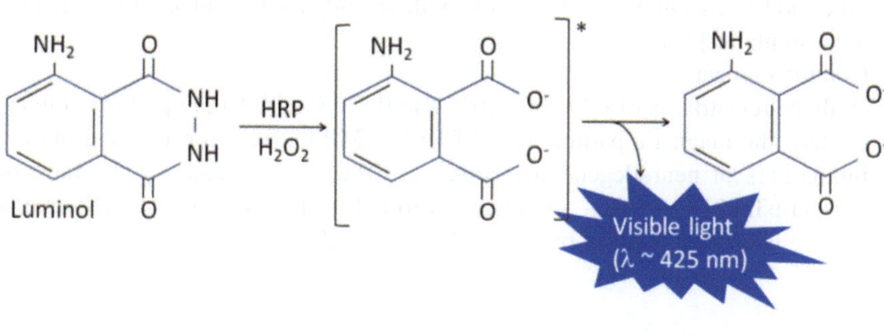

Fig. 1.2 Chemiluminescent enzyme reaction involving HRP, the oxidant H_2O_2, and luminol [4]

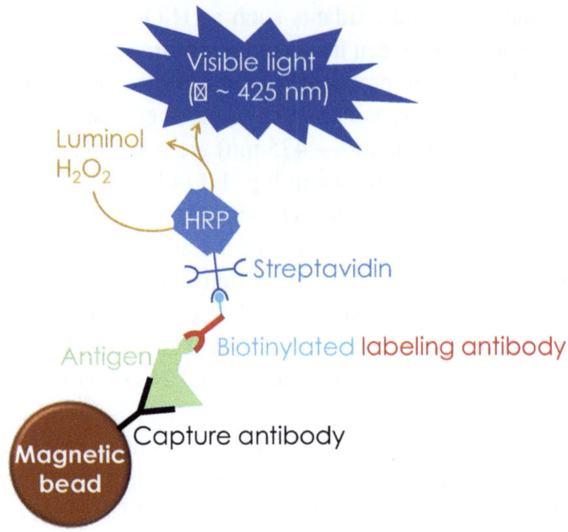

Fig. 1.3 Scheme of the magnetic particle-based CLEIA (MP-CLEIA)

- Features of the assay

  In traditional optical immunoassays, luminescence signals are generated by an optically excited enzyme. A triggering optical source is needed. However, as shown for the CLEIA mechanism in Figs. 1.1 and 1.2, light/luminescence is generated by chemical reactions. Hence, an optical source does not need in CLEIA. The complexity of the analyzers in CLEIA is significantly reduced. Moreover, the analyzers used in CLEIA show excellent low levels of background signals because they avoid possible interference by triggering light sources.

  CLEIA is a sandwich technology. At least two types of antibodies are needed for one biomarker. Once magnetic beads are utilized, one additional antibody is needed.
- Plasma preparation

  Blood is drawn in a lavender-top ethylenediaminetetraacetic acid (EDTA) tube and inverted to mix with preservatives. The blood is centrifuged at 2000 rpm for 10 min at 4 °C, after which the plasma is transferred to a labeled plastic transport tube and frozen. Dilutions of 1–2-fold with serum/plasma samples are needed for measurements [17].
- Commercialization

  With the registration of ISO 9001:2015 and ISO 13485:2016, Fujirebio commercialized the magnetic particle-based CLEIA (MP-CLEIA) for assaying plasma biomarkers of neurodegenerative diseases. The kits for assaying the ratio of plasma pTau217 to $A\beta_{1-42}$ are approved to aid in the diagnosis of Alzheimer's disease in clinical practice by the US FDA in 2025.

Table 1.2 Low limit of detection (LLoD), dynamic range, and utility for research use only (RUO) or diagnostic use (Dx) of CLEIAkits

Biomarker	LLoD	Dynamic range	RUO	Dx
$A\beta_{1-40}$	2.8 pg/ml	7.8–100 pg/ml	○	
$A\beta_{1-42}$	65 pg/ml	225–1452 pg/ml	○	
T-tau	25 pg/ml	NA	○	
pTau181	15.6 pg/ml	NA	○	
pTau217	0.03 pg/ml	0.00–0.18 pg/ml	○	

Table 1.3 Low limit of detection (LLoD), dynamic range, and utility for research use only (RUO) or diagnostic use (Dx) of MP-CLEIAkits

Biomarker	LLoD	Dynamic range	RUO	Dx
$A\beta_{1-40}$	0.44 pg/ml	0.45–3404 pg/ml	○	
$A\beta_{1-42}$	0.27 pg/ml	0.83–682 pg/ml	○	○
pTau181	0.05 pg/ml	0.05–60 pg/ml	○	
pTau217	0.03 pg/ml	0.030–10 pg/ml	○	○

- Assay kits for plasma biomarkers of neurodegenerative diseases

  Assay kits are currently available for measuring $A\beta_{1-40}$, $A\beta_{1-42}$, T-tau, pTau181, and pTau217 levels in human plasma. The low limit of detection (LLoD), dynamic range, and utility for research use only (RUO) or diagnostic use (Dx) of each assay kit are listed in Tables 1.2 and 1.3 [57].

1.4 High-Sensitivity Chemiluminescence Enzyme Immunoassay (HISCL)

- Mechanism

  The measurement principle of a high-sensitivity chemiluminescence enzyme immunoassay (HISCL) is based on a chemiluminescence enzyme immunoassay (CLEIA). The most popular enzyme used in traditional CLEIA is horseradish peroxidase (HRP). In HISCL, the most frequently used enzyme is alkaline phosphatase (ALP) [58]. The substrate is, for example, 3-chloro-2,4-difluorobenzoic acid (CDP-Star) [59]. A typical reaction of ALP–oxidant–CDP-Star in HISCL is shown in Fig. 1.4. After each cycle of the reaction, light at a wavelength of 466 nm, i.e., blue light, is emitted.

The advantages of using ALP instead of HRP are its stability and sensitivity [60]. The activity of HRP is easily inhibited by sodium azide and thimerosal, which are common in microorganisms and antibiotics. The activity of HRP is also easily inhibited by sulfides (-S), cyanides ($-CN_2$), and azides ($-N_3$). However, ALP activity is not reduced by antibiotics or other agents, such as thimerosal or sodium azide. Thus, ALP is much more stable than HRP. In addition, the reaction rates of ALP are linear, which are easy to control and record. In practice, a long reaction period is

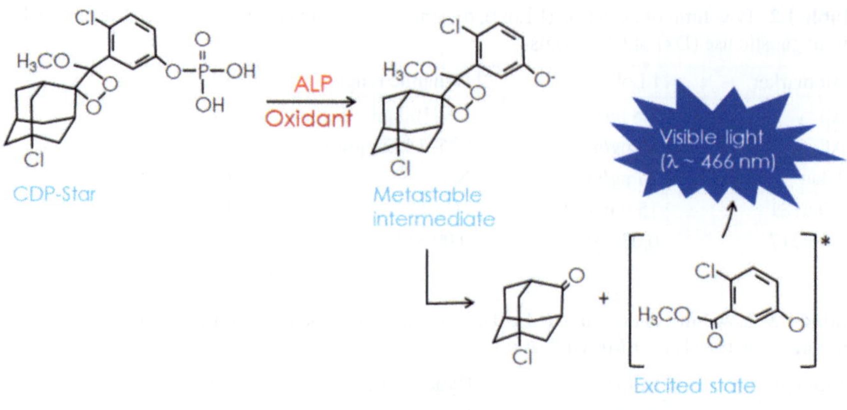

Fig. 1.4 Chemiluminescent enzyme reaction involving ALP, CDP-Star, and an oxidant

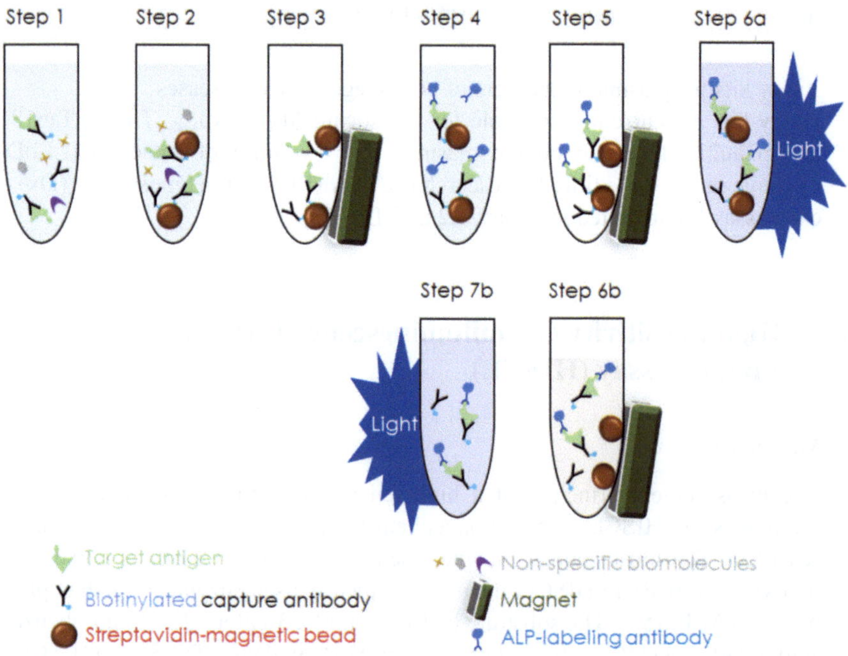

Fig. 1.5 Assay of high-sensitivity chemiluminescence enzyme immunoassay

usually designed to collect more intense emitted light. Hence, the assay sensitivity is greater when ALP is used than when HRP is used.

In addition to ALP, HISCL utilizes magnetic beads to achieve a homogeneous assay, which increases the detection sensitivity [61, 62]. For example, a sample with a target antigen such as Aβ was first mixed with a biotinylated capture antibody against the N-terminus of Aβ, as shown in Step 1 in Fig. 1.5. After the formation of the capture antibody–antigen complex, magnetic beads coated with streptavidin are added to form the capture antibody–antigen–magnetic bead complex as shown in Step 2. A magnet is used to extract the magnetic beads, and the solution with unbound antigen and nonspecific biomolecules is eluted, as shown in Step 3. The solution with the ALP-conjugated labeling antibody against the C-terminus of Aβ is added to form the capture antibody–antigen–magnetic bead–labeling antibody–ALP complex, as shown in Step 4. Using magnetic separation, as shown in Step 5, the unbound labeling antibody is removed. Finally, by adding the CDP-Star reagent and oxidant, blue light is emitted, as shown in Step 6a. Notably, the entire reaction can be performed at 40–45 °C to speed up the interactions between the antibody and antigen.

The presence of magnetic beads in Step 6a in Fig. 1.5 can cause significant background noise due to the dark brown color of the magnetic beads. Hence, sometimes, a reagent is added to disrupt the associations between the capture antibody and magnetic beads, as shown in Step 6b in Fig. 1.5. After magnetic separation, the solution containing the capture antibody–antigen–labeling antibody–ALP mixture is eluted, followed by the addition of CDP-Star and an oxidant for luminescence detection, as shown in Step 7b. By reducing the background noise in HISCL, the sensitivity of detection can be increased by 2 orders of magnitude in concentration.

- Features of the assay
  In contrast to the traditional CLEIA, the entire reaction in HISCL is performed at 40–45 °C, resulting in a rapid assay.
- Plasma preparation
  A 6-ml K_2EDTA tube is used for blood collection, followed by centrifuging the blood at 2000 × g for 10 min. After centrifugation, 0.5 ml of plasma is transferred into a 2-ml protein-free tube. The separated plasma samples are frozen at temperatures below −70 °C within 30 min of plasma separation [5].
- Commercialization
  Sysmex Corp. has commercialized HSICLs to measure plasma $A\beta_{1\text{-}40}$ and $A\beta_{1\text{-}42}$ levels for assessing Alzheimer's disease, with registration under ISO 13485:2016. The HISCL kits developed by Sysmex Corp. are approved for clinical use by the Pharmaceuticals and Medical Devices Agency (Japan).
- Assay kits for plasma biomarkers of neurodegenerative diseases
  Assay kits are currently available for measuring $A\beta_{1\text{–}40}$ and $A\beta_{1\text{–}42}$ levels in human plasma. The detailed information about the low limit of detection (LLoD), dynamic range, and utility for research use only (RUO) or diagnostic use (Dx) of the assay kits is listed in Table 1.4 [5].

Table 1.4 Low limit of detection (LLoD), dynamic range, and utility for research use only (RUO) or diagnostic use (Dx) of HISCLkits

Biomarker	LLoD	Dynamic range	RUO	Dx
$A\beta_{1\text{–}40}$	1.31 pg/ml	1.31–24,000 pg/ml	○	○
$A\beta_{1\text{–}42}$	0.13 pg/ml	0.13–40,000 pg/ml	○	○

1.5 Electrochemiluminescence (ECL)

- Mechanism

  Electrochemiluminescence (ECL) is a sandwich immunoassay that involves electrochemical reactions and luminesce, as illustrated in Fig. 1.6 [63, 64]. Capture antibodies are immobilized at the anode. A liquid sample is injected into an assay well. After the association between the antigen and capture antibody is terminated, the unbound antigen is washed out. Labeling antibodies, which are linked to luminophores such as $Ru(bpy)_3^{2+}$ (where bpy = 2,2′-bipyridine), are added to the well to form a capture antibody–antigen–labeling antibody–luminophore complex. After the unbound labeling antibodies are washed out, tripropylamine (TPA) is added to the wells of the assay plate. By applying an electric voltage to the solution, the luminophore $Ru(bpy)_3^{2+}$ releases an electron to become $Ru(bpy)_3^{3+}$. Moreover, tripropylamine (TPA) around the anode releases an electron to become the radical cation TPA+, followed by the immediate release of a proton to become the free radical TPA·. The oxidation–reduction reaction between $Ru(bpy)_3^{3+}$ and TPA· occurs to form excited $^*Ru(bpy)_3^{2+}$, which decays to ground-state $Ru(bpy)_3^{2+}$ and emits a photon with a wavelength of 620 nm (red light).

After the cycles of the reactions shown in Fig. 1.6 are complete, TPA and $Ru(bpy)_3^{2+}$ still exist. Multiple cycles of reactions can be triggered by continuously applying an electric voltage. Hence, photons are continuously emitted, and the optical signals are amplified. By integrating the emitted optical signals, high sensitivity can be achieved with ECL. The low limit of detection (LLoD) of ECL is approximately 10^{-16} mol/ml [65, 66].

ECL luminophores play a role in ECL [67–69]. Researchers have been attempting to develop highly efficient, low-cost, high-stability, versatile color luminophores for ECL immunoassays. An increasing number of new types of luminophores are currently available. In the review by Wu et al. [70], luminophores are categorized into three groups in terms of their composition. The first is small organic molecules [71]. The second is metal complexes [72]. The third type is nanoparticles [73]. The detailed interpretations of the design of luminophores and their chemical structures and emission properties are provided in published articles [70, 74, 75].

- Features of the assay

  The ECL immunoassay has several merits compared to traditional optical emission immunoassays. For example, the luminophores used in ECL are small and are able to label molecules with wide ranges of molecular weights. Similar to CLEIA, a light module does not need to be triggered in ECL. The background interference in ECL is very limited. Furthermore, multiple cycles of light emission in ECL are achieved, and the detection sensitivity of ECL is high (~10^{-16} mol/ml).

  Electric transfers are involved in ECL, and the ion concentration in the solution might affect the light emission efficiency, implying that the sample matrix effect

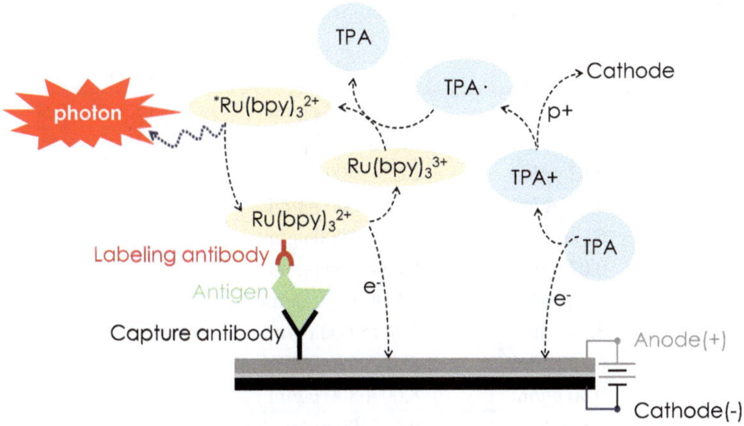

Fig. 1.6 Illustration of electrochemiluminescence (ECL)

could be significant. A sample is usually diluted with a buffer solution 1–2-fold to ensure a consistent matrix quality and control the electrical susceptibility of the solution for each ECL measurement. ECL is a type of sandwich immunoassay in which at least two different antibodies are needed for one biomarker. Instead of an optical triggering module, an electric module is needed to trigger the reactions shown in Fig. 1.6.

- Plasma preparation
  Please review the current literature and protocols, such as those proposed by the Alzheimer's Disease Neuroimaging Initiative (ADNI) [76], in which blood is collected in a uniform manner using EDTA as an anticoagulant and then centrifuged at room temperature at 3000 rpm for 15 min. Important issues for blood collection, centrifugation, sample dilution, reagent mixing, and ECL measurements are described in a previously published paper [77]. In particular, for plasma samples, all solid material should be removed by centrifugation [78]. In addition, samples are usually diluted with buffer solution 1–2-fold to ensure a consistent matrix quality for each ECL measurement.
- Commercialization
  Meso Scale Discovery (MSD) Inc. has commercialized an ECL immunoassay for assaying the plasma levels of biomarkers of neurodegenerative diseases with the registration of ISO 9001:2015 and ISO 13485:2016.
- Assay kits for plasma biomarkers of neurodegenerative diseases
  Assay kits are currently available to measure the levels of $A\beta_{1-40}$, sAPPα, sAPPβ, T-tau, pTau181, pTau231, pTau217, α-synuclein, NfL, BDNF, and GFAP in human plasma. The low limit of detection (LLoD), dynamic range, and utility for research use only (RUO) or diagnostic use (Dx) of each assay kit are listed in Table 1.5 [79].

Table 1.5 Low limit of detection (LLoD), dynamic range, and utility for research use only (RUO) or diagnostic use (Dx) of ECLassaykits

Biomarker	LLoD	Dynamic range	RUO	Dx
$A\beta_{1-40}$	5.41 pg/ml	5.41–6000 pg/ml	O	
$A\beta_{1-42}$	0.516 pg/ml	0.516–1270 pg/ml	O	
$A\beta_{1-38}$	22.2 pg/ml	22.2–7500 pg/ml	O	
sAPPα	0.12 ng/ml	0.12–1000 ng/ml	O	
sAPPβ	52 pg/ml	0.052–1000 ng/ml	O	
T-tau	0.07 pg/ml	0.63–2000 pg/ml	O	
pTau181	77 fg/ml	77–990,000 fg/ml	O	
pTau231	3.1 pg/ml	3.1–15,000 pg/ml	O	
pTau217	0.88 pg/ml	880–3,590,000 fg/ml	O	
α-Synuclein	0.90 pg/ml	8.00–6800 pg/ml	O	
NfL	2.6 pg/ml	9.6–5300 pg/ml	O	
BDNF	0.72 pg/ml	0.72–2000 pg/ml	O	
GFAP	8.8 pg/ml	52–10,400 pg/ml	O	

1.6 Magnetically Enhanced Electrochemiluminescence (ME-ECL)

- Mechanism

  For immunoassays of biomarkers in liquid samples, the biomarkers are homogeneously dispersed in a liquid sample. As the liquid sample is injected into an assay well in an ECL immunoassay, in which the capture antibodies are bound to the surface of the anode, time is usually needed for homogeneously dispersed biomarkers to diffuse to the surface of the anode to achieve an interaction between the biomarker and capture antibody on the surface of the anode. In fact, many biomarkers are dispersed in liquid and do not reach the surface of the anode. Thus, the detection sensitivity could be improved if the uncaptured biomarkers could interact with antibodies on the surface of the anodes. Magnetically enhanced electrochemiluminescence (ME-ECL) is an enhanced ECL-based immunoassay that uses a magnet to attract homogeneously dispersed biomarkers to capture antibodies on the surface of anodes in a well [80–82]. The scheme of the ME-ECL immunoassay is shown in Fig. 1.7.

In ME-ECL, target biomarkers are labeled with biotinylated antibodies, followed by an interaction with streptavidin-functionalized magnetic beads that are several hundred nanometers in diameter. The biomarker–antibody–biotin–streptavidin–magnetic bead complex is formed homogeneously in a liquid sample. The most common core material of magnetic beads is Fe_3O_4 [83], which is usually coated with polyethylene glycol (PEG) or dextran [84–86]. Streptavidin is covalently bound to PEG or dextran [87].

A magnet is placed under the electrodes in the wells of the ME-ECL plate [88, 89]. Once the magnet is activated, magnetically labeled biomarkers that are

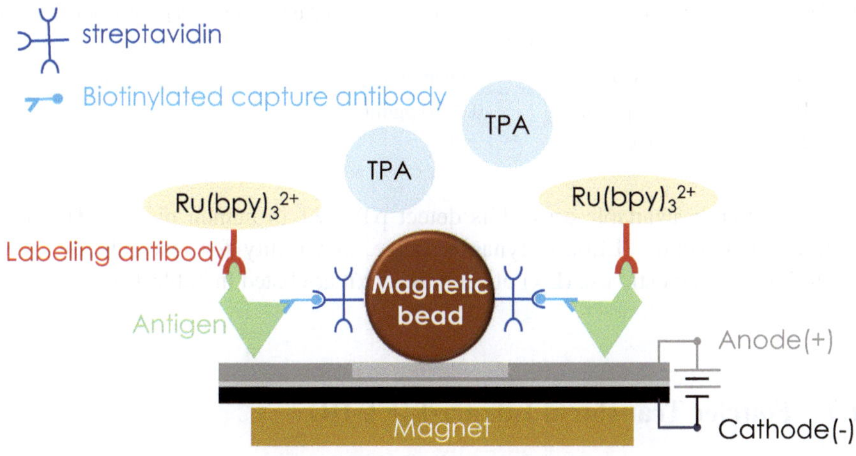

Fig. 1.7 Scheme of magnetically enhanced electrochemiluminescence (ME-ECL)

homogeneously dispersed in a liquid sample can be rapidly attracted to the surface of the anode. Thus, the interaction between the magnetically labeled biomarkers and capture antibodies on the surface of anodes becomes faster and more efficient. With the aid of the magnetic labeling of biomarkers, the detection sensitivity of the ECL immunoassay is enhanced, and the assay becomes faster.

- Features of the assay
  ECL is an inhomogeneous assay in which only antigen that reaches the bottom of the wells can be assayed. In contrast, ME-ECL is a homogeneous assay in which magnetic beads can capture antigens throughout the liquid sample and deliver the captured antigen to the bottom of the wells. The combination of ECL with magnetic beads results in fast assays with very high sensitivity and a broad measurement range. Similar to ECL, a 1–2-fold dilution of the serum/plasma sample is needed to suppress matrix effects. An electrically triggered luminophore is needed. In addition to labeling antibodies, biotinylated capture antibodies that interact with streptavidin-coated magnetic beads are needed in ME-ECL immunoassays.
- Plasma preparation
  No official suggestions are provided. The following protocol is usually used [90]: blood is collected in EDTA–plasma tubes and centrifuged (2000xg, 4 °C) for 10 min. After centrifugation, the plasma is transferred into one 50-ml tube and mixed, and 1 ml is aliquoted into polypropylene tubes and stored at −80 °C within 30–60 min of collection.
- Commercialization
  Roche Diagnostics has been commercializing fully automated ME–ECL immunoassays to measure plasma levels of biomarkers of neurodegenerative diseases with the registration of ISO 13485:2016.
- Assay kits for plasma biomarkers of neurodegenerative diseases

Table 1.6 Low limit of detection (LLoD), dynamic range, and utility for research use only (RUO) or diagnostic use (Dx) of the ME-ECLassaykit

Biomarker	LLoD	Dynamic range	RUO	Dx
pTau181	0.3 pg/ml	0.3–10 pg/ml	O	O
pTau217	NA	NA	O	

The currently available assay kits detect pTau181 in human plasma. The low limit of detection (LLoD), dynamic range, and utility for research use only (RUO) or diagnostic use (Dx) of each assay kit are listed in Table 1.6.

1.7 Fourier Transform Infrared (FT-IR)

- Mechanism

Fourier transform (FT) spectroscopy is a technology involving Fourier transformation to analyze the wavelengths absorbed by materials [91, 92]. In FT-IR spectroscopy, infrared (IR) light is used to illuminate the sample. In traditional FT-IR spectroscopy, light of various wavelengths illuminates a material. The light transmitted through materials is analyzed to observe the absorption spectrum of the material. The FT-IR absorption spectrum is associated with the composition of the material. Hence, FT-IR spectroscopy is frequently used for qualitative or quantitative measurements of material compositions.

In immunoassays, the capture antibodies are coated on a solid substrate, which is not transparent to light. Hence, transmitted FT-IR spectroscopy is not applicable for immunoassays. Instead, the attenuated total reflectance (ATR) becomes the dominant method, as illustrated in Fig. 1.8(a).

The infrared (IR) light is normally incident on the side face of the ATR crystal, such as zinc selenide, germanium, silicon, diamond, thallium iodide, or thallium bromide [93], and total reflection occurs as the light travels to the interface between the top face of the ATR crystal and the liquid, as indicated by the total reflection points in Fig. 1.8(a). The light is completely reflected back to the ATR crystal from the top face of the ATR crystal and is transmitted out of the ATR crystal, which is referred to as the reflected light in Fig. 1.8(a). In fact, the light at the total reflection point crosses the interface between the top face of the ATR crystal and the liquid. The intensity of the light decays exponentially with increasing propagation depth in the liquid. This light in the liquid at the total reflection point is called an evanescent wave [94]. The penetration depths of evanescent waves range from 30 to 300 nm. The depth depends on several factors, such as the composition of the materials and the wavelength of the light. The materials present in the region of the penetration depth can absorb light. Absorption can be obtained by analyzing the FT-IR spectrum of the reflected light, as shown in Fig. 1.8(a). As the compositions of materials change, the pen-

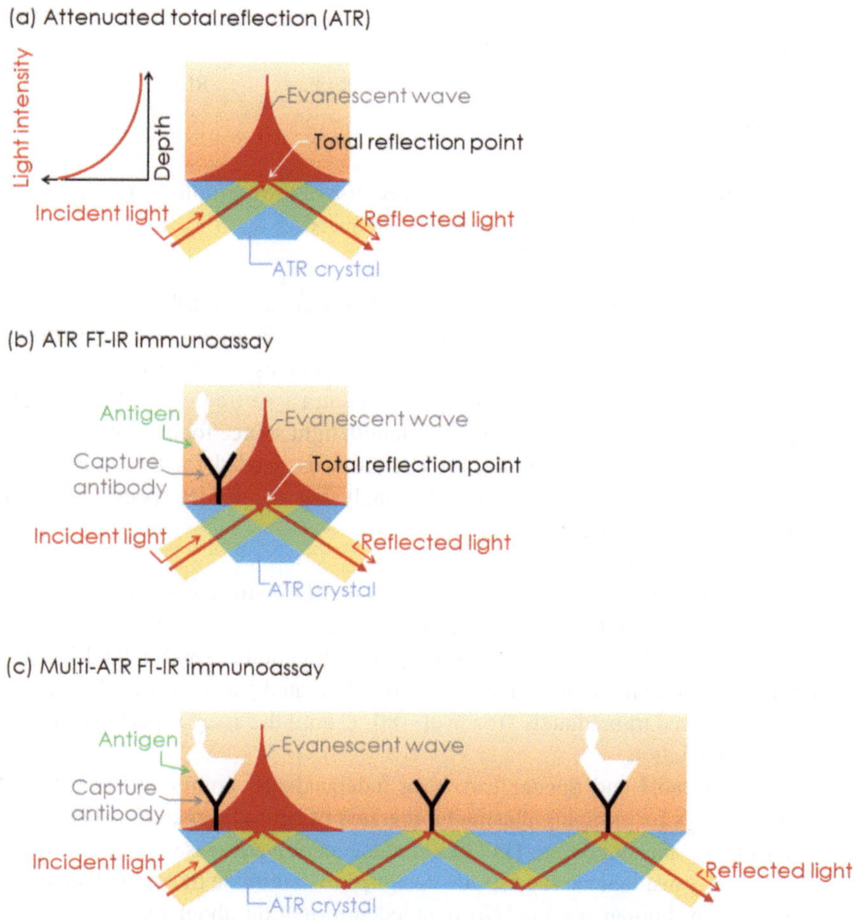

Fig. 1.8 Scheme of (**a**) attenuated total reflection (ATR), (**b**) ATR FT-IR, and (**c**) multi-ATR FT-IR

etration depth varies. The measured FT-IR spectrum of the reflected light shifts. For example, the FT-IR spectrum decreases in frequency from 1650 cm^{-1} to approximately 1622 cm^{-1} as α-helices and random coils transition to β-sheets of the amide I band [95, 96].

In the ATR FT-IR immunoassay, the capture antibody is immobilized on the top face of the ATR crystal, as shown in Fig. 1.8(b), which has a region of evanescent waves. Thus, the captured antigen can absorb evanescent waves, resulting in a characteristic FT-IR spectrum for the reflected light. As the capture antibodies interact with the antigen, the composition changes. The measured FT-IR spectrum of the reflected light shifts. By calculating the absorption at a fixed wavelength specific to the antigen in the FT-IR spectrum, the antigen can be quantified.

The propagation path is lengthened to create more total reflection points in ATR FT-IR spectroscopy, as illustrated in Fig. 1.8(c) [97]. Hence, the absorption of

Table 1.7 Low limit of detection (LLoD), dynamic range, and utility for research use only (RUO) or diagnostic use (Dx) of the FT-IRassaykit

Biomarker	LLoD	Dynamic range	RUO	Dx
Secondary structure of Aβ	NA	NA	O	

incident IR light by the antigen is enhanced because of the multiple points of attenuated total reflection. The assay sensitivity is enhanced.

- Features of the assay
  FT-IR is an inhomogeneous assay. The detected signals are attributed to analyte–probe reactions on a crystal surface functionalized with probes, such as antibodies. Dilutions of 1–2-fold of serum/plasma samples are needed. Only one antibody is needed for one biomarker. In contrast to traditional luminescent detection methods that use a single-wavelength light source for excitation, a light source with various wavelengths for excitation is needed for this method. In addition, a mathematical algorithm for analyzing the FT-IR spectrum of the reflected light is needed.
- Plasma preparation
  The standard operating procedures for sampling EDTA-treated blood plasma follow the guidelines of the German Competence Net Dementias. EDTA Monovettes are used for blood collection, followed by centrifugation at 1600xg for 15 min at room temperature to obtain plasma. The EDTA-treated plasma is transferred into a matrix vial and immediately frozen at −80 °C until the assay is performed [98].
- Commercialization
  BetaSense GmbH and Specac Ltd. have independently commercialized FT-IR immunoassays for assaying plasma biomarkers of neurodegenerative diseases.
- Assay kits for plasma biomarkers of neurodegenerative diseases
  The currently available assay kits are designed to determine the secondary structure of Aβ in human plasma. No detailed information about the low limit of detection (LLoD), dynamic range, and utility for research use only (RUO) or diagnostic use (Dx) of the assay kit is available, as listed in Table 1.7 [99].

1.8 Multimer Detection System (MDS)

- Mechanism

  The multimer detection system (MDS) is based on a sandwich enzyme-linked immunosorbent assay (ELISA). In traditional ELISA, capture antibodies and labeling antibodies are against separate epitopes of the antigen, as shown for the monomer in Fig. 1.9(a). Once one of the epitopes is inhibited or blocked due to antigen aggregation or truncation, as illustrated in the multimeric and truncated antigen in Fig. 1.9(a), these antigens cannot simultaneously interact with the capture antibody and labeling antibody and are ignored in traditional ELISA. Hence, traditional ELISA mainly assays monomers of antigens.

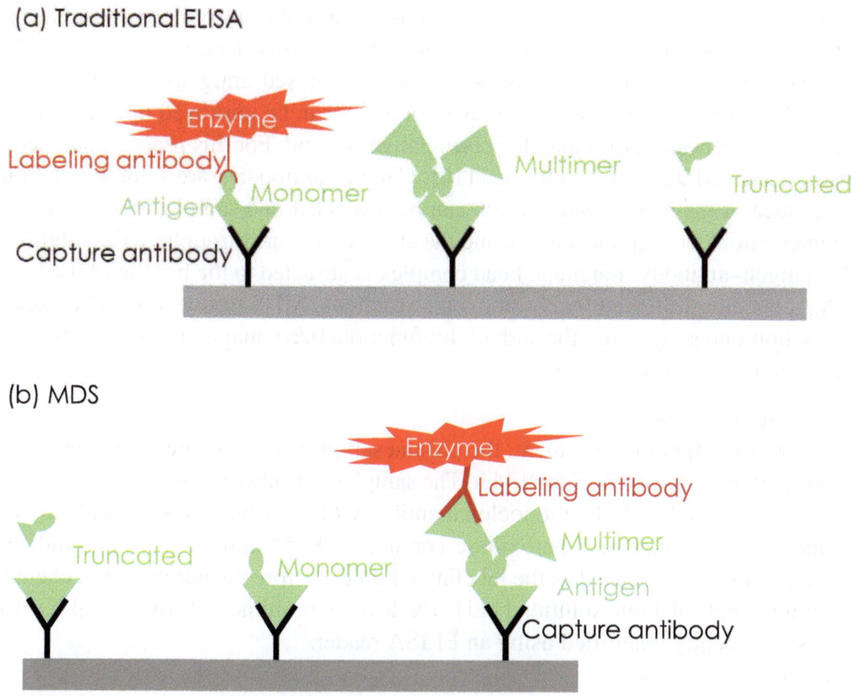

Fig. 1.9 Scheme of (a) the traditional ELISA and (b) MDS

In contrast, in MDS, the capture antibody and labeling antibody are against the same epitope of the antigen, as illustrated in Fig. 1.9(b) [100, 101]. In this case, only aggregated antigens, i.e., multimers, can have two identical epitopes. The monomeric antigen and truncated antigen do not have two identical epitopes. Thus, only multimers of antigens are assayed in MDS. Specifically, MDS can be used to detect multimers of antigens quantitatively. This property explains why it is called a multimer detection system. For example, capture antibodies and labeling antibodies used in MDS are specific for the N-terminus of amyloid-β peptides (Aβ) and are used to assay the multimeric form of Aβ (OAβ) [101–104].

As clearly shown in Fig. 1.9(b), the optical detection method is based on a sandwich ELISA, for which the low limit of detection (LLoD) is ng/ml. It is not sufficiently sensitive to assay Aβ levels in human plasma, whose levels are approximately pg/ml [105, 106]. Synthetic Aβ is added to the plasma sample and incubated with OAβ in the plasma sample to use the sandwich ELISA for MDS detection [101]. After the incubation, the AD group presents higher levels of OAβ than the normal controls. Thus, the oligomerization of Aβ in plasma from AD patients is more significant than that in plasma from normal controls. These results might be due to the abundant misfolding Aβ or biological environments in the blood of AD patients [107–109].

In MDS, the antibody–antigen interactions occur at the bottom of an assay well. It is an inhomogeneous assay. Many antigens are still dispersed in solution and do not reach the bottom of an assay well. These dispersed antigens are ignored in MDS. Therefore, the detection sensitivity of MDS can be improved once the interactions between antibodies and dispersed antigens occur. For this process, magnetic bead-assisted MDS was developed [110]. Capture antibodies are immobilized on the surface of magnetic beads, which can be dispersed in a sample solution. Thus, the interaction between the antigen and the antibody occurs throughout the solution. The antigen–antibody–magnetic bead complex is attracted to the bottom of the well using a magnet, followed by chromogenic reactions. MDS has become a homogeneous immunoassay with the aid of biofunctionalized magnetic beads and has increased detection sensitivity.

- Features of the assay
  Before the MDS measurement, the plasma samples are incubated with synthetic $A\beta_{42}$ (10 ng/mL) at 37 °C for 144 h. The samples are subsequently added to assay wells precoated with the monoclonal antibody 6E10, which captures Aβ at residues 3–8. A horseradish peroxidase-conjugated FF51 antibody, which binds to Aβ residues 1–4, is used as the labeling antibody before the addition of a chemiluminescent substrate solution [111]. The level of multimeric forms of Aβ (OAβ) is subsequently quantified using an ELISA reader.
- Plasma preparation
  Blood samples are collected in heparin-containing tubes and centrifuged at 850 × g for 30 min at room temperature [112]. The plasma (supernatant) is divided into aliquots and stored at −80 °C until analysis [101].
- Commercialization
  PeopleBio Inc. has been commercializing MDS immunoassays to measure the plasma levels of biomarkers of neurodegenerative diseases with the registration of ISO 13485:2016. The MDS assay kits developed by PeopleBio Inc. are approved for clinical use by the Ministry of Food and Drug Safety, South Korea.
- Assay kits for plasma biomarkers of neurodegenerative diseases
  The currently available assay kits are designed to detect oligomers of Aβ in human plasma. The low limit of detection (LLoD), dynamic range, and utility for research use only (RUO) or diagnostic use (Dx) of each assay kit are listed in Table 1.8.

Table 1.8 Low limit of detection (LLoD), dynamic range, and utility for research use only (RUO) or diagnostic use (Dx) of the MDSassaykit

Biomarker	LLoD	Dynamic range	RUO	Dx
Oligomers of aβs	244 pg/ml	0.244–5 ng/ml	○	○

1.9 Single Molecule Array (SIMOA)

- Mechanism

  Single molecule array (SIMOA) is based on a sandwich enzyme-linked immunosorbent assay (ELISA), but the fluorescent signals are digitized [113–115]. The flow chart of the SIMOA is illustrated in Fig. 1.10. Magnetic beads with a diameter of 2.7 μm and a functionalized capture antibody are mixed with a liquid sample containing the target antigen. A solution containing a labeling antibody conjugated with a fluorescent marker is added to the liquid sample to form a magnetic bead–capture antibody–antigen–labeling antibody complex, as shown in Fig. 1.10(a). Magnetic separation is utilized to remove nonspecifically bound molecules from the sample, as shown in Figs. 1.10(b) and (c).

The complexes are then transferred to a cartridge. The cartridge is a two-dimensional array of 235,000 microwells. Each microwell can hold only one bead. A cross-sectional view of the microwells holding beads is shown in Fig. 1.10(d). A substrate solution is added to the microwells, as shown in Fig. 1.10(e). Oil is applied onto the cartridge to form a liquid-tight seal, as indicated by the gray block in

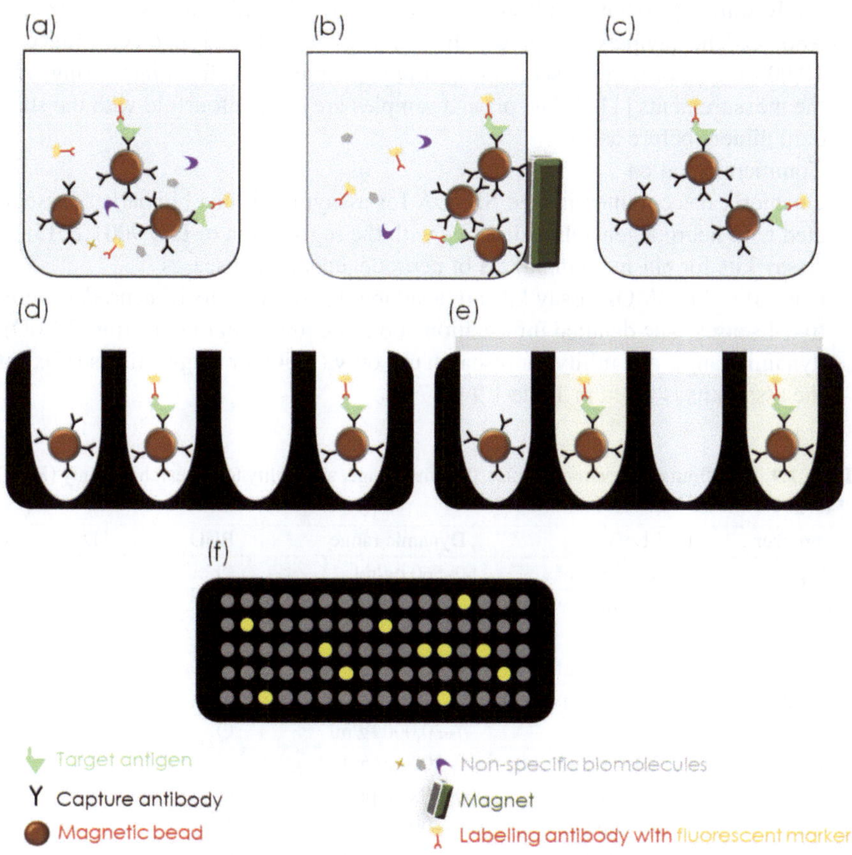

Fig. 1.10 Flow chart of the single molecule assay (SIMOA)

Fig. 1.10(e). Typically, especially at low concentrations, only one molecule of antigen is associated with one bead. One fluorescent spot on the cartridge denotes one molecule of antigen, as shown in Fig. 1.10(f). Thus, the technology is called a single molecule assay. Since each well can hold only one bead, the number of microwells with fluorescence is proportional to the amount of target antigen in the sample. A fluorescence microscope is used to detect the total number of two-dimensional fluorescent spots on the cartridge. SIMOA can detect antigens at subfemtomolar concentrations, i.e., 10^{-16} M, which is more sensitive than a traditional ELISA (~ 10^{-12} M).

- Features of the assay
  SIMOA is a homogeneous and ultrasensitive assay. Instead of the photodetector used in optical-based immunoassays, the SIMOA utilizes a microscope to analyze two-dimensional fluorescence information. The cost of the optical detection module is costly. A 2–4-fold dilution of the plasma sample is needed in SIMOA.
- Plasma preparation
  No official protocol is suggested. However, according to reported methods, the protocol of plasma preparation described below is usually used. Venous blood is collected in tubes containing EDTA and then centrifuged at 4 °C and 2000 × g for 10 min. The obtained plasma is divided into ~500 μl aliquots and frozen at −80 °C. The samples are rapidly thawed at 22 ± 2 °C and then centrifuged at 10,000 × g prior to analysis to prevent any sample debris from interfering with the measurements [116]. The plasma samples are diluted fourfold with the standard diluent before assay.
- Commercialization
  Quanterix has commercialized SIMOA for assaying plasma biomarkers associated with neurodegenerative diseases with the registration of ISO 9001:2015.
- Assay kits for plasma biomarkers of neurodegenerative diseases
  Currently, 11 SIMOA assay kits are available for biomarkers of neurodegenerative diseases. The detailed information about the low limit of detection (LLoD), dynamic range, and utility for research use only (RUO) or diagnostic use (Dx) of the assay kits is listed in Table 1.9 [117].

Table 1.9 Low limit of detection (LLoD), dynamic range, and utility for research use only (RUO) or diagnostic use (Dx) of the SIMOAassaykits

Biomarker	LLoD	Dynamic range	RUO	Dx
$A\beta_{1-40}$	0.196 pg/ml	0–560 pg/ml	O	
$A\beta_{1-42}$	45 fg/ml	0–240 pg/ml	O	
T-tau	19 fg/ml	0–400 pg/ml	O	
pTau181	0.620 pg/ml	0–1280 pg/ml	O	
pTau231	0.621 pg/ml	N/A	O	
pTau217	5.2 fg/ml	7–30,000 fg/ml	O	
NfL	0.062 pg/ml	0–2000 pg/ml	O	
BDNF	4.2 fg/ml	0–60,000 pg/ml	O	
GFAP	0.211 pg/ml	0–4000 pg/ml	O	
TDP-43	2.48 pg/ml	0–8000 pg/ml	O	
UCH-L1	1.05 pg/ml	0–40 ng/ml	O	

1.10 Single Molecule Counting (SMC)

- Mechanism

  A sample solution containing antigen is mixed with a reagent containing the capture antibody-functionalized magnetic beads [118, 119], as shown in Fig. 1.11(a). Through magnetic separation, the antigens and magnetic beads are extracted from the solution, as shown in Fig. 1.11(b). A solution containing antibodies labeled with fluorescent markers is added to the magnetic bead–antigen–labeling antibody complex, as shown in Fig. 1.11(c). The complexes are purified through magnetic separation, as shown in Fig. 1.11(d). A solution that is able to disrupt the interactions among magnetic beads, antigens, and labeling antibodies is added, as shown in Fig. 1.11(e). The isolated magnetic beads are removed with a magnet shown in Fig. 1.11(f). The solution containing analytes and labeling antibodies is placed into a well. The excitation light scans around the well, as shown in Fig. 1.11(g). Once the labeling antibody is scanned, a fluorescent signal is emitted, as shown in Fig. 1.11(h). One fluorescent signal emitted by a labeling antibody denotes one molecule of antigen. Thus, this assay technology is referred to as single molecule counting (SMC). The total fluorescent signals represent the total number of analyte molecules, i.e., the analyte concentration.

- Features of the assay
  SMC is a homogeneous and ultrasensitive assay. The fluorescent signals emitted by the labeling antibodies are digitized to quantify the concentrations of the antigens.
- Plasma preparation
  A K_2EDTA tube is used for blood collection. A 250-μl blood sample is placed into a filter plate well and centrifuged for ≥ 10 min at 1100xg. The plasma samples are diluted twofold with the standard diluent.
- Commercialization
  Millipore Sigma has been commercializing SMC assays for plasma biomarkers associated with neurodegenerative diseases with the registration of ISO 9001:2015 and ISO 13485:2016.
- Assay kits for plasma biomarkers of neurodegenerative diseases
  Currently, 14 SMC assay kits are available for biomarkers of neurodegenerative diseases. The detailed information about the low limit of detection (LLoD), dynamic range, and utility for research use only (RUO) or diagnostic use (Dx) of the assay kits is listed in Table 1.10 [120].

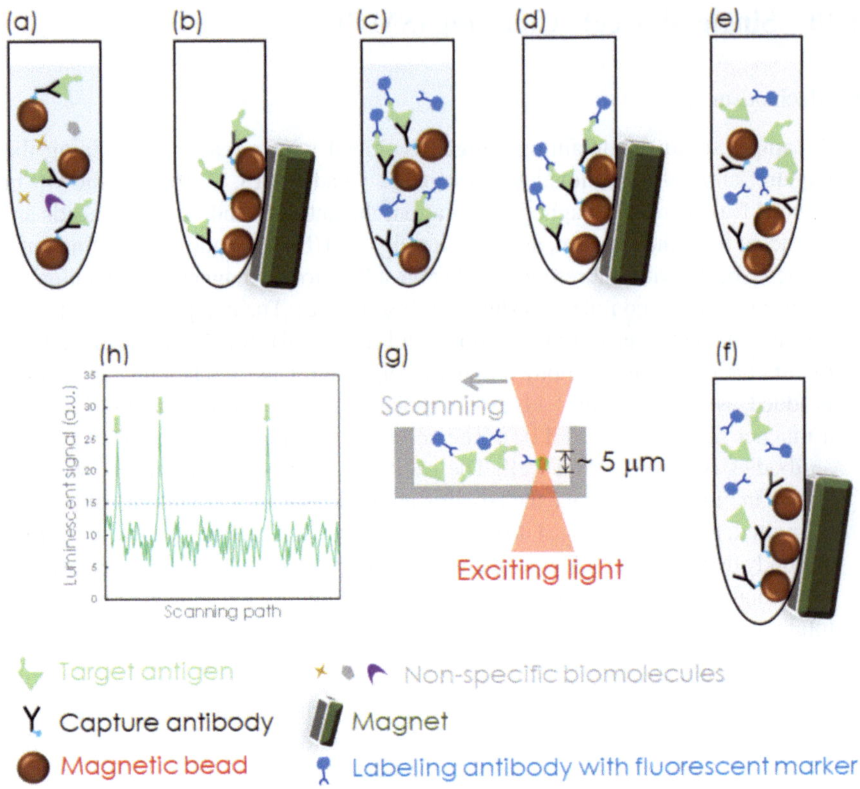

Fig. 1.11 Assay of single molecule counting (SMC) assay

Table 1.10 Low limit of detection (LLoD), dynamic range, and utility for research use only (RUO) or diagnostic use (Dx) of SMCassaykits

Biomarker	LLoD	Dynamic range	RUO	Dx
$A\beta_{1-40}$	5.86 pg/ml	NA	○	
$A\beta_{1-42}$	0.98 pg/ml	0.24–250 pg/ml	○	
T-tau	4.69 pg/ml	NA	○	
pTau181	0.94 pg/ml	NA	○	
pTau231	1.95 pg/ml	0.49–500 pg/ml	○	
pTau217	1.95 pg/ml	0.49–500 pg/ml	○	
α-Synuclein	1.95 pg/ml	NA	○	
p-α-syn129	0.1 pg/ml	NA	○	
NfL	0.87 pg/ml	NA	○	
BDNF	40 fg/ml	NA	○	
GFAP	0.23 pg/ml	NA	○	
TDP-43	5.21 pg/ml	2.6–6000 pg/ml	○	
UCH-L1	2.60 pg/ml	NA	○	
SNAP-25	0.26 pg/ml	NA	○	

1.11 Surface Plasmon Resonance (SPR)

- Mechanism

What is plasmon? Plasmons are collective oscillations of free electrons [121, 122]. For example, the free/conductive electrons in metals such as gold, silver, or copper oscillate collectively following excitation with incident light. The oscillations of free/conductive electrons in metals form plasmons. The oscillation of collective free/conduction electrons, i.e., plasmons, has resonant frequencies. For example, the resonant frequency of plasma in silver is around the band of ultraviolet light. Once the frequencies of the light are the same as the resonant frequencies of the plasmons on a metal surface, this property is called surface plasmon resonance (SPR).

As a layer of metal is coated on a dielectric material, light with suitable frequencies is incident from the dielectric material to the interface between the metal layer and the dielectric material, and plasmons are generated in the metal layer within the region of several to several tens of micrometers from the interface, which is known as surface plasmon. The surface plasmons further interact with the incident light to generate surface plasmon polaritons (SPPs). SPPs propagate along the interface and finally disappear because the energy of SPPs is absorbed by the metal layer. Obviously, the incident light interacts with the plasmons in the metal. The properties of the reflected light differ from those of the incident light.

The generation of the SPPs is illustrated in Fig. 1.12. A metal layer (Au, silver, or copper) is coated on a dielectric material such as glass. The thickness of the metal layer is several tens of nanometers. Since the refractive index of the dielectric material is higher than that of the metal layer, total reflection occurs at the interface of the metal layer and the dielectric material when the incident angle of the light is greater than the critical angle. As introduced in the description of

Fourier transform infrared (FT-IR), an evanescent wave propagates into the metal layer in the case of total reflection. The evanescent wave generates surface plasmons and SPPs, which implies that the intensities of the reflected light differ from those of the incident light. Usually, the absorption of the incident light is detected. The absorption of the incident light depends on the composition of the metal layer. Once the target antigen is captured on the metal layer with the aid of antibodies, the

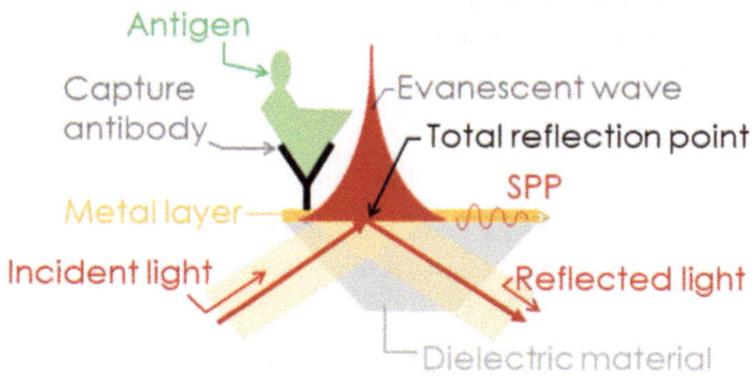

Fig. 1.12 Scheme for the generation of surface plasma polaritons (SPPs) and its application in bioassays

absorption of the incident light changes. Hence, SPPs are applied to quantify the biomolecule concentrations, as illustrated in Fig. 1.12.

The idea of localized SPR (LSPR) was proposed to enhance SPR [123]. In LSPR, the surface plasmon is localized in a nanoparticle, whose size is approximately equal to or smaller than the wavelength of incident light. Instead of coating a layer of metal, individual metal nanoparticles are fabricated on a dielectric material in LSPR. In LSPR, the electric fields of light near the nanoparticle surface are greatly enhanced, and the optical absorption by the nanoparticles reaches a maximum at the plasmon resonant frequency. Similar to SPPs, the properties of the reflected light are different from those of the incident light because of several factors, such as the nanoparticle composition, nanoparticle shape, and dielectric environment [124–126]. The changes can be manipulated with respect to the nanoparticle composition once the configuration of the LSPR is structured, as shown in Fig. 1.13. Hence, by immobilizing antibodies onto nanoparticles, the properties of reflected light can be changed once the antibodies capture target antigens. With the use of metal nanoparticles to localize SPR, the sensitivity of the assay for the target antigen is increased. Therefore, in addition to SPP, LSPR has also been shown to be a promising assay for ultralow concentrations of plasma biomarkers associated with Alzheimer's disease [127, 128].

Owing to the great progress in optical fiber technologies, an optical fiber is commonly used as the dielectric material in LSPR [129]. The cladding layer of the optical fiber is removed and seated with metal nanoparticles, followed by the immobilization of antibodies on the nanoparticles. The light transmitted through the optical fiber is detected to reveal the concentrations of the target antigen.

- Features of the assay
  SPR is an inhomogeneous assay. The detected signals are attributed to analyte–probe reactions on the metal surface functionalized with probes. Serum/plasma samples do not need to be diluted. Only one antibody is needed for one biomarker. A complicated light emitting and receiving module is needed.
- Serum preparation
  The blood samples are incubated at room temperature for 30–40 min to allow them to clot. The clotted blood is centrifuged for 20 min at 3000 rpm to separate the serum.

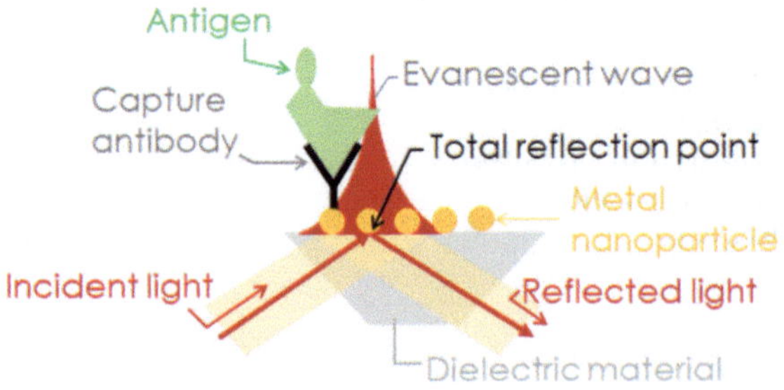

Fig. 1.13 Scheme of bioassay using localized surface plasmon resonance (LSPR)

Table 1.11 Low limit of detection (LLoD), dynamic range, and utility for research use only (RUO) or diagnostic use (Dx) of SPRassaykits

Biomarker	LLoD	Dynamic range	RUO	Dx
T-tau	2.4 pg/ml	0–2360 pg/ml	O	
pTau181	1.6 pg/ml	0–4360 pg/ml	O	
pTau217	2.4 fg/ml	0.0024–10 pg/ml	O	
NfL	24.4 fg/ml	0.04–125 pg/ml	O	
GFAP	2.81 fg/ml	0.01–1000 pg/ml	O	

- Commercialization
  Instant NanoBiosensors has commercialized LSPR with optical fibers to measure blood biomarkers associated with neurodegenerative diseases.
- Assay kits for plasma biomarkers of neurodegenerative diseases
  Currently, three SPR assay kits are commercially available for biomarkers of neurodegenerative diseases (pTau217, NfL, and GFAP). The detailed information about the low limit of detection (LLoD), dynamic range, and utility for research use only (RUO) or diagnostic use (Dx) of the assay kits is listed in Table 1.11. Information about T-tau and pTau181 was obtained from in-house measurements [129].

1.12 Multi-analyte Profiling (xMAP)

- Mechanism

  The flexible multi-analyte profiling (xMAP) system utilizes microsphere-based fluorescence and flow cytometry detection for the antigens of interest [130–133]. The microspheres are composed of polystyrene, are 5.6 μm in diameter, and are functionalized with capture antibodies. When a sample containing the antigens of interest is mixed with the reagent containing the microspheres, a complex of microspheres and antigens is formed. Through a washing step, nonspecific molecules are removed. Then, antibodies labeled with fluorescent reporters are added to form the microsphere–antigen–labeling antibody–reporter complex. After a wash step, unbound antibodies labeled with fluorescent reporters are removed. The fluorescent reporters are counted using flow cytometry.

Remarkably, the microspheres are internally labeled with various spectrally distinct fluorochromes, e.g., two spectrally distinct fluorochromes [134, 135]. Using precise amounts of each of these fluorochromes, dyes of various "colors" are obtained. Microspheres labeled with fluorochromes of various "colors" are functionalized with different capture antibodies to measure multiple different antigens, as shown in Fig. 1.14(a). Each labeled microsphere of a specific "color" is functionalized with specific capture antibodies against the antigens of interest, followed by an interaction with the antibodies with labeled fluorescent reporters. Once all the microspheres are individually labeled with various colors and rapidly separated using a flow cytometer, one laser (Laser 1, 635 nm in wavelength) emits light to

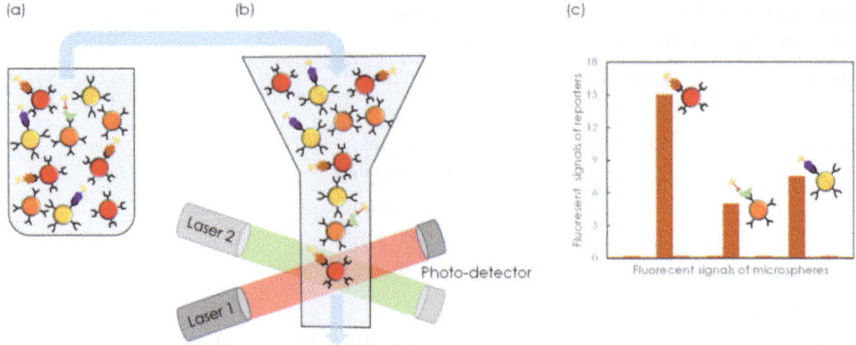

Fig. 1.14 Scheme for the generation of the flexible multi-analyte profiling (xMAP) assay

excite the fluorescence of the microspheres, as shown in Fig. 1.14(b) [136]. Another laser (Laser 2, 525–532 nm in wavelength) simultaneously emits light to excite the fluorescent reporter on the labeling antibody. A high-speed digital photodetector receiving fluorescent signals from the microspheres is used to classify the antigen samples. The photodetector receiving the fluorescent signals from the reporter on the labeling antibody quantifies the concentrations of the antigen, as shown in Fig. 1.14(c). Thus, multiple antigens can be assayed simultaneously with the received fluorescent signals emitted by microspheres and reporters on labeling antibodies in one test using xMAP.

Microspheres embedding magnetic beads are used to reduce the noise of detected optical signals with the aid of magnetic separation at wash steps in the xMAP assay [137, 138]. Microspheres with magnetic beads become slightly larger, at 6.5 μm in diameter. Thus, high-sensitivity and multiple assays are features of xMAP.

- Features of the assay
  xMAP is a homogeneous and sandwich-based assay. Two antibodies are needed for one biomarker. Microspheres labeled with fluorochromes are prepared. Two lasers that excite the fluorescent signals generated by microspheres and fluorescent reporters are needed. A complicated light-emitting and light-receiving module is equipped with an analyzer. xMAP is a powerful multiplexed assay. A 2-fold dilution of a plasma sample is needed.
- Plasma preparation
  After the blood is drawn, the blood samples are centrifuged at 1000 × g for 10 min at 4 °C within 30 min of collection. The plasma samples are frozen at −80 °C. Multiple (> 2) freeze/thaw cycles are avoided before the xMAP assay.
- Commercialization
  Luminex has been commercializing xMAP assays to measure the levels of blood biomarkers associated with neurodegenerative diseases with the registration of ISO 9001:2015 and ISO 13485:2016.

Table 1.12 Low limit of detection (LLoD), dynamic range, and utility for research use only (RUO) or diagnostic use (Dx) of xMAPassaykits

Biomarker	LLoD	Dynamic range	RUO	Dx
$A\beta_{1-40}$	10.2 pg/ml	21–15,000 pg/ml	O	
$A\beta_{1-42}$	2.5 pg/ml	2–2000 pg/ml	O	
T-tau	14.2 pg/ml	11–8000 pg/ml	O	
pTau181	1.5 pg/ml	0.7–500 pg/ml	O	
α-Synuclein	140.7 pg/ml	244–1000,000 pg/ml		
GFAP	3.1 pg/ml	12–50,000 pg/ml	O	
UCH-L1	8.2 pg/ml	20–80,000 pg/ml	O	

- Assay kits for plasma biomarkers of neurodegenerative diseases
  Currently, six xMAP assay kits are commercially available for biomarkers of neurodegenerative diseases. The detailed information about the low limit of detection (LLoD), dynamic range, and utility for research use only (RUO) or diagnostic use (Dx) of the assay kits is listed in Table 1.12 [139].

1.13 Immunomagnetic Reduction (IMR)

- Mechanism
  Instead of light signals, the immunomagnetic reduction (IMR) assay generates magnetic signals to quantify the target antigen, i.e., the so-called magnetically labeling immunoassay [87]. In IMR, the reagent consists of magnetic nanoparticles that are 55 nm in diameter and are functionalized with capture antibodies [140, 141], as shown in Fig. 1.15. Antibody-functionalized magnetic nanoparticles are well suspended in a pH 7.2 phosphate buffered saline (PBS) solution.

Under external alternating-current (AC) magnetic fields, magnetic nanoparticles in the reagents oscillate with AC magnetic fields, resulting in AC magnetic susceptibility, χ_{ac} [142, 143]. The reagent is mixed with a sample containing the target antigen. Portions of magnetic nanoparticles in the reagent interact with the target antigen via the antibody on the particle surface. Due to the enlargement of bound nanoparticles, the oscillations of magnetic nanoparticles with external AC magnetic fields are suppressed. Thus, the AC magnetic susceptibility of the reagent is reduced, as shown in Fig. 1.15. In principle, for samples with higher concentrations of the target antigens, more magnetic nanoparticles are associated with the target antigen. The reduction in the AC magnetic susceptibility of the reagent is enhanced. Therefore, the greater reduction in signals for the AC magnetic susceptibility of the reagents results in samples with more concentrated target antigens. Once the relationship between the reduction in the AC magnetic susceptibility of the reagent and the concentration of antigen is determined, the antigen in a sample can be quantitatively detected via IMR [144].

In IMR, antibody-functionalized magnetic nanoparticles are homogeneously dispersed in a reagent solution. IMR is clearly a homogeneous assay that is able to

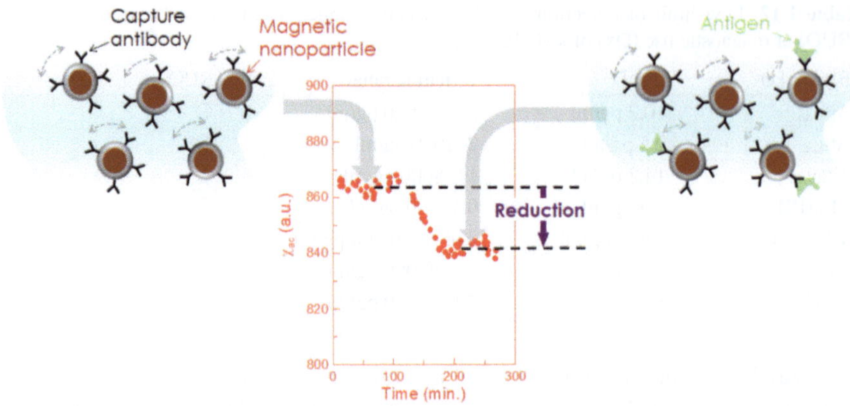

Fig. 1.15 Schematic illustration of the immunomagnetic reduction (IMR) assay

capture antigens throughout the solution, resulting in high sensitivity. In addition, the total area of the binding surface on magnetic nanoparticles is large. This property also contributes to the high sensitivity of the IMR.

For ultralow-concentration samples, the reduction in the AC magnetic susceptibility of the reagent will be extremely small. An ultrasensitive sensor that is able to detect extremely tiny reductions in the AC magnetic susceptibility of reagents precisely is needed. A high-temperature superconducting quantum interference device (high-T_c SQUID) magnetometer is one of the candidates [145, 146]. With the aid of a high-T_c SQUID magnetometer, the low limits of detection of the IMR assay can be as low as fg/ml or lower [147–150].

A SQUID magnetometer is a superconducting loop consisting of two Josephson junctions in parallel. The Josephson junctions are the macroscopic phenomena of quantum tunneling, in which superconducting electrons tunnel through a resistive barrier. The magnetic flux thought the superconducting loop of a SQUID is quantized. Once a non-quantized magnetic flux goes through the loop, the superconducting electrons will be driven to generate an additional magnetic flux to quantize the magnetic flux through the loop. Once the superconducting electrons are driven, they will tunnel the resistive barriers in the loop to generate voltage cross the SQUID. Hence, the voltage of the SQUID will be manipulated by the magnetic signals through the loop of the SQUID. In IMR, the changes in the magnetic signals generated by nanoparticles owing to the association with target antigens is guided to a SQUID, resulting in changes in the voltage across the SQUID. Therefore, the reduction in the magnetic signals in IMR is converted to the decrease in the voltage of a SQUID. Notably, the theory of macroscopic quantum tunneling was awarded the Nobel Prize in Physics in 1973 (Leo Esaki, Ivar Giaever, and Brian David Josephson). The experimental milestone of macroscopic quantum tunneling was awarded the Nobel Prize in Physics in 2025 (John Clarke, Michel H. Devoret, and John M. Martinis). Obviously, SQUID technologies are well recognized as highly influential discovery and inventions.

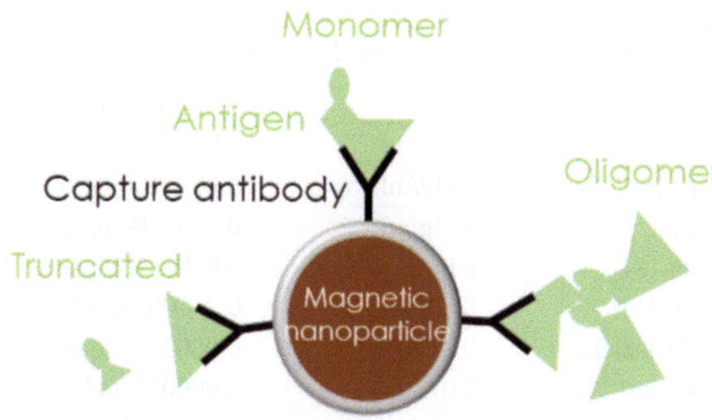

Fig. 1.16 IMR assays of monomers, oligomers, and truncated forms of antigens

Unlike sandwich immunoassay technologies, only capture antibodies are needed in IMR, and they are called single-antibody immunoassays. A labeling antibody is not necessary in IMR. Antibody-functionalized magnetic nanoparticles can interact with the target antigen once the epitope of the antigen is exposed. Hence, IMR can assay monomers, oligomers, and truncated forms of antigens, as illustrated in Fig. 1.16.

Notably, IMR is a one-step assay without wash steps to remove unbound antibody-functionalized magnetic nanoparticles. In addition, neither enzymes nor labeling antibodies are used in IMR. The measurement processes are very simple. IMR benefits from its friendly use.

For sandwich technologies, the use of labeling antibodies can increase the specificity of the assay. However, no labeling antibody is used for IMR. How can the IMR be a highly specific assay?

As shown in Fig. 1.15, the magnetic nanoparticles oscillate with external AC magnetic fields. The antigen captured on a nanoparticle is subjected to a centrifugal force due to oscillation. Nanoparticles can interact with both nonspecific biomolecules and specific antigens. The binding force between antibodies and nonspecific biomolecules is weaker than that between antibodies and specific antigens. Once the centrifugal force is stronger than the nonspecific binding force but weaker than the specific binding force, nonspecific associations are prohibited, which is the so-called spin-wash technology. Thus, a highly specific assay is achieved for IMR. Considering the size of the nanoparticles, the viscosity of the reagent and the binding force between the antibody and antigen, the suitable frequency of oscillation is approximately 20 kHz. This method to prohibit nonspecific binding in IMR is named spin-wash technology [151].

- Features of the assay

  IMR is a homogeneous assay. Magnetic signals are detected. The reader in the IMR system is a magnetic sensor, not an optical sensor. IMR is a single-antibody immunoassay with the aid of spin-wash technology and has high specificity. IMR is user-friendly because of the one-step assay. The low limits of detection of the IMR immunoassay are fg/ml or lower. No dilution is needed for the plasma samples.

Table 1.13 Low limit of detection (LLoD), dynamic range, and utility for research use only (RUO) or diagnostic use (Dx) of IMRassaykits

Biomarker	LLoD	Dynamic range	RUO	Dx
$A\beta_{1-40}$	64 fg/ml	0.064–200 pg/ml	O	O
$A\beta_{1-42}$	80 fg/ml	0.080–100 pg/ml	O	O
T-tau	35 fg/ml	0.035–100 pg/ml	O	O
pTau181	8 fg/ml	0.0196–100 pg/ml	O	O
pTau217	66 fg/ml	0.1–100 pg/ml	O	O
α-Synuclein	0.21 fg/ml	0.00021–100 pg/ml	O	O
p-α-syn129	0.072 fg/ml	0.000072–0.48 pg/ml	O	O
NfL	0.18 fg/ml	0.00018–100 pg/ml	O	O
BDNF	3.0 fg/ml	0.003–1000 pg/ml	O	O
GFAP	31.4 fg/ml	0.00314–100 pg/ml	O	O
TDP-43	0.68 fg/ml	0.00068–100 pg/ml	O	O
UCH-L1	3.3 fg/ml	0.0033–100 pg/ml	O	O

- Plasma preparation
  A 6-ml or 9-ml K_2EDTA or K_3EDTA tube is used for blood collection. The blood samples are placed at room temperature, not on wet ice, followed by centrifuging the blood at 1500–2000 × g at room temperature (20–25 °C) for 15 mins. The blood tubes are not centrifuged at 4 °C [152]. After centrifugation, 0.5–1 ml of plasma is transferred into a 1.5-ml microcentrifuge tube. The separated plasma samples are frozen at −80 °C within 4.5 h after the blood is drawn [153]. Plasma samples stored for more than 1 year are not recommended for IMR immunoassays [154].
- Commercialization
  MagQu Co. Ltd. has commercialized IMR assays to detect plasma biomarkers associated with Alzheimer's disease, Parkinson's disease, frontotemporal dementia, etc., with the registration of ISO 13485:2016. The IMR assay kits developed by MagQu Co. Ltd. are approved for clinical use by the CE IVD in Europe and the Ministry of Health and Welfare, Taiwan.
- Assay kits for assaying plasma biomarkers of neurodegenerative diseases
  Currently, 12 IMR assay kits are available for biomarkers of neurodegenerative diseases. The detailed information about the low limit of detection (LLoD), dynamic range, and utility for research use only (RUO) or diagnostic use (Dx) of the assay kits is listed in Table 1.13 [155].

1.14 Proximity Ligation Assay (PLA)

- Mechanism

Two independent binders with individual free single-stranded oligonucleotides with 3′ and 5′ ends [156]. The binders could be aptamers or antibodies, able to specifically bind epitopes of target molecules [157–160]. The two independent

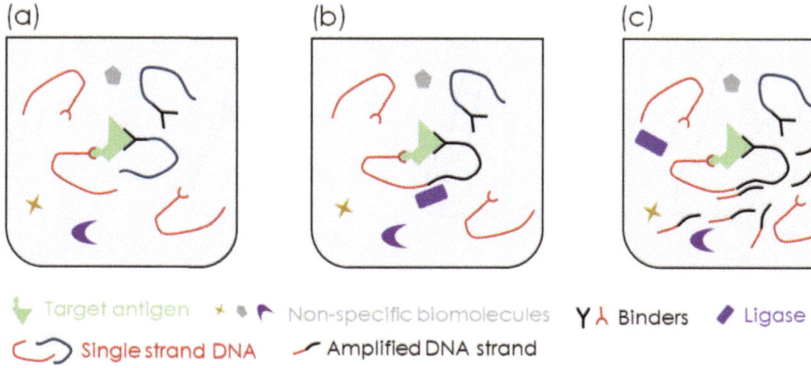

Fig. 1.17 Illustration of the proximity ligation assay (PLA)

binders associate simultaneously with the molecule of interest at two epitopes, as shown in Fig. 1.17(a). Due to the proximity between the two free single-stranded oligonucleotides (20–50 nm), the 3′ and 5′ ends of single-stranded oligonucleotides are ligated with the aid of a ligase enzyme, as shown in Fig. 1.17(b). The ligated sector is designed with a specific sequence of deoxyribonucleic acid (DNA) or ribonucleic acid (RNA), which acts as a primer for the following real-time polymerase chain reaction, as shown in Fig. 1.17(c). According to the cycle threshold in real-time polymerase chain reaction, the molecule of interest can be quantified.

The reliance of the polymerase chain reaction on heating (80–95 °C) and cooling (50–60 °C) cycles might destroy biological molecules of interest [161]. Isothermal approaches to DNA or RNA amplification in the polymerase chain reaction have been developed. One convenient isothermal approach could be the rolling circle amplification (RCA) [162], which amplifies DNA or RNA strands in the range of 13–60 °C [163–166]. In RCA, two linear DNA or RNA strands, which will form a very small circular oligonucleotide (25–100 nucleotides), are formed independently by the two free single-stranded oligonucleotides linked to binders, as shown in Fig. 1.18(a) and (b). These two linear DNA or RNA strands are ligated to form the small circular oligonucleotide with the aid of a ligase enzyme, as shown in Fig. 1.18(c). Since the sequence of the two free single-stranded oligonucleotides is specifically designed, the DNA or RNA sequence of the small circle is well defined. The small circle of oligonucleotide acts as a template for a DNA or an RNA polymerase, as shown in Figs. 1.18(d) and (e). Linearly repeating DNA or RNA strands, copying the sequence of the small circle, are produced, as shown in Figs. 1.18(f) and (g). Fluorescent dyes binding to the linearly repeating DNA or RNA strands emit optical densities to show the cycle threshold in RCA. The cycle threshold corresponds to the quantity of the molecule of interest.

In order to enhance the sensitivity of PLA, biofunctionalized magnetic beads, which are able to specifically extract molecules of interest through magnetic separation, are frequently used before adding paired binders to the solution [167–169].

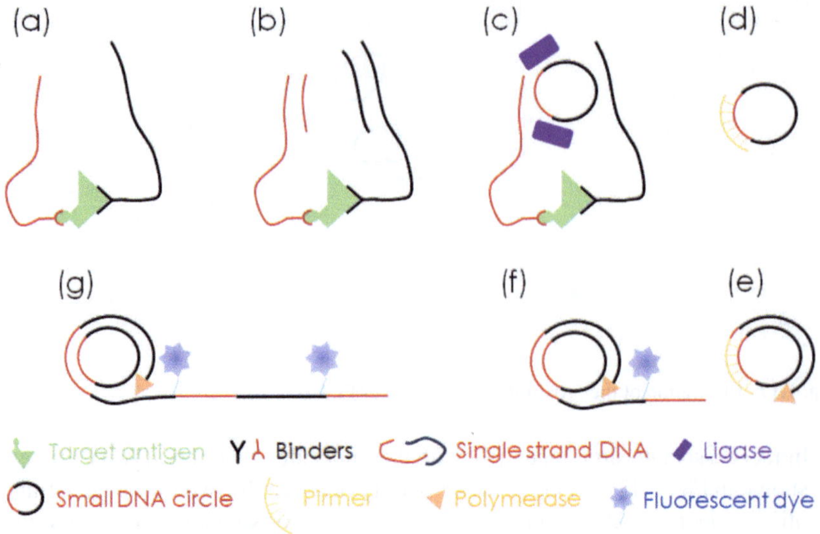

Fig. 1.18 Illustration of the rolling circle amplification (RCA) in the proximity ligation assay (PLA)

Table 1.14 Low limit of detection (LLoD), dynamic range, and utility for research use only (RUO) or diagnostic use (Dx) of magnetic bead-assisted PLAkit

Biomarker	LLoD	Dynamic range	RUO	Dx
pTau217	NA	0.2–200 pg/ml	○	

- Features of the assay
  PLA is a homogeneous assay, involving paired binders to the molecule of interest to achieve high specificity. With the ability to amplify DNA or RNA, ultrasensitive assays are achieved. However, the entire process of the assay is complicated and includes the binding of paired binders to the molecule of interest, a ligation reaction, and qPCR or RCA.
- Plasma preparation
  Blood is collected in K_2EDTA–plasma tubes and centrifuged at 2000xg for 10 min at 4 °C. The plasma is aliquoted into 1.5-ml polypropylene tubes (1 ml per tube) and stored at −80 °C until analysis.
- Commercialization
  Taudia has been commercializing magnetic bead-assisted PLA for plasma biomarkers of neurodegenerative diseases.
- Assay kits for plasma biomarkers of neurodegenerative diseases
  Currently, only one magnetic bead-assisted PLA kit is available: pTau217. The detailed information about the low limit of detection (LLoD), dynamic range, and utility for research use only (RUO) or diagnostic use (Dx) of the assay kit is listed in Table 1.14.

1.15 Nucleic Acid-Linked Immuno-sandwich Assay (NULISA)

- Mechanism

Nucleic acid-linked immuno-sandwich assay (NULISA) is a technology that combines antibody–antigen interactions, ligation reactions, and polymerase chain reaction [170]. In antigen–antibody interactions, the capture antibody is linked to a double-stranded deoxyribonucleic acid (DNA), which consists of a poly-A tail and a target-specific molecular identifier (TMI). The labeling antibody is linked to another double-stranded DNA, which contains biotin and a matching TMI. When both antibodies bind to the antigen of interest, an immunocomplex is formed, as shown in Fig. 1.19(a). Magnetic beads coated with oligo(dT) are mixed with the solution. Oligo(dT)–poly(A) hybridization occurs on the bead surface, as shown in Fig. 1.19(b). The immunocomplexes and unbound capture antibodies bind to the magnetic beads. Through magnetic separation with the aid of a magnet, immunocomplexes and unbound capture antibodies are selected, as shown in Fig. 1.19(c) and (d). A low-salt solution is added to the solution to dissociate the oligo(dT)–poly(A) hybridized on the bead surface. The immunocomplexes and unbound capture antibodies are independently dissolved in the solution, as shown in Fig. 1.19(e). Through magnetic separation, the supernatant containing the immunocomplexes and unbound capture antibody is transferred to a tube, as shown in Fig. 1.19(f). Magnetic beads functionalized with streptavidin are added to the solution, as shown in Fig. 1.19(f). The biotin

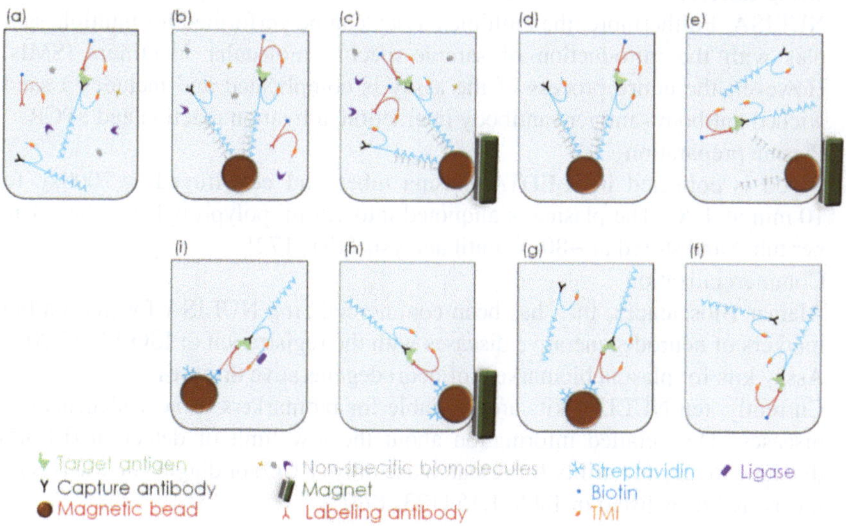

Fig. 1.19 Illustration of the nucleic acid-linked immuno-sandwich assay (NULISA)

in the immunocomplexes binds to streptavidin on the magnetic beads, as shown in Fig. 1.19(g). However, unbound capture antibodies do not bind to the magnetic beads. Through magnetic separation, unbound capture antibodies are washed away, as shown in Fig. 1.19(h). The antibody–antigen interaction in NULISA is terminated, followed by a ligation reaction.

A solution containing T4 DNA ligase and a specific DNA ligator sequence is added to the solution to achieve the ligation reaction in NULISA, as shown in Fig. 1.19(h). The proximal ends of the DNA attached to the capture and labeling antibodies interact with each other, resulting in the generation of a DNA reporter molecule containing unique pairs of TMIs, as shown in Fig. 1.19(i). The DNA reporter is quantified by quantitative real-time polymerase chain reaction (qPCR). Thus, the concentration of the target antigen is quantified.

Different capture/labeling antibodies combined with specific paired TMIs are used along with next-generation sequencing (NGS) for a multiplex analysis of one sample. Notably, NULISA is able to perform a multiplex assay not only in one sample but also in multiple samples. Sample-specific molecular identifiers (SMIs) are introduced to identify samples, as shown in Fig. 1.19(h). Thus, the SMIs are ligated between paired TMIs. Next-generation sequencing (NGS) is used to generate antigen- and sample-specific counts.

- Features of the assay
  NULISA is a homogeneous assay. Both the use of capture/labeling antibodies and paired target-specific molecular identifiers (TMIs) result in the high specificity of NULISA. With the ability to amplify DNA, ultrasensitive immunoassays are achieved with NULISA. The low limit of detection is approximately fg/ml. Using different paired TMIs, a multiplex assay in one sample is available with NULISA. Furthermore, the multiplex assay can be performed for multiple samples with the introduction of sample-specific molecular identifiers (SMIs). However, the entire process of the assay is complicated and includes a sandwiched antibody–antigen–antibody interaction, a ligation reaction and qPCR.
- Plasma preparation
  Blood is collected in K_2EDTA–plasma tubes and centrifuged at 2000xg for 10 min at 4 °C. The plasma is aliquoted into 1.5-ml polypropylene tubes (1 ml per tube) and stored at −80 °C until analysis [171, 172].
- Commercialization
  Alamar Biosciences, Inc., has been commercializing NULISA for plasma biomarkers of neurodegenerative diseases with the registration of ISO 13485:2016.
- Assay kits for plasma biomarkers of neurodegenerative diseases
  Currently, ten NULISA kits are available for biomarkers of neurodegenerative diseases. The detailed information about the low limit of detection (LLoD), dynamic range, and utility for research use only (RUO) or diagnostic use (Dx) of the assay kits is listed in Table 1.15 [173, 174].

Table 1.15 Low limit of detection (LLoD), dynamic range, and utility for research use only (RUO) or diagnostic use (Dx) of NULISAkits

Biomarker	LLoD	Dynamic range	RUO	Dx
$A\beta_{1-40}$	NA	NA	O	
$A\beta_{1-42}$	NA	NA	O	
T-tau	NA	NA	O	
pTau181	27 fg/ml	NA	O	
pTau231	NA	NA	O	
pTau217	7 fg/ml	NA	O	
α-Synuclein	NA	NA	O	
NfL	9 fg/ml	NA	O	
GFAP	NA	NA	O	
UCH-L1	NA	NA	O	

1.16 Successive Proximity Extension Amplification Reaction (SPEAR)

- Mechanism

Successive proximity extension amplification reaction (SPEAR) is a next-generation proximity extension assay (PEA) [175–179]. Two successive interactions between colocalized probes are utilized in SPEAR to reduce false signals due to the synthesis of the complete deoxyribonucleic acid (DNA) sequence by the interaction between free-floating probes, which occurs frequently in conventional PEA [176]. This process significantly reduced the background compared with that of conventional PEA. The details of SPEAR are illustrated in Fig. 1.20.

The SPEAR reagent consists of a pair of antibodies linked with specific-sequence double-stranded DNA (Probe A) and single-stranded DNA (Probe B) probes, as shown in Fig. 1.20(a). The first extension of SPEAR occurs as the pair of antibodies colocalize at the antigen of interest, as shown in Fig. 1.20(b). The end of Probe B hybridizes to Probe A with the aid of DNA polymerase until the polymerase encounters a chemical stopper sequence, which terminates the extension, as shown in Figs. 1.20(c) and (d). Probe B is extended with a segment of single-stranded DNA, depicted as the orange segment in Fig. 1.20(d). The extended Probe B disassociates from Probe A, as shown in Fig. 1.20(e), followed by the second extension of SPEAR.

The second extended region of Probe B (orange segment) hybridizes to the other end of Probe A to execute the second extension with the aid of DNA polymerase, as shown in Figs. 1.20(f) and (g). After completing the second extension, Probe B is extended with a segment of single-stranded DNA, depicted as the dark blue segment in Fig. 1.20(h), and dissociates from Probe A. The resulting sequence of Probe B can be amplified and quantified through quantitative real-time polymerase chain reaction (qPCR).

Notably, the interactions of free-floating probes or transient nonspecific binding are not able to undergo the second extension of SPEAR, as illustrated in Fig. 1.20.

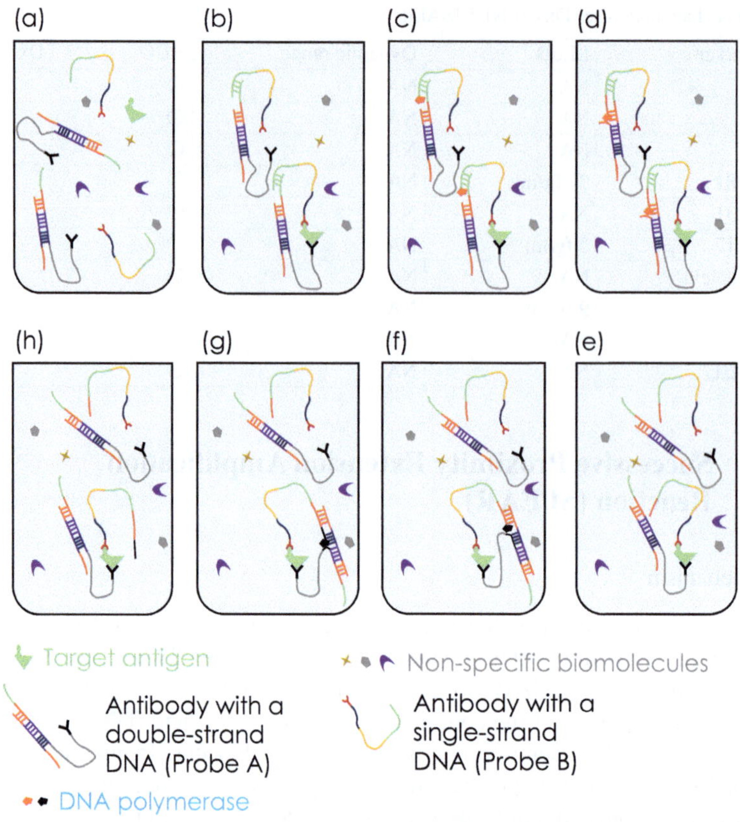

Fig. 1.20 Illustration of the successive proximity extension amplification reaction (SPEAR)

If only a single extended Probe B is generated, it can be amplified by qPCR. As a result, the second extension in SPEAR to produce a specific sequence significantly reduces the background signal compared with the conventional PEA.

- Features of the assay
  SPEAR is a homogeneous assay. By utilizing qPCR, SPEAR benefits from very high sensitivity. The sample volume is as small as 1 μl [175]. In addition, SPEAR is a wash-free, multiplex assay. However, the entire process of the assay is complicated, including a sandwiched antibody–antigen–antibody interaction, several DNA ligations, hybridization reactions, and qPCR.
- Plasma preparation
  Blood samples are collected in K_2EDTA tubes and centrifuged at 2000 × g for 5 min at room temperature. The plasma samples are frozen within 2 h of collection and stored at −80 °C until use [180].
- Commercialization
  Spear Bio Inc. has commercialized SPEAR assays for plasma biomarkers of neurodegenerative diseases with the registration of ISO 13485:2016.

Table 1.16 Low limit of detection (LLoD), dynamic range, and utility for research use only (RUO) or diagnostic use (Dx) of SPEARassaykits

Biomarker	LLoD	Dynamic range	RUO	Dx
pTau231	17.7 fg/ml	NA	O	
pTau217	2.5 pg/ml	NA	O	
NfL	10 fg/ml	NA	O	

- Assay kits for plasma biomarkers of neurodegenerative diseases
  Currently, three SPEAR assay kits are available for biomarkers of neurodegenerative diseases. The detailed information about the low limit of detection (LLoD), dynamic range, and utility for research use only (RUO) or diagnostic use (Dx) of the assay kits is listed in Table 1.16 [181].

1.17 Real-Time Quaking-Induced Conversion (RT-QuIC)

- Mechanism

  The conformational changes of unfolded proteins associated with neurodegenerative diseases into β-sheets play a role in the pathology of diseases [182]. The β-sheet misfolded protein molecules can act as templates, inducing the aggregation of unfolded protein molecules and finally generate fibrils, which further form insoluble aggregates like amyloid plaques or Lewy bodies in the brain [183]. As compared to unfolded proteins, the fibrils of misfolded protein are easily analyzed with technologies such as Western blot or fluorescent labeling. However, it usually takes a few years to form enough quantities of fibrils for detection. In order to speed up the formation of fibrils, misfolded protein molecules are externally added to a sample as seeds to generate templates. Sufficient quantities of fibrils can be formed within a few hours for analysis. This is the so-called seed amplification assay (SAA) [184]. The details of SAA are illustrated in Fig. 1.21.

A sample containing native unfolded proteins is prepared, as shown in Fig. 1.21(a). Misfolded proteins, usually oligomers, are added to the sample as seeds. It takes time for the diffusion of native unfolded protein molecules and seeds to collide with each other. This is referred to as the log phase, in which no signal related to the aggregation of native unfolded protein is detected. After the associations between seeds and unfolded protein molecules, the unfolded protein molecules become folded, such as β-sheet folding, as shown in Figs. 1.21(c) and (d). The signals related to the aggregation of native unfolded proteins are detectable. More native unfolded protein molecules bind to the chains of folded protein, as shown in Fig. 1.21(e). The chains of folded protein become longer and longer, which is referred to as the elongation phase. Meanwhile, the signals related to the aggregation of native unfolded protein increase.

In order to speed up the aggregation of native unfolded proteins, the chains of folded proteins are broken down into several segments. Thus, more seeds are

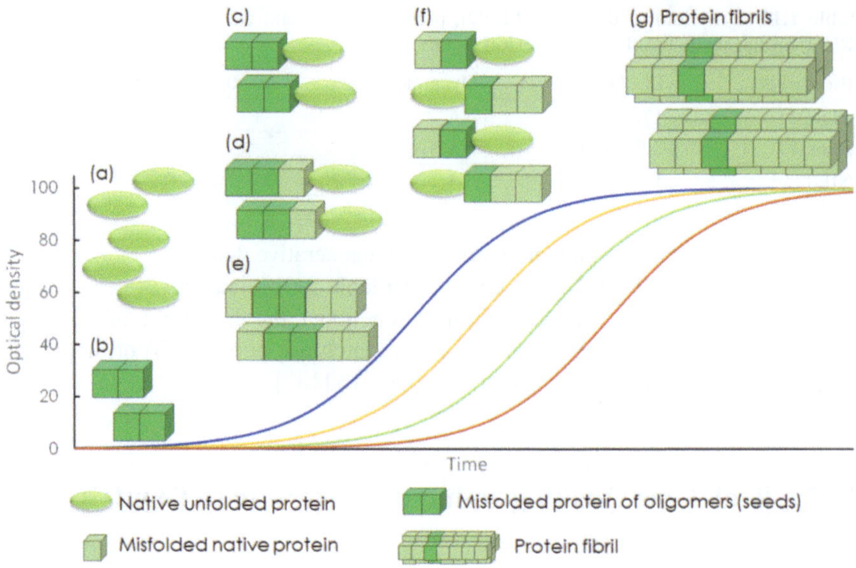

Fig. 1.21 Illustration of the seed amplification assay (SAA) or protein misfolding cyclic amplification (PMCA) or real-time quaking-induced conversion (RT-QuIC)

created to speed up the aggregation of native unfolded protein, as shown in Fig. 1.21(f). Sonication is used for seed fragmentation. Hereafter, the SAA is renamed as protein misfolding cyclic amplification (PMCA). These chains of folded proteins form fibrils until the aggregation reactions finish, which is referred to as the stationary phase, as shown in Fig. 1.21(g).

The first species of the misfolded protein seeds was brain homogenate [185]. In order to increase the production and lower the cost of misfolded protein seeds, the seeds of recombinant folded proteins were used instead of brain homogenate. The conversion rates of native unfolded protein molecules to aggregates significantly increase by using seeds of recombinant folded proteins [186]. Thus, the analytical sensitivity of PMCA is also improved as compared to SAA.

In both SAA and PMCA, the method to detect signals related to the aggregation of native unfolded proteins is Western blot. By replacing Western blot with fluorescent detections using thioflavin T (ThT), which specifically binds to fibrils, the real-time signals during the aggregation reactions can be monitored, as shown with the solid lines in Fig. 1.21. Hereafter, PMCA is renamed as real-time quaking-induced conversion (RT-QuIC) [187]. With the real-time signals, it is not necessary to detect the signals at the stationary phase. Instead, the threshold cycle (Ct) in the real-time signals is used as an index to reveal the concentration of native unfolded proteins. In principle, the higher the concentration, the lower the Ct.

- Features of the assay

  The most studied protein associated with neurodegenerative disease assayed with RT-QuIC is α-synuclein. Several groups have reported the discriminations

Table 1.17 Low limit of detection (LLoD), dynamic range, and utility for research use only (RUO) or diagnostic use (Dx) of LC/MS/MSassaykits

Biomarker	LLoD	Dynamic range	RUO	Dx
α-Synuclein	110 fg/ml	NA	O	

between healthy controls and patients with Parkinson's disease [188–192]. According to these reports, the operation of RT-QuIC is very tricky. The aggregation reactions in α-synuclein RT-QuIC depend on several factors, such as the concentration and quality of seeds, the ion concentration of samples, the pH value of the reaction buffer, the incubation time, etc. [193–195]. It is difficult to standardize the protocols of α-synuclein RT-QuIC. Most of α-synuclein RT-QuIC in the reported results were performed in-house.

- Plasma preparation
  In fact, RT-QuIC is frequently used for assaying proteins in cerebrospinal fluid (CSF), and seldom for blood biomarkers associated with neurodegenerative diseases. Hence, the protocol of plasma preparation for optimizing RT-QuIC performance is not available. In common, blood is collected in K_2EDTA tubes, which are centrifuged at 500–2300xg at 4 °C to obtain plasma. The plasma is aliquoted into 2.0-ml polypropylene vials, frozen on dry ice, and stored at −65 to −80 °C until measurement.
- Commercialization
  Amprion Inc. has been commercializing RT-QuIC for assaying CSF α-synuclein to assess Parkinson's disease (PD), dementia with Lewy bodies (DLB), Alzheimer's disease with Lewy body pathology (AD-LB), and multiple system atrophy (MSA). Amprion is using the assay to provide testing services for customers. Amprion Clinical Laboratory owns the certificates of Clinical Laboratory Improvement Amendments (CLIA) and College of American Pathologists (CAP).
- Assay kits for plasma biomarkers of neurodegenerative diseases
  The detailed information about the low limit of detection (LLoD), dynamic range, and utility for research use only (RUO) or diagnostic use (Dx) of the RT-QuIC kit for CSF α-synuclein is listed in Table 1.17.

1.18 Liquid Chromatography Tandem Mass Spectrometry (LC/MS/MS)

- Mechanism
  Mass spectrometry (MS) is designed to measure the mass-to-charge ratio (m/z) in spectra to determine the molecules present in a sample [196–199]. The fundamental components of MS include an ionization source, a mass analyzer, and an ion detection system. With the ion source, molecules are nebulized, desolvated, and ionized, creating charged particles. The charged particles are driven through a mass analyzer (magnetic quadrupole) by electric and magnetic fields under

high vacuum. The charged particles are separated according to the mass-to-charge (m/z) ratio, the so-called mass spectrum. Each peak in a mass spectrum shows a component with a unique m/z in the sample, and the heights of the peaks denote the relative abundance of the various components in the sample. With the aid of a data system with abundant patterns of mass spectra, the molecules in a sample can be determined and quantified. Detailed introductions to the designs and operations of MS are available in many review articles.

Tandem mass spectrometry (MS/MS) was designed to increase the selectivity or resolution of the mass spectrum, as shown in Fig. 1.22. In MS/MS, the charged particles, referred to as parent ions, pass through the first magnetic quadrupole to select specific m/z precursor charged particles. The selected charged particles are fragmented by collision with an inert gas in a collision cell. The second magnetic quadrupole is used to further separate fragmented charged particles, referred to as product ions. The separated product ions are quantified with an electron multiplier. The advanced selection of product ions from precursor charged particles in MS/MS is highly specific to the structure of the molecules of interest. Therefore, MS/MS has a high degree of selectivity.

Mass spectrometry is commonly used in molecular analyses of composition and for quantification. The composition of a sample can be analyzed according to the m/z spectra. For quantification, a stable isotope-labeled target analyte (SILTA) of known concentration is added to the sample. By comparing the area of peaks in the m/z spectrum of the analyte of interest in a sample to that of SILTA, the analyte of interest in a sample can be quantified. In reality, a sample solution contains various types of molecules. The molecules of interest must be roughly selected from the sample before MS/MS. Since biological samples are usually liquid, liquid chromatography (LC) is popularly used for the pretreatment of samples for MS/MS. This method is the so-called LC/MS/MS [200, 201].

The working principle of liquid chromatography (LC) is described in Fig. 1.23 [202, 203]. A liquid biological sample with various analytes is injected into an LC column, which is packed with both a solid stationary phase and a liquid mobile phase, as shown in Fig. 1.23(a). Different types of analytes have different affinities for the solid stationary phase, resulting in various flow rates in the LC column for

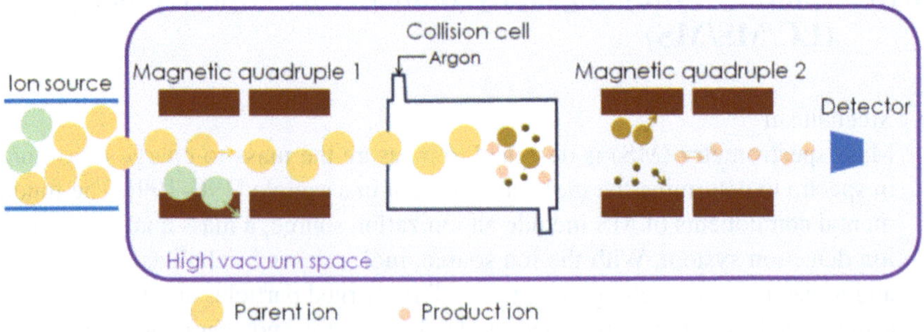

Fig. 1.22 Architecture of tandem mass spectrometry (MS/MS)

Fig. 1.23 Working principle of liquid chromatography (LC)

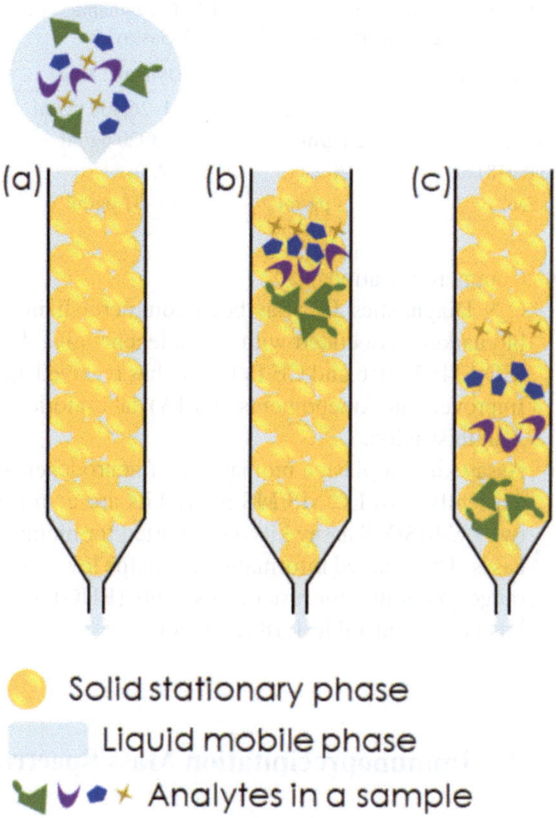

different analytes, as shown in Fig. 1.23(b). To this end, different analytes are separated and eluted from the LC column at different time points, as shown in Fig. 1.23(c). The affinity between the solid stationary phase and analytes can be manipulated based on the molecular size, analyte charge, hydrophobicity, or specific binding interactions. LC is able to select the analyte of interest and is helpful for achieving high specificity for MS/MS.

- Features of the assay
  LC/MS/MS is a two-step homogeneous assay. LC is used to isolate the target analyte in a sample. The featured mass-to-charge (m/z) spectrum is patterned to identify and quantify the specific analyte. A stable isotope-labeled target analyte is needed as a standard sample to quantify the analyte to be detected. LC might cause analyte loss and plays a role in analytical sensitivity. LC/MS/MS is a delicate and complicated system. A well-trained technician is needed.
- Plasma preparation
  Blood is collected in K_2EDTA tubes (Becton Dickinson, Franklin Lakes, NJ), placed on ice, and centrifuged at 500–2300xg at room temperature to obtain plasma. The plasma is aliquoted into 2.0-ml polypropylene vials, frozen on dry ice, and stored at −65 to −80 °C until measurement [204, 205].

Table 1.18 Low limit of detection (LLoD), dynamic range, and utility for research use only (RUO) or diagnostic use (Dx) of LC/MS/MSassaykits

Biomarker	LLoD	Dynamic range	RUO	Dx
$A\beta_{1-40}$	11 pg/ml	11–26,400 pg/ml	O	O
$A\beta_{1-42}$	2 pg/ml	2–3920 pg/ml	O	O
pTau181	NA	NA	O	
pTau217	1.3 pg/ml	0.1–81 pg/ml	O	

- Commercialization
  C_2N Diagnostics Inc. has been commercializing LC/MS/MS assays for blood biomarkers associated with neurodegenerative diseases, with the registration of ISO 13485:2016 and CE IVD, and has received approval for Clinical Laboratory Improvement Amendments (CLIA) laboratories from the US Food and Drug Administration.
- Assay kits for plasma biomarkers of neurodegenerative diseases
  Currently, two LC/MS/MS assay kits are commercially available, and another two LC/MS/MS assay kits are reported for biomarkers of neurodegenerative diseases. The detailed information about the low limit of detection (LLoD), dynamic range, and utility for research use only (RUO) or diagnostic use (Dx) of the assay kits is listed in Table 1.18 [205, 206].

1.19 Immunoprecipitation Mass Spectrometry (IP/MS)

- Mechanism

In addition to liquid chromatography, an alternative method to select samples by the analyte of interest is immunoprecipitation (IP). IP is a technique for enriching or isolating target analytes from biological samples. The enriched or isolated target analyte is quantitatively detected using mass spectrometry (MS) [207–210]. The flow chart of IP/MS is illustrated in Fig. 1.24. A stable isotope-labeled target antigen (SILTA) is added to the sample, as shown in Fig. 1.24(a). The SILTA is a reference analyte for MS.

A solid matrix such as magnetic beads with capture antibodies against the target antigen is mixed with the sample to perform IP [211, 212]. Both the SILTA and the target antigen bind to the magnetic beads, as shown in Fig. 1.24(b). Through the magnetic separation shown in Fig. 1.24(c), the SILTA and the target antigen bound to the magnetic beads are selected, as shown in Fig. 1.24(d). By adding a separating buffer to the sample, the SILTA and the target antigen dissociate from the magnetic beads, as shown in Fig. 1.24(e). Through another magnetic separation, the SILTA and the target antigen are eluted, as shown in Fig. 1.24(f). The eluted target antigen is quantitatively analyzed with MS.

The steps in Figs. 1.24(b)-(f) are IP. IP is good for the small-scale enrichment of analytes in biological samples. This method saves time and is user-friendly

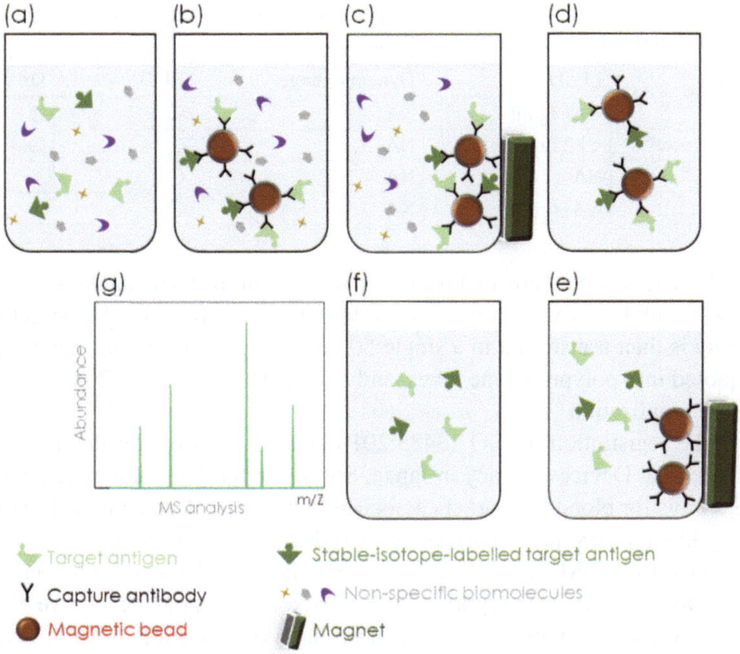

Fig. 1.24 Flow chart of immunoprecipitation mass spectrometry (IP/MS)

compared with affinity chromatography, such as liquid chromatography, which is often time-consuming and involves cycles of binding and washing. However, non-specific analytes can be isolated together with specific analytes via IP. Sometimes, a two-cycle IP procedure is performed to reduce nonspecific contamination [213, 214]. Furthermore, the molecular structures of nonspecific analytes differ from those of specific analytes. MS is a good approach for differentiating nonspecific analytes from specific analytes. Therefore, IP/MS is able to achieve high-sensitivity and high-specificity assays for target analytes. Recently, IP has been integrated with LC/MS, referred to as IP-LC/MS, to achieve high-speed, high-sensitivity, and high-specificity assays [204, 215].

- Features of the assay
  IP/MS is a two-step, homogeneous assay. Solid matrices, such as magnetic beads, and functionalized specific antibodies, are used to enrich or isolate target analytes via IP. With the aid of a stable isotope-labeled target analyte, the featured mass/charge (m/z) spectrum is obtained to identify and quantify the specific analyte. IP might cause analyte loss and play a role in analytical sensitivity. IP/MS is a delicate and complicated assay. A well-trained technician is needed.
- Plasma preparation
  Blood is drawn into two 10-ml syringes precoated with 0.5 M EDTA and then transferred to two 15-ml polypropylene tubes containing 120 μl of 0.5 M

Table 1.19 Low limit of detection (LLoD), dynamic range, and utility for research use only (RUO) or diagnostic use (Dx) of IP/MSassaykits

Biomarker	LLoD	Dynamic range	RUO	Dx
$A\beta_{1-40}$	< 41 fg/ml	NA	O	O
$A\beta_{1-42}$	< 82 fg/ml	NA	O	O
pTau181	NA	NA	O	
pTau217	NA	NA	O	

EDTA. The samples are incubated on wet ice until centrifugation (< 2 h) at 2500xg and 4 °C for 10 min to separate the plasma from the blood cells. The plasma is then transferred to a single 50-mL polypropylene tube, gently mixed, aliquoted into polypropylene tubes, and stored at −80 °C [204, 208].

- Commercialization
  With the registration of ISO 13485:2016 and approval by the Pharmaceuticals and Medical Devices Agency in Japan, Shimadzu has been commercializing IP/MS assays for blood biomarkers associated with neurodegenerative diseases.
- Assay kits for plasma biomarkers of neurodegenerative diseases
  Currently, two IP/MS assay kits are commercially available, and another two IP/MS assay kits have been reported for biomarkers of neurodegenerative diseases. The detailed information about the low limit of detection (LLoD), dynamic range, and utility for research use only (RUO) or diagnostic use (Dx) of the assay kits is listed in Table 1.19 [210, 215].

1.20 Key Notes of the Assays

All the details of each assay technology are described in the Assay Technology section. These 17 assay technologies were developed based on versatile working principles. However, these technologies are not absolutely different from each other. Some factors are shared among the technologies. Through inspections of the key notes of these technologies, we can clearly understand the consistencies and differences among technologies, which would be helpful for the readers to further understand these 17 assay technologies. The key notes inspected here include blood collection tubes, centrifugation/clotting temperature of blood, low limit of detections, and analytical specificity.

1.21 Blood Collection Tubes

Two types of blood collection tubes are used in the reported assay technologies: ethylenediaminetetraacetic acid (EDTA)- and heparin-coated tubes. Both EDTA and heparin are anticoagulants that prevent the blood from forming blood clots or becoming larger.

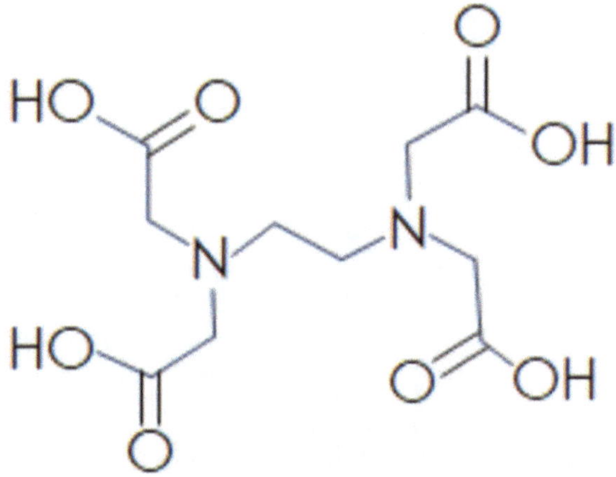

Fig. 1.25 Chemical structure of EDTA [216]

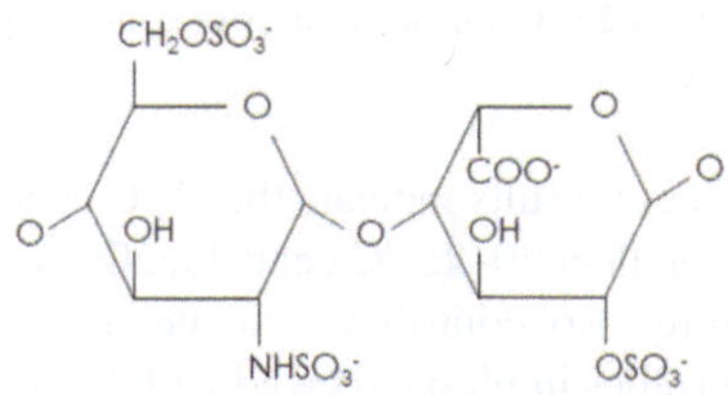

Fig. 1.26 Major repeating disaccharide unit in heparin [219]

The chemical structure of EDTA is shown in Fig. 1.25 [216]. EDTA can chelate Ca^{2+} in the blood, causing Ca^{2+} to lose its coagulation effect and prevent blood coagulation [217]. EDTA is typically used in the form of a sodium salt or potassium salt. The EDTA tubes are also referred to as lavender tubes.

Heparin is a complex, linear, acidic polysaccharide that is a member of the glycosaminoglycan (GAG) family [218]. The chemical structure of the major repeating disaccharide unit in heparin is shown in Fig. 1.26 [219]. Heparin increases the activity of antithrombin, preventing the formation of thrombin and the aggregation of platelets. Heparin sodium is the most common form used.

In plasma separation using reported assay technologies, EDTA tubes are recommended for CLEIA, ECL, ME-ECL, FT-IR, HISCL, IMR, NILISA, SPEAR, SIMOA, SMC, LC/MS/MS, and IP/MS. Heparin tubes are recommended for MDS. A blood collection tube has not been recommended for SPR and xMAP.

1.22 Centrifugation/Clotting Temperature of Blood

The preanalytical steps used to prepare blood samples play a role in ensuring an adequate sample quality. One of the critical factors is the centrifugation/clotting temperature of blood. The effects of the centrifugation temperature of the blood on coagulation were investigated via enzyme-linked immunosorbent assay (ELISA).

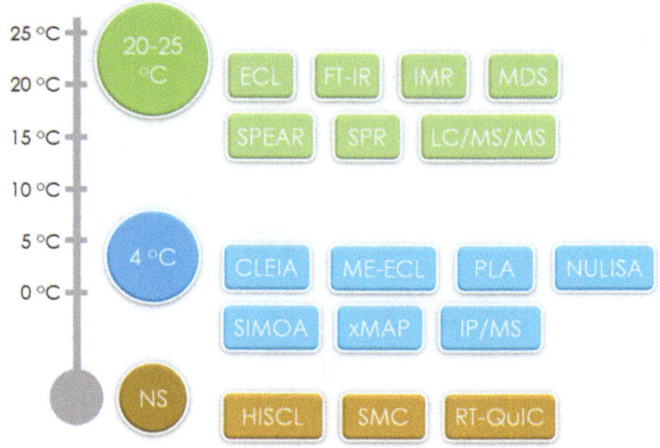

Fig. 1.27 Centrifugation or clotting temperatures used in various assay technologies (NS: not specified)

These results indicate that 4 °C centrifugation results in a longer period of coagulation than 20–25 °C centrifugation of blood [220, 221]. The centrifugation temperature also contributes to the measured concentrations of metabolites or protein profiles in blood assayed with LC/MS. [222, 223] Chang et al. utilized an immunomagnetic reduction (IMR) assay and showed that the measured concentrations of total α-synuclein in plasma are lower when the plasma is centrifuged at 4 °C than at room temperature [152]. Lee et al. reported that lower levels of $A\beta_{1-42}$ and T-tau are measured in plasma samples centrifuged at 4 °C than at room temperature, resulting in a high degree of overlap in biomarker levels between normal controls and patients with Alzheimer's disease [224]. These results reveal that suitable temperatures exist for the centrifugation/clotting of blood samples. Thus, the centrifugation/clotting temperature has to be clearly described for plasma preparation using various assay technologies. The centrifugation/clotting temperature could be 4 °C or room temperature (20–25 °C) for the 15 assay technologies, as shown in Fig. 1.27.

Users must prepare plasma/serum samples at the correct centrifugation/clotting temperatures for specific assay technologies. Otherwise, the integrity of plasma/serum could vary, resulting in distorted measurements of the concentrations of the analytes of interest.

1.23 Low Limit of Detection

The low limit of detection (LLoD) is an important specification of an assay, especially for assaying ultralow concentrations of biomarkers in bio-liquid samples. Most plasma biomarkers associated with neurodegenerative diseases are present at concentrations of several tens of pg/ml or lower. The LLoD should be 1–10 pg/ml

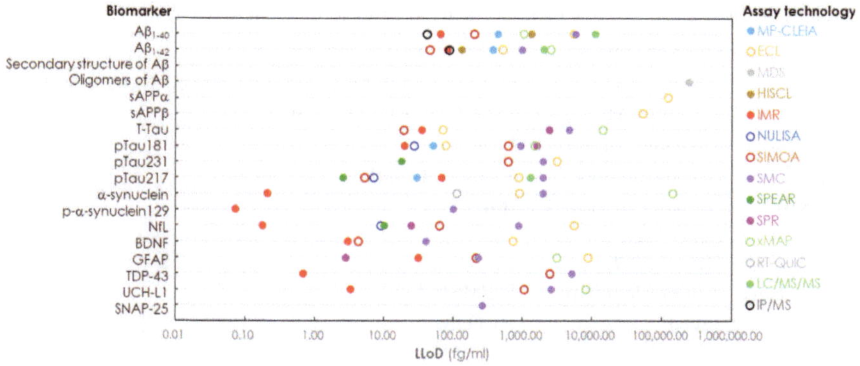

Fig. 1.28 Reported low limit of detections (LLoDs) of assays of biomarkers using various technologies

to achieve a precise assay. The LLoD values listed in the section describing assay kits for plasma biomarkers of neurodegeneration for each technology are summarized in Fig. 1.28. Most LLoD values are below 10,000 fg/ml, i.e., 10 pg/ml. Some of them are 1 pg/ml or lower. Obviously, these assay technologies are so sensitive that they are able to detect ultralow concentrations of biomarkers in plasma samples precisely. Notably, IMR (•) has LLoD values below 1 fg/ml for the kits that are available. The assay technology is supersensitive.

1.24 Analytical Specificity

For high-sensitivity assays, specificity is an important issue because of the limited value of false-positives. Mechanisms to achieve high specificity of the assay with these 15 technologies are introduced. Briefly, the mechanisms are categorized into three groups, two antibodies, a single antibody, and no antibody, as shown in Table 1.20.

The most common method to increase analytical specificity is the use of both a capture antibody and a secondary antibody (or labeling antibody), referred to as two antibodies in Table 1.19. Only the analyte that simultaneously has the two epitopes individually targeted by the capture and labeling antibodies is assayed. The technologies for two antibodies include CLEIA, HISCL, ECL, ME-ECL, MDS, SIMOA, SMC, xMAP, SPEAR, and NULISA.

Instead of two antibodies, some technologies utilize only one antibody, referred to as a single antibody in Table 1.20. These technologies include FT-IR, SPR, IP/MS, and IMR. In the group of single antibody, the specificity of the antibody to the analyte of interest is very demanding because only one antibody is used. Otherwise, nonspecific binding between the capture antibody and nontarget analytes frequently occurs, resulting in a high value of false-positives and low accuracy.

Table 1.20 Mechanisms for the high specificity of assay technologies

Mechanism for high specificity		Technology
Two antibodies	Capture antibody + Labeling antibody	CLEIA, HISCL, ECL, ME-ECL, MDS, SIMOA, SMC, xMAP, SPEAR, NULISA
	2 capture antibodies	PLA
Single antibody	Capture antibody	FT-IR, SPR, IP/MS
	Capture antibody + Spin-wash technology	IMR
No antibody	Chromatography	LC/MS/MS
	Seed	RT-QuIC

In addition to the use of high-specificity antibodies for capturing target analytes, IMR involves so-called spin-wash technology to achieve high-specificity assays [151]. In IMR, magnetic nanoparticles are immobilized with capture antibodies, which might bind to nonspecific analytes and target analytes. The binding force between the capture antibody and nonspecific analytes is weaker than that between the capture antibody and the target analyte. After binding, the magnetic nanoparticles oscillate with external alternating-current magnetic fields. Thus, the bound analytes experience a centrifugal force, which is proportional to the square of the oscillating frequency of the nanoparticles. The oscillating frequency of nanoparticles is equal to that of external alternating-current magnetic fields. By effectively manipulating the frequency of magnetic fields, the centrifugal force acting on the bound analytes can be adjusted. At suitable frequencies, the centrifugal force can be stronger than the binding force between the capture antibody and nonspecific analytes but weaker than the binding force between the capture antibody and target analyte. Consequently, nonspecific binding is eliminated, and specific binding still occurs. Therefore, IMR shows high specificity in assaying analytes of interest with the aid of high-specificity capture antibodies and spin-wash technology.

LC/MS/MS is the only technology that does not use antibodies to achieve high specificity of the assay. The target analyte is selected with liquid chromatography. The selection of target analytes can be executed by controlling the molecular size, analyte charge, hydrophobicity, or specific binding interactions. LC is very popular in analytical chemistry and biomolecular assays.

References

1. Hardy, J., & Allsop, D. (1991). Amyloid deposition as the central event in the aetiology of Alzheimer's disease. *Trends in Pharmacological Sciences, 12*, 383–388.
2. Iqbal, K., Alonso, A. D. C., Chen, S., Chohan, M. O., El-Akkad, E., Gong, C. X., Khatoon, S., Li, B., Liu, F., Rahman, A., Tanimukai, H., & Grundke-Iqbal, I. (2005). Tau pathology in Alzheimer disease and other tauopathies. *Biochimica et Biophysica Acta (BBA) – Mol Basis Dise, 1739*, 198–210.

3. Wakabayashi, K., Tanji, K., Mori, F., & Takahashi, H. (2007). The lewy body in Parkinson's disease: Molecules implicated in the formation and degradation of α-synuclein aggregates. *Neuropathology, 27*, 494–506.
4. Arseni, D., Hasegawa, M., Murzin, A. G., Kametani, F., Arai, M., Yoshida, M., & Ryskeldi-Falcon, B. (2022). Structure of pathological TDP-43 filaments from ALS with FTLD. *Nature, 601*, 139–143.
5. Chu, D., & Liu, F. (2019). Pathological changes of tau related to Alzheimer's disease. *ACS Chemical Neuroscience, 10*, 931–944.
6. Wu, M., Zhang, M., Yin, Y., Chen, K., Hu, Z., Zhou, Q., Cao, X., Chen, Z., & Liu, D. (2021). The role of pathological tau in synaptic dysfunction in Alzheimer's diseases. *Translational Neurodegeneration, 10*, 45-1-11.
7. Zhang, H., Wei, W., Zhao, M., Ma, L., Jiang, X., Pei, H., Cao, Y., & Li, H. (2021). Interaction between aβ and tau in the pathogenesis of Alzheimer's disease. *International Journal of Biological Sciences, 17*, 2181–2192.
8. Luk, K. C., Kehm, V., Carroll, J., Zhang, B., O'Brien, P., Trojanowski, J. Q., & Lee, V. M. Y. (2012). Pathological α-synuclein transmission initiates Parkinson-like neurodegeneration in nontransgenic mice. *Science, 338*, 949–953.
9. Calabresi, P., Mechelli, A., Natale, G., Volpicelli-Daley, L., Lazzaro, G. D., & Ghiglieri, V. (2023). Alpha-synuclein in Parkinson's disease and other synucleinopathies: From overt neurodegeneration back to early synaptic dysfunction. *Cell Death and Dis, 14*, 176-1-16.
10. Cairns, N. J., Neumann, M., Bigio, E. H., Holm, I. E., Troost, D., Hatanpaa, K. J., Foong, C., White, C. L., 3rd., Schneider, J. A., Kretzschmar, H. A., Carter, D., Taylor-Reinwald, L., Paulsmeyer, K., Strider, J., Gitcho, M., Goate, A. M., Morris, J. C., Mishra, M., Kwong, L. K., Stieber, A., Xu, Y., Forman, M. S., Trojanowski, J. Q., Lee, V. M., & Mackenzie, I. R. (2007). TDP-43 in familial and sporadic frontotemporal lobar degeneration with ubiquitin inclusions. *The American Journal of Pathology, 171*, 227–240.
11. Jo, M., Lee, S., Jeon, Y. M., Kim, S., Kwon, Y., & Kim, H. J. (2020). The role of TDP-43 propagation in neurodegenerative diseases: Integrating insights from clinical and experimental studies. *Exper Mollar Med, 52*, 1652–1662.
12. Gaetani, L., Blennow, K., Calabresi, P., Di Filippo, M., Parnetti, L., & Zetterberg, H. (2019). Neurofilament light chain as a biomarker in neurological disorders. *Journal of Neurology, Neurosurgery, and Psychiatry, 90*, 870–881.
13. Arslan, B., & Zetterberg, H. (2023). Neurofilament light chain as neuronal injury marker—What is needed to facilitate implementation in clinical laboratory practice? *Clinical Chemistry and Laboratory Medicine, 61*, 1140–1149.
14. Karve, I. P., Taylor, J. M., & Crack, P. J. (2016). The contribution of astrocytes and microglia to traumatic brain injury. *British Journal of Pharmacology, 173*, 692–702.
15. Kim, K. Y., Shin, K. Y., & Chang, K. A. GFAP as a potential biomarker for Alzheimer's disease: A systematic review and meta-analysis. *Cells, 12*, 1309-1-13.
16. Jack, C. R., Jr., Bennett, D. A., Blennow, K., Carrillo, M. C., Dunn, B., Haeberlein, S. B., Holtzman, D. M., Jagust, W., Jessen, F., Karlawish, J., Liu, E., Molinuevo, J. L., Montine, T., Phelps, C., Rankin, K. P., Rowe, C. C., Scheltens, P., Siemers, E., & Snyder, H. M. (2018). NIA-AA research framework: Toward a biological definition of Alzheimer's disease. *Alzheimers Dement, 14*, 535–562.
17. Hampel, H., & Blennow, K. (2004). CSF tau and β-amyloid as biomarkers for mild cognitive impairment. *Dialogues in Clinical Neuroscience, 6*, 379–390.
18. Kasai, T., Tokuda, T., Ishigami, N., Sasayama, H., Foulds, P., Mitchell, D. J., Mann, D. M., Allsop, D., & Nakagawa, M. (2009). Increased TDP-43 protein in cerebrospinal fluid of patients with amyotrophic lateral sclerosis. *Acta Neuropathologica, 117*, 55–62.
19. Hansson, O., Lehmann, S., Otto, M., Zetterberg, H., & Lewczuk, P. (2019). Advantages and disadvantages of the use of the CSF amyloid β (aβ) 42/40 ratio in the diagnosis of Alzheimer's disease. *Alzheimer's Research & Therapy, 11*, 34-1-15.

20. Agnello, L., Sasso, B. L., Vidali, M., Scazzone, C., Gambino, C. M., Piccoli, T., Bivona, G., Ciaccio, A. M., Giglio, R. V., Bella, V. L., & Ciaccio, M. (2022). Evaluation of alpha-synuclein cerebrospinal fluid levels in several neurological disorders. *Journal of Clinical Medicine, 11*, 3139-1-7.
21. Evans, R. W. (1998). Complications of lumbar puncture. *Neurologic Clinics, 16*, 83–105.
22. Kim, K. T. (2022). Lumbar puncture: Considerations, procedure, and complications. *L'Encéphale, 2*, 93–97.
23. Yang, S. Y., Chiu, M. J., Chen, T. F., & Horng, H. E. (2017). Detection of plasma biomarkers using immunomagnetic reduction: A promising method for the early diagnosis of Alzheimer's disease. *Neurology and Therapy, 6*, 37–56.
24. Lin, C. H., Yang, S. Y., Horng, H. E., yang, C. C., Chieh, J. J., Chen, H. H., Liu, B. H., & Chiu, M. J. (2017). Plasma α-synuclein predicts cognitive decline in Parkinson's disease. *Journal of Neurology, Neurosurgery & Psychiatry, 88*, 818–824.
25. De Meyer, S., Schaeverbeke, J. M., Verberk, I. M. W., Gille, B., De Schaepdryver, M., Luckett, E. S., Gabel, S., Bruffaerts, S., Mauroo, K., Thijssen, E. H., Stoops, E., Vanderstichele, H. M., Teunissen, C. E., Vandenberghe, R., & Poesen, K. (2020). Comparison of ELISA-and SIMOA-based quantification of plasma aβ ratios for early detection of cerebral amyloidosis. *Alzheimer's Research & Therapy, 12*, 162-1-16.
26. Brand, A. L., Lawler, P. E., Bollinger, J. G., Li, Y., Schindler, S. E., Li, M. E., Lopez, S., Ovod, V., Nakamura, A., Shaw, L. M., Zetterberg, H., Hansson, O., & Bateman, R. J. (2022). The performance of plasma amyloid beta measurements in identifying amyloid plaques in Alzheimer's disease: A literature review. *Alzheimer's Research & Therapy, 14*, 195-1-15.
27. Chiu, P. Y., Yang, F. C., Chiu, M. J., Lin, W. C., Lu, C. H., & Yang, S. Y. (2022). Relevance of plasma biomarkers to pathologies in Alzheimer's disease, Parkinson's disease and frontotemporal dementia. *Scientific Reports, 12*, 17919-1-9.
28. Jack, C. R., Jr., Andrews, J. S., Beach, T. G., Buracchio, T., Dunn, B., Graf, A., Hansson, O., Ho, C., Jagust, W., McDade, E., Molinuevo, J. L., Okonkwo, O. C., Pani, L., Rafii, M. S., Scheltens, P., Siemers, E., Snyder, H. M., Sperling, R., Teunissen, C. E., & Carrillo, M. C. (2024). Revised criteria for diagnosis and staging of Alzheimer's disease: Alzheimer's association workgroup. *Alzheimers Dement, 20*, 5143–5169.
29. Baird, A. L., Westwood, S., & Lovestone, S. (2015). Blood-based proteomic biomarkers of Alzheimer's disease pathology. *Frontiers in Neurology, 6*, 236-1-16.
30. Prete, E. D., Beatino, M. F., Campese, N., Giampietri, L., Siciliano, G., Ceravolo, R., & Baldacci, F. (2020). Fluid candidate biomarkers for Alzheimer's disease: A precision medicine approach. *Journal of Personalized Medicine, 10*, 221-1-34.
31. Chen, Y. R., Liang, C. S., Chu, H., Voss, J., Kang, X. L., O'Connell, G., Jen, H. J., Liu, D., Hsiao, S. T. S., & Chou, K. R. (2021). Diagnostic accuracy of blood biomarkers for Alzheimer's disease and amnestic mild cognitive impairment: A meta-analysis. *Ageing Research Reviews, 71*, 101446-1-9.
32. Zubelzu, M., Morera-Herreras, T., Irastorza, G., Gomez-Esteban, J. C., & Murueta-Goyena, A. (2022). Plasma and serum alpha-synuclein as a biomarker in Parkinson's disease: A meta-analysis. *Parkinsonism Related Disorers, 99*, 107–115.
33. Tsao, H. H., Huang, C. G., & Wu, Y. R. (2022). Detection and assessment of alpha-synuclein in Parkinson disease. *Neurochemistry International, 158*, 105358-1-9.
34. Leuzy, A., Mattsson-Carlgren, N., Palmqvist, S., Janelidze, S., Dage, J. L., & Hansson, O. (2022). Blood-based biomarkers for Alzheimer's disease. *EMBO Molecular Medicine, 14*, e14408-1-15.
35. Cheng, L., Li, W., Chen, Y., Lin, Y., Wang, B., Guo, Q., & Miao, Y. (2022). Plasma aβ as a biomarker for predicting aβ-PET status in Alzheimer's disease:A systematic review with meta-analysis. *Journal of Neurology, Neurosurgery, and Psychiatry, 93*, 513–520.
36. Chen, L., Niu, X., Wang, Y., Lv, S., Zhou, X., Yang, Z., & Peng, D. (2022). Plasma tau proteins for the diagnosis of mild cognitive impairment and Alzheimer's disease: A systematic review and meta-analysis. *Frontiers in Aging Neuroscience, 14*, 942629-1-10.

37. Paisa, M. V., Forlenza, O. V., & Diniz, B. S. (2023). Plasma biomarkers of Alzheimer's disease: A review of available assays, recent developments, and implications for clinical practice. *Journal of Alzheimer's Disease Reports, 7*, 355–380.
38. Zabala-Findlay, A., Penny, L. K., Lofthouse, R. A., Porter, A. J., Palliyil, S., Harrington, C. R., Wischik, C. M., & Arastoo, M. Utility of blood-based tau biomarkers for mild cognitive impairment and Alzheimer's disease: Systematic review and meta-analysis. *Cells, 12*, 1184-1-16.
39. Richardson, B., Gunputh, U., & Wood, P. Could a non-Invasive biosensor be used to diagnose Alzheimer's disease? *American Journal of Biomedical Science and Research, 21*, 541–542.
40. Zhang, Y., Bi, K., Zhou, L., Wang, J., Huang, L., Sun, Y., Peng, G., & Wu, W. (2024). Advances in blood biomarkers for Alzheimer's disease: Ultra-sensitive detection technologies and impact on clinical diagnosis. *Degenerative Neurological Neuromuscular Disease, 14*, 85–102.
41. Halmann, M., Velan, B., & Sery, T. (1977). Rapid identification and quantitation of small numbers of microorganisms by a chemiluminescent immunoreaction. *Applied and Environmental Micobiology, 34*, 473–477.
42. Thorpe, G. H., Kricka, L. J., Moseley, S. B., & Witehead, T. P. (1985). Phenols as enhancers of the chemiluminescent horseradish peroxidase-luminol-hydrogen peroxide reaction: Application in luminescence-monitored enzyme immunoassays. *Clinical Chemistry, 31*, 1335–1441.
43. https://www.lsbio.com/products/elisakits/clia
44. Zhu, H., Jia, Z., Trush, M. A., & Li, Y. R. (2016). A highly sensitive chemiluminometric assay for real-time detection of biological hydrogen peroxide formation. *React Oxyg Species (Apex), 1*, 216–227.
45. Yang, X. L., Wang, S. W., Jiang, Z. H., Bi, J. J., & Ma, L. R. (1995). The development of enhanced chemiluminescence enzyme immunoassay and applications in biology and medicine. *Basic Medical Sciences Clinics, 15*, 241–246.
46. Hai, Z., Li, J., Wu, J., Xu, J., & Liang, G. (2017). Alkaline phosphatase-triggered simultaneous hydrogelation and chemiluminescence. *Journal of the American Chemical Society, 139*, 1041–1044.
47. Gnaim, S., Green, O., & Shabat, D. (2018). The emergence of aqueous chemiluminescence: New promising class of phenoxy 1,2-dioxetane luminophores. *Chemical Communications, 54*, 2073–2085.
48. Fu, X., Liu, Y., Qiu, R., Fo, M. F., Zhang, Y., Wang, T., & Li, J. (2018). The fabrication of magnetic particle-based chemiluminescence immunoassay for human epididymis protein-4 detection in ovarian cancer. *Biochemistry and Biophysics Reports, 13*, 73–77.
49. Sakharov, I. Y., Demiyanova, A. S., Gribas, A. V., Uskova, N. A., Efremov, E. E., & Vdovenko, M. M. (2013). 3-(10'-Phenothiazinyl) propionic acid is a potent primary enhancer of peroxidase-induced chemiluminescence and its application in sensitive ELISA of methylglyoxal-modified low density lipoprotein. *Talanta, 115*, 414–417.
50. Vidyasagar, G. V. C., Reddy, P. V. J., Suravajhala, P., Suravajhala, R., Kiran V, U., Kishor PB, K., Venkateswarulu, T. C., & Polavarapu, R. (2023). Performance evaluation of in-house developed covid-19 IgG/IgM antibody rapid diagnostic kit. *AMB Express, 13*, 116-1-8.
51. Easton, P. M., Simmonds, A. C., Rakishev, A., Egorov, A. M., & Candeias, L. P. (1996). Quantitative model of the enhancement of peroxidase-induced luminol luminescence. *Journal of the American Chemical Society, 118*, 6619–6624.
52. Lind, J., Merenyi, G., & Eriksen, T. E. (1983). Chemiluminescence mechanism of cyclic hydrazides such as luminol in aqueous solutions. *Journal of the American Chemical Society, 105*, 7655–7661.
53. Vlasenko, S. B., Arefyev, A. A., Klimov, A. D., Kim, B. B., Gorovits, E. L., Osipov, A. P., Gavrilova, A. M., & Yegorov, A. M. (1989). An investigation on the catalytic mechanism of enhanced chemiluminescence: Immunochemical applications of this reaction. *Journal of Bioluminescence and Chemiluminescence, 4*, 164–176.

54. Dotsikas, Y., & Loukas, Y. L. (2004). Employment of 4-(1-imidazolyl)phenol as a luminol signal enhancer in a competitive-type chemiluminescence immunoassay and its comparison with the conventional antigen–horseradish peroxidase conjugate-based assay. *Analytica Chimica Acta, 509*, 103–109.
55. Bao, D. Q., Liu, Y. B., Guan, G. Y., Wang, Y. X., Han, S. Q., & Luo, Z. F. (2004). Chemiluminescent characteristic of a new type of enhanced chemiluminescent substrate solution and its application in immunoassay. *Atom Energy Science and Technology, 38*, 125–130.
56. Yang, S. J., Chen, G. R., Zhang, W., Zhang, P. Y., Sun, C. W., Chen, D., & Cheng, J. (1994). Experimental study of enhancement effect of p-iodophenol on luminol chemiluminescence. *Chinese Journal of Luminescence, 15*, 249–252.
57. Arranz, J., Zhu, N., Rubio-Guerra, S., Rodriguez-Baz, I., Ferrer, R., Carmona-Iragui, M., Barroeta, I., Illan-Gala, I., Santos-Santos, M., Fortea, J., Lleo, A., Tondo, M., & Alcolea, D. (2024). Diagnostic performance of plasma pTau217, pTau181, Aβ1-42 and Aβ1-40 in the LUMIPULSE automated platform for the detection of Alzheimer disease. *Alzheimer's Research & Therapy, 16*, 139-1-15.
58. https://www.sysmex.co.jp/en/rd/technologies/protein.html
59. Hu, L., Hu, X., Huang, T., Wang, M., & Xu, G. (2021). Chemiluminescence of the Ce(IV)/CDP-Star system based on the phosphatase-like activity of Ce(IV) ions. *ACS Omega., 6*, 6379–6384.
60. https://michigandiagnostics.squarespace.com/use-of-horseradish- peroxidase-and-alkaline-phosphatase-in-chemiluminescence-the-pros-and-cons.
61. Goto, M., Chamoto, K., Higuchi, K., Yamashita, S., Noda, K., Iino, T., Miura, M., Yamasaki, T., Ogawa, O., Sonobe, M., Date, H., Hamanishi, J., Mandai, M., Tanaka, Y., Chikuma, S., Hatae, R., Muto, M., Minamiguchi, S., Minato, N., & Honjo, T. (2019). Analytical performance of a new automated chemiluminescent magnetic immunoassays for soluble PD-1, PD-L1, and CTLA-4 in human plasma. *Scientific Reports, 9*, 10144-1-9.
62. Yamashita, K., Watanabe, S., Ishiki, K., Miura, M., Irino, Y., Kubo, T., Matsui, J., Hagino, K., Iwanaga, S., & Yoshida, T. (2021). Fully automated chemiluminescence enzyme immunoassays showing high correlation with immunoprecipitation mass spectrometry assays for β-amyloid (1–40) and (1–42) in plasma samples. *Biochemical and Biophysical Research Communications., 576*, 22–26.
63. Hercules, D. M., & Lytle, F. E. (1966). Chemiluminescence from reduction reactions. *Journal of the American Chemical Society, 88*, 4745–4746.
64. Richter, M. M. (2004). Electrochemiluminescence (ECL). *Chemical Reviews, 104*, 3003–3036.
65. Yoo, S. M., Jeon, Y. M., & Heo, S. Y. (2022). Electrochemiluminescence systems for the detection of biomarkers: Strategical and technological advances. *Biosensors-Basel, 12*, 738-1-25.
66. Zhao, W., Chen, H. Y., & Xu, J. J. (2021). Electrogenerated chemiluminescence detection of single entities. *Chemical Science, 12*, 5720–5736.
67. Li, L., Chen, Y., & Zhu, J. J. (2017). Recent advances in electrochemiluminescence analysis. *Analytical Chemistry, 89*, 358–371.
68. Ma, C., Cao, Y., Gou, X., & Zhu, J. J. (2020). Luminol electrochemiluminescence for the analysis of active cholesterol at the plasma membrane in single mammalian cells. *Analytical Chemistry, 92*, 431–454.
69. Valenti, G., Scarabino, S., Goudeau, B., Lesch, A., Jovic, M., Villani, E., Sentic, M., Rapino, S., Arbault, S., Paolucci, F., & Sojic, N. (2017). Single cell electrochemiluminescence imaging: From the proof-of-concept to disposable device-based analysis. *Journal of the American Chemical Society, 139*, 16830–16837.
70. Wu, K., Zheng, Y., Zhang, Y., Chen, R., Zhou, Z., Liu, S., & Shen, Y. (2023). Advances in electrochemiluminescence luminophores based on small organic molecules for biosensing. *Biosensors and Bioelectronics, 223*, 115031-1-18.

71. Wei, X., Zhu, M. J., Yan, H., Lu, C., & Xu, J. J. (2019). Recent advances in aggregation-induced electrochemiluminescence. *Chemistry – A European Journal, 25*, 12671–12683.
72. Haghighatbin, M. A., Laird, S. E., & Hogan, C. F. (2018). Electrochemiluminescence of cyclometalated iridium (III) complexes. *Current Opinion in Electrochemistry, 7*, 216–223.
73. Chen, Y., Zhou, S., Li, L., & Zhu, J. (2017). Nanomaterials-based sensitive electrochemiluminescence biosensing. *Nano Today, 12*, 98–115.
74. O'Connor, S., Dennany, L., & O'Reilly, E. (2023). Evolution of nanomaterial electrochemiluminescence luminophores towards biocompatible materials. *Bioelectrochemistry, 149*, 108286.
75. Chu, K., Ding, Z., & Zysman-Colman, E. (2023). Materials for electrochemiluminescence: TADF, hydrogen bonding, and aggregation- and crystallization-induced emission luminophores. *Chemistry – A European Journal, 29*, e202301504-1-13.
76. http://www.adni-info.org/Scientists/ProceduresManuals.aspx
77. Premnath, S. M., & Zubair, M. (2024). *Electrochemiluminescence method.* StatPearls Publishing.
78. https://www.mesoscale.com/~/media/files/product%20inserts/ human%206e10%20abeta%20us.pdf
79. https://www.mesoscale.com/en/applications/neurodegenerative
80. Erler, K. (1998). Elecsys immunoassay systems using electrochemiluminescence detection. *Wiener Klinische Wochenschrift, 110*(Suppl 3), 5–10.
81. Namba, Y., Usami, M., & Suzuki, O. (1999). Highly sensitive electrochemiluminescence immunoassay using ruthenium chelate-labeled antibody bound on the magnetic micro beads. *Analytical Sciences, 15*, 1087–1093.
82. Ding, C., Wei, S., & Liu, H. (2012). Electrochemiluminescent determination of cancer cells based on aptamers, nanoparticles, and magnetic beads. *Chem – A European Journal, 18*, 7263–7268.
83. Zhang, Y., Ge, S., Wang, S., Yan, M., Yu, J., Song, X., & Liu, W. (2012). Magnetic beads-based electrochemiluminescence immunosensor for determination of cancer markers using quantum dot functionalized PtRu alloys as labels. *Analyst, 137*, 2176–2182.
84. Pregibon, D. C., Toner, M., & Doyle, P. S. (2006). Magnetically and biologically active bead-patterned hydrogels. *Langmuir, 22*, 5122–5128.
85. Philippova, O. E., Barabanova, A. I., Molchanov, V., & Khokhlov, A. (2011). Magnetic polymer beads: Recent trends and developments in synthetic design and applications. *European Polymer Journal, 47*, 542–559.
86. Yang, S. Y., Jian, Z. F., Horng, H. E., Hong, C. Y., Yang, H. C., Wu, C. C., & Lee, Y. H. (2008). Dual immobilization and magnetic manipulation of magnetic nanoparticles. *Journal of Magnetism and Magnetic Materials, 320*, 2688–2691.
87. Hong, C. Y., Wu, C. C., Chiu, Y. C., Yang, S. Y., Horng, H. E., & Yang, H. C. (2006). Magnetic susceptibility reduction method for magnetically labeled immunoassay. *Applied Physics Letters, 88*, 212512-1-3.
88. Wei, Y., Zhang, J., Yang, X., Wang, Z., Wang, J., Qi, H., & Zhang, C. (2023). Performance enhancement of electrochemiluminescence magnetic microbiosensors by using double magnetic field actuation for cancer biomarkers and exosomes. *Talanta, 259*, 124485.
89. Fracassa, A., Santo, C. I., Kerr, E., Knežević, S., Hayne, D. J., Francis, P. S., Kanoufi, F., Sojic, N., Paolucci, F., & Valenti, G. (2024). Redox-mediated electrochemiluminescence enhancement for bead-based immunoassay. *Chemical Science, 15*, 1150-1-9.
90. Palmqvist, S., Janelidze, S., Stomrud, E., Zetterberg, H., Karl, J., Zink, K., Bittner, T., Mattsson, N., Eichenlaub, U., Blennow, K., & Hansson, O. (2019). Performance of fully automated plasma assays as screening tests for Alzheimer disease-related β-amyloid status. *JAMA Neurology, 76*, 1060–1069.
91. Jacquinot, P. (1960). New developments in interference spectroscopy. *Reports on Progress in Physics., 23*, 267–312.

92. Horlick, G. (1968). Introduction to fourier transform spectroscopy. *Applied Spectroscopy., 22*, 617–626.
93. Hammond, J. P. (2014). The use of spectrophotometry in the pharmaceutical industry. *Experimental Methods in the Physical Sciences, 46*, 409–456.
94. Oheim, M., Salomon, A., Weissman, A., Brunstein, M., & Becherer, U. (2019). Calibrating evanescent-wave penetration depths for biological TIRF microscopy. *Biophysical Journal, 117*, 795–809.
95. Nabers, A., Ollesch, J., Schartner, J., Kötting, C., Genius, J., Hafermann, H., Klafki, H., Gerwert, K., & Wiltfang, J. (2016). Amyloid-β-secondary structure distribution in cerebrospinal fluid and blood measured by an immuno-infrared-sensor: A biomarker candidate for Alzheimer's disease. *Analytical Chemistry, 88*, 2755–2762.
96. Stockmann, J., Verberk, I. M. W., Timmesfeld, N., Denz, R., Budde, B., Lange-Leifhelm, J., Scheltens, P., van der Flier, W. M., Nabers, A., Teunissen, C. E., & Gerwert, K. (2020). Amyloid-β misfolding as a plasma biomarker indicates risk for future clinical Alzheimer's disease in individuals with subjective cognitive decline. Alzheimers res. *Therapy, 12*, 169-1-13.
97. Kaur, H., Rana, B., Tomar, D., Kaur, S., & Jena, K. C. (2021). Fundamentals of ATR-FTIR spectroscopy and its role for probing in-situ molecular-level interactions. In D. K. Singh, M. Pradhan, & A. Materny (Eds.), *Modern Techniques of Spectroscopy* (Progress in optical Science and photonics) (Vol. 13). Springer.
98. Lewczuk, P., Kornhuber, J., & Wiltfang, J. (2006). The German competence net dementias: Standard operating procedures for the neurochemical dementia diagnostics. *Journal of Neural Transmission, 113*, 1075–1080.
99. https://specac.com/ftir-applications/amyloids-and-alzheimers-detected-with-ftir/
100. An, S. S. A., Lim, K. T., Oh, H. J., Lee, B. S., Zukic, E., Ju, Y. R., Yokoyama, T., Kim, S. Y., & Welker, E. (2010). Differentiating blood samples from scrapie infected and non-infected hamsters by detecting disease-associated prion proteins using multimer detection system. *Biochemical and Biophysical Research Communications, 392*, 505–509.
101. An, S. S. A., Lee, B. S., Yu, J. S., Lim, K., Kim, G. J., Lee, R., Kim, S., Kang, S., Park, Y. H., Wang, M. J., Yang, Y. S., Youn, Y. C., & Kim, S. Y. (2017). Dynamic changes of oligomeric amyloid β levels in plasma induced by spiked synthetic Aβ42. *Alzheimer's Research & Therapy, 9*, 86-1-10.
102. Youn, Y. C., Lee, B. S., Kim, G. J., Ryu, J. S., Lim, K., Lee, R., Suh, J., Park, Y. H., Pyun, J. M., Ryu, N., Kang, M. J., Kim, H. R., Kang, S., An, S. S. A., & Kim, S. (2020). Blood amyloid-β oligomerization as a biomarker of Alzheimer's disease: A blinded validation study. *Journal of Alzheimer's Disease, 75*, 493–499.
103. Lee, H. Y., Ugay, D., Hong, S., & Kim, Y. S. (2020). Alzheimer's disease diagnosis using misfolding proteins in blood. *Dementia and Neurocognitive Disorders, 19*, 1–18.
104. Pyun, J. M., Youn, Y. C., Park, Y. H., & Kim, S. (2023). Integration of amyloid-β oligomerization tendency as a plasma biomarker in Alzheimer's disease diagnosis. *Frontiers in Neurology, 13*, 1028448-1-5.
105. Verberk, I. M. W., Hendriksen, H. M. A., van Harten, A. C., Wesselman, L. M. P., Verfaillie, S. C. J., van den Bosch, K. A., Slot, R. E. R., Prins, N. D., Scheltens, P., Teunissen, C. E., & Van der Flier, W. M. (2020). Plasma amyloid is associated with the rate of cognitive decline in cognitively normal elderly: The SCIENCe project. *Neurobiology of Aging, 89*, 99–107.
106. Chiu, M. J., Yang, S. Y., Chen, T. F., Chieh, J. J., Huang, T. Z., Yip, P. K., Yang, H. C., Cheng, T. W., Chen, Y. F., Hua, M. S., & Hong, H. E. (2012). New assay for old markers-plasma beta amyloid of mild cognitive impairment and Alzheimer's disease. *Current Alzheimer Research, 10*, 1142–1148.
107. Salvadores, N., Shahnawaz, M., Scarpini, E., Tagliavini, F., & Soto, C. (2014). Detection of misfolded aβ oligomers for sensitive biochemical diagnosis of Alzheimer's disease. *Cell Reports, 7*, 261–268.

108. Sehlin, D., Söllvander, S., Paulie, S., Brundin, R., Ingelsson, M., Lannfelt, L., Pettersson, F. E., & Englund, H. (2010). Interference from heterophilic antibodies in amyloid-β oligomer ELISAs. *Journal of Alzheimer's Disease, 21*, 1295–1301.
109. Kuo, Y. M., Kokjohn, T. A., Kalback, W., Luehrs, D., Galasko, D. R., Chevallier, N., Koo, E. H., Emmerling, M. R., & Roher, A. E. (2000). Amyloid-β peptides interact with plasma proteins and erythrocytes: Implications for their quantitation in plasma. *Biochemical and Biophysical Research Communications, 268*, 750–756.
110. Lim, K., Kim, S. Y., Lee, B., Segarra, C., Kang, S., Ju, Y., Schmerr, M. J., Coste, J., Kim, S. Y., Yokoyama, K., & An, S. S. A. (2015). Magnetic microparticle-based multimer detection system for the detection of prion oligomers in sheep. *International Journal of Nanomedicine, 10*, 241–250.
111. Wang, M. J., Yi, S. H., Han, J. Y., Park, S. Y., Jang, J. W., Chun, I. K., Kim, S. E., Lee, B. S., Kim, G. J., Yu, J. S., Lim, K., Kang, S. M., Park, Y. H., Youn, Y. C., An, S. S. A., & Kim, S. Y. (2017). Oligomeric forms of amyloid-β protein in plasma as a potential blood-based biomarker for Alzheimer's disease. Alzheimers res. *Therapy, 9*, 98-1-10.
112. Meng, X., Li, T., Wang, X., Lv, X., Sun, Z., Zhang, J., Su, F., Kang, S., Kim, S. Y., An, S. S. A., Yu, X., Zhang, C., & Wang, H. (2019). *Alzheimer's Research & Therapy, 11*, 89-1-10.
113. Rissin, D. M., Kan, C. W., Campbell, T. G., Howes, S. C., Fournier, D. R., Song, L., Piech, T., Patel, P. P., Chang, L., Rivnak, A. J., Ferrell, E. P., Randall, J. D., Provuncher, G. K., Walt, D. R., & Duffy, D. C. (2010). Single-molecule enzyme-linked immunosorbent assay detects serum proteins at subfemtomolar concentrations. *Nature Biotechnology., 28*, 595–599.
114. Rissin, D. M., Kan, C. W., Song, L., Rivnak, A. J., Fishburn, M. W., Shao, Q., Piech, T., Ferrell, E. P., Meyer, R. E., Campbell, T. G., Fournier, D. R., & Duffy, D. C. (2013). Multiplexed single molecule immunoassays. *Lab on a Chip, 13*, 2902–2911.
115. Wilson, D. H., Rissin, D. M., Kan, C. W., Fournier, D. R., Piech, T., Campbell, T. G., Meyer, R. E., Fishburn, M. W., Cabrera, C., Patel, P. P., Frew, E., Chen, Y., Chang, L., Ferrell, E. P., von Einem, V., McGuigan, W., Reinhardt, M., Sayer, H., Vielsack, C., & Duffy, D. C. (2016). The SIMOA HD-1 analyzer: A novel fully automated digital immunoassay analyzer with single-molecule sensitivity and multiplexing. *Journal of Laboratory Automation, 21*, 533–547.
116. Jiao, B., Liu, H., Guo, L., Liao, X., Zhou, Y., Weng, L., Xiao, X., Zhou, L., Wang, X., Jiang, Y., Yang, Q., Zhu, Y., Zhou, L., Zhang, W., Wang, J., Yan, X., Tang, B., & Shen, L. (2021). Performance of plasma amyloid β, total tau, and neurofilament light chain in the identification of probable Alzheimer's disease in South China. *Frontiers in Aging Neuroscience, 13*, 749649-1-10.
117. https://www.quanterix.com/simoa-assay-kits/
118. Rissin, D. M., Kan, C. W., Campbell, T. G., Howes, S. C., Fournier, D. R., Song, L., Piech, T., Patel, P. P., Chang, L., Rivnak, A. J., Ferrell, E. P., Randall, J. D., Provuncher, G. K., Walt, D. R., & Duffy, D. C. (2010). Single-molecule enzyme-linked immunosorbent assay detects serum proteins at subfemtomolar concentrations. *Nature Biotechnology, 28*, 595–599. Hwang, J., Banerjee, M., Venable, A.S., Walden, Z., Jolly, J., Zimmerman, C., Adkisson, E., & Xiao, Q. (2019). Quantitation of low abundant soluble biomarkers using high sensitivity single molecule counting technolog. *Methods*, 158, 69–76.
119. Schindler, S. E., Li, Y., Todd, K. W., Herries, E. M., Henson, R. L., Gray, J. D., Wang, G., Graham, D. L., Shaw, L. M., Trojanowski, J. Q., Hassenstab, J. J., Benzinger, T. L. S., Cruchaga, C., Jucker, M., Levin, J., Chhatwal, J. P., Noble, J. M., Ringman, J. M., Graff-Radford, N. R., Holtzman, D. M., Ladenson, J. H., Morris, J. C., Bateman, R. J., Xiong, C., & Fagan, A. M. (2019). Emerging cerebrospinal fluid biomarkers in autosomal dominant Alzheimer's disease. *Alzheimer's & Dementia, 15*, 655–665.
120. https://www.sigmaaldrich.com/TW/en/campaigns/smc-neurology-immunoassays#form
121. Pines, D., & Bohm, D. (1952). A collective description of electron interactions: II. Collective vs individual particle aspects of the interactions. *Physics Review, 85*, 338–353.

122. Bohm, D., & Pines, D. (1953). Coulomb interactions in a degenerate electron gas. *Physics Review, 92*, 609–625.
123. Haes, A. J., Hall, W. P., Chang, L., Klein, W. L., & Van Duyne, R. P. (2004). A localized surface plasmon resonance biosensor: First steps toward an assay for Alzheimer's disease. *Nano Letters, 4*, 1029–1034.
124. Kelly, K. L. (2002). The optical properties of metal manoparticles: The influence of size, shape, and dielectric environment. *The Journal of Physical Chemistry. B, 107*, 668–677.
125. Rycenga, M., Cobley, C. M., Zeng, J., Li, W., Moran, C. H., Zhang, Q., Qin, D., & Xia, Y. (2011). Controlling the synthesis and assembly of silver nanostructures for plasmonic applications. *Chemical Reviews, 111*, 3669–3712.
126. Skrabalak, S. E., Au, L., Li, X., & Xia, Y. (2007). Facile synthesis of Ag nanocubes and Au nanocages. *Nature Protocols., 2*, 2182–2190.
127. Vestergaard, M., Kerman, K., Kimc, D. K., Hiep, H. M., & Tamiya, E. (2008). Detection of Alzheimer's tau protein using localized surface plasmon resonance-based immunochip. *Talanta, 74*, 1038–1042.
128. Shekhar, S., Kumar, R., Rai, N., Kumar, V., Singh, K., Upadhyay, A. D., Tripathi, M., Dwivedi, S., Dey, A. B., & Dey, S. (2016). Estimation of tau and phosphorylated tau181 in serum of Alzheimer's disease and mild cognitive impairment patients. *PLoS One, 11*, e0159099-1-10.
129. Nu, T. T. V., Tran, N. H. T., Nam, E., Nguyen, T. T., Yoon, W. J., Cho, S., Kim, J., Chang, K. A., & Ju, H. (2018). Blood-based immunoassay of tau proteins for early diagnosis of Alzheimer's disease using surface plasmon resonance fiber sensors. *RSC Advances, 8*, 7855–7862.
130. Ugozzoli, L. A. (2004). Multiplex assays with fluorescent microbead readout: A powerful tool for mutation detection. *Clinical Chemistry, 50*, 1963–1965.
131. Dunba, S. A. (2006). Applications of Luminex xMAP technology for rapid, high-throughput multiplexed nucleic acid detection. *Clinica Chimica Acta, 363*, 71–82.
132. Kang, J. H., Vanderstichele, H., Trojanowski, J. Q., & Shaw, L. M. (2012). Simultaneous analysis of cerebrospinal fluid biomarkers using microsphere-based xMAP multiplex technology for early detection of Alzheimer's disease. *Methods, 56*(4), 484–493.
133. Reslova, N., Michna, V., Kasny, M., Mikel, P., & Kralik, P. (2017). xMAP technology: Applications in detection of pathogens. *Frontiers in Microbiology, 8*, 55.
134. Dunbar, S., & Li, D. (2010). Introduction to Luminex xMAP technology and applications for biological analysis in China. *Asia Pac Biotech., 14*, 26–30.
135. Houser, B. (2012). Bio-Rad's bio-Plex suspension array system, xMAP technology overview. *Archives of Physiology and Biochemistry, 118*, 192–196.
136. Collins, C. J., Altman, A., Roth, W., & Carrano, J. C. (2008). Low cost, LED-based xMAP analyzer for multiplexed diagnosis and environmental detection of biological agents. *Proceedings of SPIE, 7116*, 711605-1-8.
137. Kim, J. S., Taitt, C. R., Ligler, F. S., & Anderson, G. P. (2010). Multiplexed magnetic microsphere immunoassays for detection of pathogens in foods. *Sensing and Instrumentation for Food Quality and Safety, 4*, 73–81.
138. Coppock, M. B., & Stratis-Cullum, D. N. (2019). A universal method for the functionalization of dyed magnetic microspheres with peptides. *Methods, 158*, 12–16.
139. https://www.sigmaaldrich.com/TW/en/product/mm/hns1mag95k
140. Jiang, W., Yang, H. C., Yang, S. Y., Horng, H. E., Hung, J. C., Chen, Y. C., & Hong, C. Y. (2004). Preparation and properties of superparamagnetic nanoparticles with narrow size distribution and biocompatible. *Journal of Magnetism and Magnetic Materials, 283*, 210–214.
141. Yang, C. C., Yang, S. Y., Chieh, J. J., Horng, H. E., Hong, C. Y., Yang, H. C., Chen, K. H., Shih, B. Y., Chen, T. F., & Chiu, M. J. (2011). Biofunctionalized magnetic nanoparticles for specifically detecting biomarkers of Alzheimer's disease in vitro. *ACS Chemical Neuroscience, 2*, 500–505.

142. Yang, S. Y., Chiu, M. J., Chen, T. F., & Horng, H. E. (2017). Detection of plasma biomarkers using immunomagnetic reduction: A promising method. *Neurology and Therapy, 6*(Suppl 1), S37–S56.
143. Yang, S. Y. (2024). Theoretical explorations of reductions in the alternative-current magnetic susceptibility of reagent in immunomagnetic reduction assay. *Journal of Physical and Chemical Research, 6*, 177-1-7.
144. Yang, C. C., Yang, S. Y., Chen, H. H., Weng, W. L., Horng, H. E., Chieh, J. J., Hong, C. Y., & Yang, H. C. (2012). Effect of molecule-particle binding on the reduction in the mixed-frequency alternating current magnetic susceptibility of magnetic bio-reagents. *Journal of Applied Physics, 112*, 24704.
145. Chieh, J. J., Yang, S. Y., Jian, Z. F., Wang, W. C., Horng, H. E., Yang, H. C., & Hong, C. Y. (2008). Hyper-high-sensitivity wash-free magnetoreduction assay on biomolecules using high-Tc superconducting quantum interference devices. *Journal of Applied Physics, 103*, 14703-1-6.
146. Chen, H. H., Hsu, M. H., Lee, K. H., & Yang, S. Y. (2022). Development of a 36-channel instrument for assaying biomarkers of ultralow concentrations utilizing immunomagnetic reduction. *ACS Measurement Science Au, 2*, 485–492.
147. Yang, S. Y., Chiu, M. J., Lin, C. H., Horng, H. E., Yang, C. C., Chieh, J. J., Chen, H. H., & Liu, B. H. (2016). Development of an ultra-high sensitive immunoassay with plasma biomarker for differentiating Parkinson disease dementia from Parkinson disease using antibody functionalized magnetic nanoparticles. *Journal of Nanobiotechnology, 14*, 41-1-8.
148. Yang, S. Y., Chiu, M. J., Chen, T. F., Lin, C. H., Jeng, J. S., Tang, S. C., Lee, Y. F., Yang, C. C., Liu, B. H., Chen, H. H., & Wu, C. C. (2017). Analytical performance of reagent for assaying tau protein in human plasma and feasibility study screening neurodegenerative diseases. *Scientific Reports, 7*, 9304-1-12.
149. Liu, H. C., Lin, W. C., Chiu, M. J., Lu, C. H., Lin, C. Y., & Yang, S. Y. (2020). Development of an assay of plasma neurofilament light chain utilizing immunomagnetic reduction technology. *PLoS One, 15*, e0234519-1-17.
150. Yang, S. Y., Liu, H. C., Lin, C. Y., Chiu, M. J., Chen, T. F., Lin, C. H., & Chen, H. H. (2020). Development of assaying plasma TDP-43 utilizing immunomagnetic reduction. *Journal of Neurological Disorders, 8*, 7-1-8.
151. Yang, S. Y., Wang, W. C., Lan, C. B., Chen, C. H., Chieh, J. J., Horng, H. E., Hong, C. Y., Yang, H. C., Tsai, C. P., Yang, C. Y., Cheng, I. C., & Chung, W. C. (2010). Magnetically enhanced high-specificity virus detection using bio-activated magnetic nanoparticles with antibodies as labeling markers. *Journal of Virological Methods, 164*, 14–18.
152. Chang, K. H., Liu, K. C., Lai, C. S., Yang, S. Y., & Chen, C. M. (2021). Assessing plasma levels of α-synuclein and neurofilament light chain by different blood preparation methods. *Frontiers in Aging Neuroscience, 13*, 759182-1-10.
153. Chang, J. F., Liu, H. C., Chen, H. H., Chen, W. P., Juang, J. L., Wang, P. N., & Yang, S. Y. (2020). Effect of times to blood processing on the stability of blood proteins associated with dementia. *Dementia and Geriatric Cognitive Disorders, 49*, 303–311.
154. Chiu, M. J., Lue, L. F., Sabbagh, M. N., Chen, T. F., Chen, H. H., & Yang, S. Y. (2019). Long-term storage effects on stability of Aβ1-40, Aβ1-42, and total tau proteins in human plasma samples measured with immunomagnetic reduction assays. *Dementia and Geriatric Cognitive Disorders Extra, 9*, 77–86.
155. http://www.magqu.com
156. Fredriksson, S., Gullberg, M., Jarvius, J., Olsson, C., Pietras, K., Gústafsdóttir, S. M., Östman, A., & Landegren, U. (2022). Protein detection using proximity-dependent DNA ligation assays. *Nature Biotechnology, 20*, 473–477.
157. Gullberg, M., Gústafsdóttir, S. M., Schallmeiner, E., Jarvius, J., Bjarnegård, M., Betsholtz, C., Landegren, U., & Fredriksson, S. (2004). Cytokine detection by antibody-based proximity ligation. *PNAS, 101*, 8420–8424.

158. Blazek, M., Betz, C., Hall, M. N., Reth, M., Zengerle, R., & Meier, M. (2013). Proximity ligation assay for high-content profiling of cell signaling pathways on a microfluidic chip. *Molecular & Cellular Proteomics, 12*, 3898–3907.
159. Zatloukal, B., Kufferath, I., Thueringer, A., Landegren, U., Zatloukal, K., & Haybaeck, J. (2014). Sensitivity and specificity of in situ proximity ligation for protein interaction analysis in a model of Steatohepatitis with Mallory-Denk bodies. *PLoS One, 9*, e96690-1-6.
160. Hegazy, M., Cohen-Barak, E., Koetsier, J. L., Najor, N. A., Arvanitis, C., Sprecher, E., Green, K. J., & Godsel, L. M. (2020). Proximity ligation assay for detecting protein-protein interactions and protein modifications in cells and tissues in situ. *Current Protocols in Cell Biology, 89*, e115-1-34.
161. Lorenz, T. C. (2012). Polymerase chain reaction: Basic protocol plus troubleshooting and optimization strategies. *Journal of Visualized Experiments, 22*, e3998-1-14.
162. Mohsen, M. G., & Kool, E. T. (2016). Discovery of rolling circle amplification and rolling circle transcription. *Accounts of Chemical Research, 49*, 2540–2550.
163. Zingg, J. M., & Daunert, S. (2018). Trinucleotide rolling circle amplification: A novel method for the detection of RNA and DNA. *Methods and Protocols, 1*, 15-1-20.
164. Joffroy, B., Uca, Y. O., Pre˘sern, D., Doye, J. P. K., & Schmid, T. L. (2018). Rolling circle amplification shows a sinusoidal template length-dependent amplification bias. *Nucleic Acids Research, 46*, 538–545.
165. Zhang, K., Zhang, H., Cao, H., Jiang, Y., Mao, K., & Yang, Z. (2021). Rolling circle amplification as an efficient analytical tool for rapid detection of contaminants in aqueous environments. *Biosensors, 11*, 352-1-30.
166. Garafutdinov, R. R., Sakhabutdinova, A. R., Gilvanov, A. R., & Chemeris, A. V. (2021). Rolling circle amplification as a universal method for the analysis of a wide range of biological targets. *Russian Journal of Bioorganic Chemistry, 47*, 1172–1189.
167. Ke, R., Nong, R. Y., Fredriksson, S., Landegren, U., & Nilsson, M. (2013). Improving precision of proximity ligation assay by amplified single molecule detection. *PLoS One, 8*, e69813-1-5.
168. Chen, H., Crum, M., Chavan, D., Vu, B., Kourentzi, K., & Willson, R. C. (2018). Nanoparticle-based proximity ligation assay for ultra-sensitive, quantitative detection of protein biomarkers. *ACS Applied Materials & Interfaces, 10*, 31845–31849.
169. Gleichenhagen, J., Arndt, C., Casjens, S., Topfer, C., Gerullis, J., Raiko, I., Taeger, D., Ecke, T., Bruning, T., & Johnen, G. (2022). Exploring solid-phase proximity ligation assay for survivin detection in urine. *PLoS One, 17*, e0270535-1-17.
170. Feng, W., Beer, J. C., Hao, Q., Ariyapala, I. S., Sahajan, A., Komarov, A., Cha, K., Moua, M., Qiu, X., Xu, X., Iyengar, S., Yoshimura, T., Nagaraj, R., Wang, L., Yu, M., Engel, K., Zhen, L., Xue, W., Lee, C. J., Park, C. H., Peng, C., Zhang, K., Grzybowski, A., Hahm, J., Schmidt, S. V., Odainic, A., Spitzer, J., Buddika, K., Kuo, D., Fang, L., Zhang, B., Chen, S., Latz, E., Yin, Y., Luo, Y., & Ma, X. J. (2023). NULISA: A proteomic liquid biopsy platform with attomolar sensitivity and high multiplexing. *Nature Communications, 14*, 7238-1-14.
171. Warmenhoven, N., Salvadó, G., Janelidze, S., Mattsson-Carlgren, N., Bali, D., Dolado, A.O., Kolb, H., Triana-Baltzer, G., Barthélemy, N.R., Schindler, S.E., Aschenbrenner, A.J., Raji, C.A., Benzinger, T.L.S., Morris, J.C., Ibanez, L., Timsina, J., Cruchaga, C., Bateman, R.J., Ashton, N., Arslan, B., Zetterberg, H., Blennow, K., Binette, A.P., & Hansson, O.. (submitted).
172. Ibanez, L., Liu, M., Beric, A., Timsina, J., Kholfeld, P., Bergmann, K., Lowery, J., Sykora, N., Sanchez-Montejo, B., Brock, W., Budde, J.P., Bateman, R.J., Barthelemy, N., Schindler, S.E., Holtzman, D.M., Benzinger, T.L.S., Xiong, C., Tarawneh, R., Moulder, K., Morris, J.C., Sung, Y.J., & Cruchaga, C.. (submitted).
173. https://alamarbio.com/products-and-services/nulisa-singleplex-assays/
174. https://alamarbio.com/products-and-services/nulisaseq-cns-disease-panel/
175. Assarsson, E., Lundberg, M., Holmquist, G., Bjorkesten, J., Thorsen, S. B., Ekman, D., Eriksson, A., Dickens, E. R., Ohlsson, S., Edfeldt, G., Andersson, A. C., Lindstedt, P., Stenvang, J., Gullberg, M., & Fredriksson, S. (2014). Homogenous 96-Plex PEA immu-

noassay exhibiting high sensitivity, specificity, and excellent scalability. *PLoS One, 9*, e95192-1-11.
176. Gullberg, M., Gustafsdottir, S. M., Schallmeiner, E., Jarvius, J., Bjarnegård, M., Betsholtz, C., Landegren, U., & Fredriksson, S. (2004). Cytokine detection by antibody-based proximity ligation. *PNAS, 101*, 8420–8424.
177. Schaus, T. E., Woo, S., Xuan, F., Chen, X., & Yin, P. (2017). A DNA nanoscope via autocycling proximity recording. *Nature Communications, 8*, 696-1-9.
178. Kishi, J. Y., Schaus, T. E., Gopalkrishnan, N., Xuan, F., & Yin, P. (2018). Programmable autonomous synthesis of single-stranded DNA. *Nature Chemistry, 10*, 155–164.
179. Schueder, F., Lara-Gutirrez, J., Haas, D., Beckwith, K. S., Yin, P., Ellenberg, J., & Jungmann, R. (2024). Super-resolution spatial proximity detection with proximity-PAINT. *Angewandte Chemie, International Edition, 60*, 716–720.
180. Kivisäkk, P., Fatima, H. A., Cahoon, D. S., Otieno, B., Chacko, L., Minooei, F., Demos, C., Stengelin, M., Sigal, G., Wohlstadter, J., & Arnold, S. E. Clinical evaluation of a novel plasma pTau217 electrochemiluminescence immunoassay in Alzheimer's disease. *Scientific Reports, 14*, 629-1-8.
181. https://spear.bio/biomarker-research-applications/immunoassays-neurology/
182. Soto, C., & Pritzkow, S. (2018). Protein misfolding, aggregation, and conformational strains in neurodegenerative diseases. *Nature Neuroscience, 21*, 1332–1340.
183. Jarrett, J. T., & Lansbury, P. T. (1993). Seeding 'one-dimensional crystallization' of amyloid: A pathogenic mechanism in Alzheimer's disease and scrapie? *Cell, 73*, 1055–1058.
184. Kocisko, D. A., Come, J. H., Priola, S. A., Chesebro, B., Raymond, G. J., Lansbury, P. T., & Caughey, B. (1994). Cell-free formation of protease-resistant prion protein. *Nature, 370*, 471–474.
185. Saborio, G. P., Permanne, B., & Soto, C. (2001). Sensitive detection of pathological prion protein by cyclic amplification of protein misfolding. *Nature, 411*, 810–813.
186. Atarashi, R., Moore, R. A., Sim, V. L., Hughson, A. G., Dorward, D. W., Onwubiko, H. A., Priola, S. A., & Caughey, B. (2007). Ultrasensitive detection of scrapie prion protein using seeded conversion of recombinantprion protein. *Nature Methods, 4*, 645–650.
187. Atarashi, R., Wilham, J. M., Christensen, L., Hughson, A. G., Moore, R. A., Johnson, L. M., Onwubiko, H. A., Priola, S. A., & Caughey, B. (2008). Simplified ultrasensitive prion detection by recombinant PrP conversion with shaking. *Nature Methods, 5*, 211–212.
188. Iranzo, A., Fairfoul, G., Ayudhaya, A. C. N., Serradell, M., Gelpi, E., Vilaseca, I., Sanchez-Valle, R., Gaig, C., Santamaria, J., Tolosa, E., Riha, R. L., & Green, A. J. E. (2021). Detection of α-synuclein in CSF by RT-QuIC in patients with isolated rapid-eye-movement sleep behaviour disorder: A longitudinal observational study. *Lancet Neurology, 20*, 203–212.
189. Manne, S., Kondru, N., Jin, H., Serrano, G. E., Anantharam, V., Kanthasamy, A., Adler, C. H., Beach, T. G., & Kanthasamy, A. G. (2020). Blinded RT-QuIC analysis of α-synuclein biomarker in skin tissue from Parkinson's disease patients. *Movement Disorders, 35*, 2230–2239.
190. Kuzkina, A., Bargar, C., Schmitt, D., Rößle, J., Wang, W., Schubert, A. L., Tatsuoka, C., Gunzler, S. A., Zou, W. Q., Volkmann, J., Sommer, C., Doppler, K., & Chen, S. G. (2023). Diagnostic value of skin RT-QuIC in Parkinson's disease: A two-laboratory study. *Parkinson's Related Disorders, 117*, 105807-1-11.
191. Gomes, B. F., Farris, C. M., Ma, Y., Concha-Marambio, L., Lebovitz, R., Nellgård, B., Dalla, K., Constantinescu, J., Constantinescu, R., Gobom, J., Andreasson, U., Zetterberg, H., & Blennow, K. (2023). α-synuclein seed amplification assay as a diagnostic tool for parkinsonian disorders. *Parkinsons's Related Disorders, 117*, 105807-1-6.
192. Li, J., Duan, S., Yang, J., Zheng, H., Yuan, Y., Tang, M., Wang, Y., Liu, Y., Xia, Z., Luo, H., & Xu, Y. (2024). Detection of skin α-synuclein using RT-QuIC as a diagnostic biomarker for Parkinson's disease in the Chinese population. *European Journal of Medical Research, 29*, 114-1-10.
193. Paciotti, S., Bellomo, G., Gatticchi, L., & Parnetti, L. (2018). Are we ready for detecting α-synuclein prone to aggregation in patients? The case of "protein-misfolding cyclic ampli-

fication" and "real-time quaking-induced conversion" as diagnostic tools. *Frontiers in Neurology, 9*, 415-1–415-9.
194. Janarthanam, C., Clabaugh, G., Wang, Z., Melvin, B. R., Scheibe, I., Jin, H., Anantharam, V., Urbauer, R. J. B., Urbauer, J. L., Ma, J., Kanthasamy, A., Huang, X., Donadio, V., Zou, W., & Kanthasamy, A. G. (2024). High-yield α-synuclein purification and ionic strength modification pivotal to seed amplification assay performance and reproducibility. *International Journal of Molecular Sciences, 25*, 5988-1-123.
195. Hatano, T., Okuzumi, A., Matsumoto, G., Tsunemi, T., & Hattori, N. (2024). α-Synuclein: A promising biomarker for Parkinson's disease and related disorders. *Journal of Movement Disorders, 17*, 127–137.
196. Potts, P. J. (1987). Mass spectrometry: Principles and instrumentation. In *A Handbook of Silicate Rock Analysis*. Springer.
197. Ho, C. S., Lam, C. W. K., Chan, M. H. M., Cheung, R. C. K., Law, L. K., Lit, L. C. W., Ng, K. F., Suen, M. W. M., & Tai, H. L. (2003). Electrospray ionisation mass spectrometry: Principles and clinical applications. *Clinical Biochemist Reviews, 24*, 1–10.
198. McLafferty, F. W. (2011). A century of progress in molecular mass spectrometry. *Annual Review of Analytical Chemistry, 4*, 1–22.
199. Vadakedath, S., Kandi, V., Godishala, V., Pinnelli, V. B. K., Alkafaas, S. S., & Have, S. S. E. (2022). The principle, types, and applications of mass spectrometry: A comprehensive review. *Biomedical and Biotechnology, 7*, 6–22.
200. Pitt, J. J. (2009). Principles and applications of liquid chromatography-mass spectrometry in clinical biochemistry. *Clinical Biochemist Reviews, 30*, 19–34.
201. Thomas, S. N., French, D., Jannetto, P. J., Rappold, B. A., & Clarke, W. A. (2022). Liquid chromatography-tandem mass spectrometry for clinical diagnostics. *Nature Review Methods Primers, 2*, 96-1-14.
202. Nawrocki, J. (1997). The silanol group and its role in liquid chromatography. *Journal of Chromatography A, 779*, 29–71.
203. Kitagawa, S. (2019). Liquid chromatography. *Analytical Sciences, 35*, 949–950.
204. Schindler, S. E., Bollinger, J. G., Ovod, V., Mawuenyega, K. G., Li, Y., Gordon, B. A., Holtzman, D. M., Morris, J. C., Benzinger, T. L. S., Xiong, C., Fagan, A. M., & Bateman, R. J. (2019). High-precision plasma β-amyloid 42/40 predicts current and future brain amyloidosis. *Neurology, 93*, E1647–e1659.
205. Eastwood, S. M., Meyer, M. R., Kirmess, K. M., Wente-Roth, T. L., Irvin, F., Holubasch, M. S., Verghese, P. B., West, T., Braunstein, J. B., Yarasheski, K. E., & Contois, J. H. (2024). PrecivityAD2™ blood test: Analytical validation of an LC-MS/MS assay for quantifying plasma phospho-tau217 and non-phospho-tau217 peptide concentrations that are used with plasma amyloid-β42/40 in a multianalyte assay with algorithmic analysis for detecting brain amyloid pathology. *Diagnostics, 14*, 1739-1-18.
206. Kirmess, K. M., Meyer, M. R., Holubasch, M. S., Knapik, S. S., Hu, Y., Jackson, E. N., Harpstrite, S. E., Verghese, P. B., West, T., Fogelman, I., Braunstein, J. B., Yarasheski, K. E., & Contois, J. H. (2021). The PrecivityAD™ test: Accurate and reliable LC-MS/MS assays for quantifying plasma amyloid beta 40 and 42 and apolipoprotein E proteotype for the assessment of brain amyloidosis. *Clinica Chimica Acta, 519*, 267–275.
207. Have, S. T., Boulon, S., Ahmad, Y., & Lamond, A. I. (2011). Mass spectrometry-based immunoprecipitation proteomics—The user's guide. *Proteomics, 11*, 1153–1159.
208. Pannee, J., Törnqvist, U., Westerlund, A., Ingelsson, M., Lannfelt, L., Brinkmalm, G., Persson, R., Gobom, J., Svensson, J., Johansson, P., Zetterberg, H., Blennow, K., & Portelius, E. (2014). The amyloid-β degradation pattern in plasma-A possible tool for clinical trials in Alzheimer's disease. *Neuroscience Letters, 573*, 7–12.
209. Nakamura, A., Kaneko, N., Villemagne, V. L., Kato, T., Doecke, J., Doré, V., Fowler, C., Li, Q. X., Martins, R., Rowe, C., Tomita, T., Matsuzaki, K., Ishii, K., Ishii, K., Arahata, Y., Iwamoto, S., Ito, K., Tanaka, K., Masters, C. L., & Yanagisawa, K. (2018). High performance plasma amyloid-β biomarkers for Alzheimer's disease. *Nature, 554*, 249–254.

210. Kaneko, N., Yamamoto, R., Sato, T. A., & Tanaka, K. (2014). Identification and quantification of amyloid beta-related peptides in human plasma using matrix-assisted laser desorption/ionization time-of-flight mass spectrometry. *Proceedings of the Japan Academy, Series B, Physical and Biological Sciences, 90*, 104–117.
211. Sato, C., Barthelemy, N. R., Mawuenyega, K. G., Patterson, B. W., Gordon, B. A., Jockel-Balsarotti, J., Sullivan, M., Crisp, M. J., Kasten, T., Kirmess, K. M., Kanaan, N. M., Yarasheski, K. E., Baker-Nigh, A., Benzinger, T. L. S., Miller, T. M., Karch, C. M., & Bateman, R. J. (2018). Tau kinetics in neurons and the human central nervous system. *Neuron, 97*, 1284–1298.
212. van Andel, E., Roosjen, M., van der Zanden, S., Lange, S. C., Weijers, D., Smulders, M. M. J., Savelkoul, H. F. J., Zuilhof, H., & Tijhaar, E. J. (2022). Highly specific protein identification by immunoprecipitation–mass spectrometry using antifouling microbeads. *ACS Applied Materials & Interfaces, 14*, 23102–23116.
213. Doolittle, M. H., Martin, D. C., Davis, R. C., Reuben, M. A., & Elovson, J. (1991). A two-cycle immunoprecipitation procedure for reducing nonspecific protein contamination. *Analytical Biochemistry, 195*, 364–368.
214. Zeng, X., Zeng, W. H., Zhou, J., Liu, X. M., Huang, G., Zhu, H., Xiao, S., Zeng, Y., & Cao, D. (2022). Removal of nonspecific binding proteins is required in co-immunoprecipitation with nuclear proteins. *BioTechniques, 73*, 289–296.
215. Barthelemy, N. R., Horie, K., Sato, C., & Bateman, R. J. (2020). Blood plasma phosphorylated-tau isoforms track CNS change in Alzheimer's disease. *The Journal of Experimental Medicine, 217*, e20200861-1-11.
216. Paolieri, M. (2017). Ferdinand Münz: EDTA and 40 years of inventions. *Bulletin for the History of Chemistry, 42*, 133–140.
217. Banfi, G., Salvagno, G. L., & Lippi, G. (2007). The role of ethylenediamine tetraacetic acid (EDTA) as in vitro anticoagulant for diagnostic purposes. *Clinical Chemistry and Laboratory Medicine, 45*, 565–576.
218. Lima, M., Rudd, T., & Yates, E. (2017). New applications of heparin and other glycosaminoglycans. *Molecules, 22*, 749-1-11.
219. Shriver, Z., Capila, I., Venkataraman, G., & Sasisekharan, R. (2012). Heparin and heparan sulfate: Analyzing structure and microheterogeneity. *Handbook of Experimental Pharmacology, 207*, 159–176.
220. Lippi, G., Salvagno, G. L., Montagnana, M., Poli, G., & Guidi, G. C. (2206). Influence of centrifuge temperature on routine coagulation testing. *Clinical Chemistry, 52*, 537–538.
221. Yazar, H., Özdemir, F., & Köse, E. (2018). Effect of centrifuge temperature on routine coagulation tests. *Acta Haematologica, 139*, 158–163.
222. Zheng, R., Brunius, C., Shi, L., Zafar, H., Paulson, L., Landberg, R., & Naluai, Å. T. (2021). Prediction and evaluation of the effect of pre-centrifugation sample management on the measurable untargeted LC-MS plasma metabolome. *Analytica Chimica Acta, 1182*, 338968-1-11.
223. Si, J., Gao, Y., Yan, B., Shen, X., Zhu, C., & Yao, Q. (2024). Evaluating the impact of delayed centrifugation on protein profiles analyzed by LC/MS in serum and plasma samples. *Advanced Sensor and Energy Materials, 3*, 100101-1-8.
224. Lee, P. J., Tsai, C. L., Liang, C. S., Peng, G. S., Lee, J. T., Tsai, C. K., Lin, Y. K., & Yang, F. C. (2021). Biomarkers with plasma amyloid β and tau protein assayed by immunomagnetic reduction in patients with amnestic mild cognitive impairment and Alzheimer's disease. *Acta Neurologica Taiwanica, 30*, 1–9.

Chapter 2
Roles of Plasma Preparation in Biomarker Measurements

Abstract Owing to the ultralow concentrations of biomarkers in human blood, the sample (plasma) preparation significantly contributes to the measured results. Different concentrations could be obtained by using different recipes of plasma preparation, probably resulting in false positives or negatives. Hence, it is necessary to realize the contributions of recipes to measured results and standardize the plasma preparation. In this chapter, the effects of factors usually faced in clinical practice, such as blood collecting tubes, storage temperature of whole blood, centrifugation temperatures of whole blood, time period to centrifuge blood samples after blood draw, lifetime of biomarkers in frozen plasma, and freeze/thaw cycles of plasma, on the measured concentrations of blood-based biomarkers associated with neurodegenerative diseases are reviewed. The precautions for subjects undergoing blood draw are suggested. The contents in this chapter are really helpful for developers, laboratory technicians, and physicians to establish standard operating processes for plasma preparation in assaying blood-based biomarkers of neurodegenerative diseases.

Keywords Plasma preparation · Lifetime · Standard operating processes · Blood draw · Centrifugation · Freeze/thaw cycles

2.1 Factors Affecting the Assay Results of Biomarkers During Sample Preparation

The samples used for the assay of biomarkers associated with neurodegenerative diseases are typically plasma or serum. Plasma or serum preparation is necessary before biomarker levels are measured. However, the procedures for preparing plasma and serum are different, probably leading to different measured results of biomarker levels.

Kac et al. utilized SIMOA assay to measure pTau181 and pTau231 levels in human plasma and serum (n = 32) [1]. They reported that pTau181 and pTau231

C. S.-Y. Yang, M. N. Sabbagh, *Assessing Neurodegenerative Diseases: High-Sensitivity Technologies That Assay Pathological Biomarkers in Blood*,
https://doi.org/10.1007/978-3-032-23863-4_2

concentrations were twice as high in plasma as in serum. Chang et al. used IMR assay and reported 40-fold higher levels of plasma α-synuclein than serum α-synuclein in cognitively normal subjects (n = 40) [2]. Bibl et al. investigated the serum and plasma samples of 19 cognitively normal subjects [3]. The panels of $A\beta_{1-37/38/39/40/42}$ and $A\beta_{2-40}$ in both serum and plasma were analyzed via one--dimensional Aβ–SDS–PAGE/immunoblotting, either immediately after or after storage at room temperature for 24 h. In both serum and plasma, sample storage led to a significant decrease in measurable peptide concentrations. This effect was stronger in serum than in plasma. These results imply that plasma is more appropriate than serum for analyzing blood biomarkers associated with neurodegenerative diseases.

To prepare plasma samples, many recipes can be manipulated. In particular, for ultralow concentrations of plasma biomarkers associated with neurodegenerative diseases, the recipes used for sample preparation affect the measured concentrations of biomarkers. All the recipes should be optimized and standardized to control variations in measured concentrations among assays. In this chapter, the effects of recipes in plasma/serum preparation, starting from blood draw to plasma/serum storage, on the measured concentrations of biomarkers are discussed. The recipes include the following items:

Blood collecting tubes
Storage temperature of whole blood
Centrifugation temperatures of whole blood
Time period to centrifuge blood samples after blood draw
Lifetime of biomarkers in frozen plasma
Freeze/thaw cycles of plasma

We used the assay technology, immunomagnetic reduction (IMR), as an example to discuss the effects of these recipes on the measured concentrations of plasma biomarkers associated with neurodegenerative diseases. Notably, the effects of these recipes on the measured concentrations of biomarkers could differ among assay technologies. Detailed studies should be performed to clarify the effects of various assay technologies.

In addition to recipes, the precautions for collecting blood samples for the assay of biomarkers in neurodegenerative diseases are described to prevent abnormal results.

2.2 Blood Collecting Tubes

For tests and biochemical assays, serum or plasma is prepared. Serum is a clear-colored liquid produced by clotting a blood sample followed by centrifuging the clotted blood sample [4]. Serum does not contain blood cells, coagulants or fibrin clots but contains components such as electrolytes, antibodies, hormones, and other proteins. For example, serum is frequently used in assays of antibody levels due to infections or vaccinations [5, 6]. Thus, serum is usually prepared for detecting exposure to pathogens such as viruses and bacteria.

Red top and gold top tubes were used to collect blood for the production of serum. Red top tubes do not contain any additives or anticoagulants and are essential for conducting tests that require the separation of serum (the liquid part of the blood) or clots (the solid part of the blood), such as basic metabolic panels and comprehensive metabolic panels. Gold top tubes are used for serum preparation. The gold top tube, usually referred to as a serum separator tube (SST), contains a gel separator at the bottom. This gel separator acts as a barrier between the serum and the clot after blood has been centrifuged.

Plasma is the straw-colored liquid portion of blood and is obtained by centrifuging the blood sample. Anticoagulants are added to blood samples to prevent the formation of fibrin, which plays a role in the clotting process [7, 8]. Thus, plasma is rich in clotting factors, blood cells, and other proteins. Owing to its rich composition, plasma is regularly used in proteomic and metabolomic assays. There are several types of blood collecting tubes, depending on the anticoagulants used, for producing plasma in clinical practice. These tubes are briefly introduced as follows.

The yellow, pink, and blue top tubes were used for blood culture. The anticoagulant in these tubes is mainly polyanethole sulfonate ($C_{10}H_{12}NaO_4S$), which inhibits the activation of complement function [9].

Light blue top tubes containing sodium citrate ($Na_3C_6H_5O_7$) as an anticoagulant were used for coagulation studies [10].

Green top tubes contain heparin, which is an anticoagulant that inhibits thrombin formation. Green top tubes are used for hematology, chemistry, and virology tests.

Lavender tubes, referred to as purple top tubes, contain ethylenediaminetetraacetic acid (EDTA), which achieves anticoagulation by chelating calcium ions in the blood to prevent coagulation. Lavender tubes are also called EDTA tubes and are commonly used for hematologic assays, such as proteins, peptides, or cytokines [11].

Gray top tubes contain potassium oxalate ($K_2C_2O_4$), which binds to calcium to inhibit coagulation. Gray top tubes also contain sodium fluoride (NaF), which acts as an antiglycolytic agent [12]. Thus, gray top tubes are used for plasma glucose and lactic acid testing.

EDTA tubes are frequently used to collect plasma for the detection of biomarkers of neurodegenerative diseases. There are two types of EDTA tubes: K2 and K3. The K_2EDTA tubes contain two chelated potassium ions, whereas K_3EDTA tubes contain three chelated potassium ions.

Owing to the different compositions of anticoagulants in the K_2EDTA and K_3EDTA tubes, the measured concentrations of biomarkers may differ [13]. It would be better to know the potential differences in order to select the suitable type of EDTA for standardizing the blood collection. Figure 2.1 shows the correlations between the measured concentrations of the plasma biomarkers $A\beta_{1-40}$, $A\beta_{1-42}$, T-Tau, and α-synuclein via the K2 (y-axis) and K3 (x-axis) EDTA tubes for blood collection. The assay technology used is immunomagnetic reduction (IMR). There were strong correlations in the measured concentrations of biomarkers between the K_2EDTA and K_3EDTA tubes. This explains why both K_2EDTA and K_3EDTA are

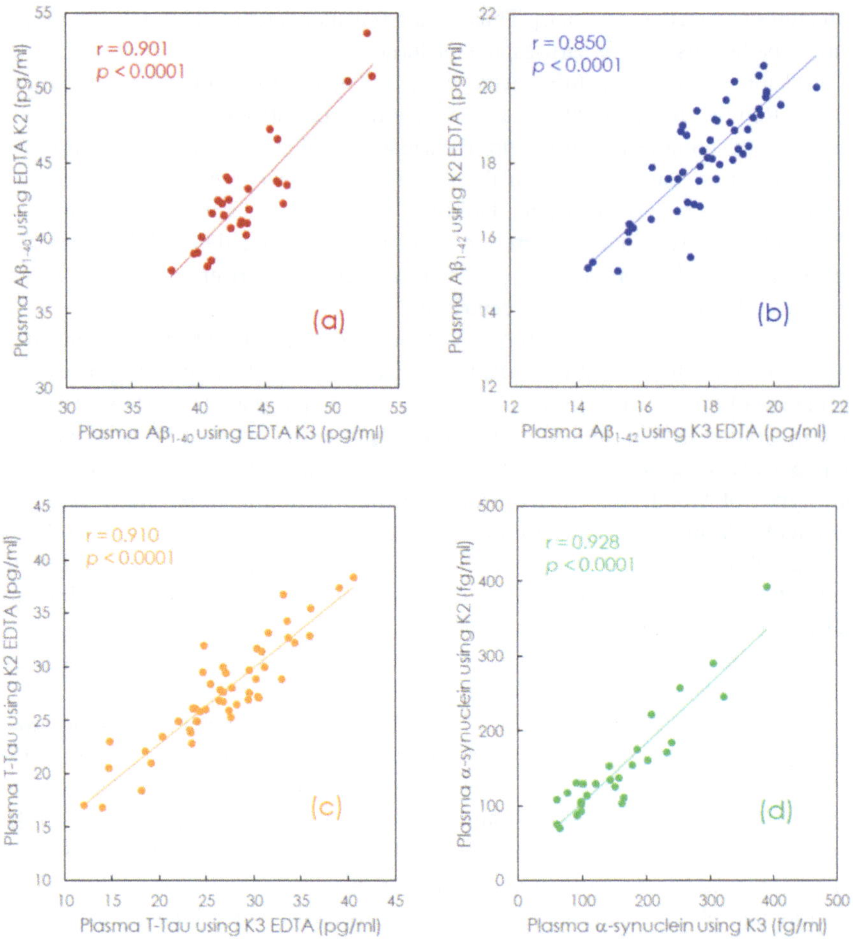

Fig. 2.1 Correlations in the measured concentrations of biomarkers (**a**) $A\beta_{1-40}$, (**b**) $A\beta_{1-42}$, (**c**) T-Tau, and (**d**) α-synuclein using K_2EDTA and K_3EDTA tubes for blood collection

currently used. However, the slopes of the linear fitting curves in Figs. 2.1(a)-(b) are 0.93, 0.81, 0.71, and 0.80, respectively. This implies that the measured concentrations of biomarkers are not identical between the K_2EDTA and K_3EDTA tubes. For example, in the case of T-Tau, as the measured concentration with the K_3EDTA tube is 30 pg/ml, the measured concentration with the K_2EDTA tube would be slightly less than 30 pg/ml. In contrast, as the measured concentration with the K_3EDTA tube is 20 pg/ml, the measured concentration with the K_2EDTA tube would be greater than 20 pg/ml. Thus, only one type of EDTA is selected to standardize blood drawn in clinical practice.

Ashton et al. utilized a SIMOA assay to investigate the differences in the measured concentrations of dementia biomarkers in plasma when different blood collecting tubes were used [14]. The results revealed that the levels of plasma $A\beta_{1-40}$,

$A\beta_{1-42}$, T-Tau, and pTau181 significantly differed between EDTA tubes and heparin tubes. Therefore, in both clinical practice and research, the use of only one type of tube for blood collection is strongly suggested. At present, the most frequently used tube for assaying plasma biomarkers in neurodegenerative diseases is the EDTA tube.

2.3 Storage Temperature of Whole Blood

After the blood is drawn, the blood collecting tubes should be placed at suitable temperatures to maintain sample quality; otherwise, the degradation of proteins and enzymes in the blood, changes in the concentrations of certain components in the blood, the loss of sample integrity, or inaccurate test results can occur, any of which leads to incorrect assay results [15–18].

Whole-blood samples contain versatile components, which are affected differently by storage temperature. Thus, suitable storage temperatures for blood collecting tubes depend on the assay parameters. The temperature range 2–8 °C for blood collecting tubes is generally utilized for most assays. This temperature range is good for stabilizing analytes or retarding metabolism, which involves the consumption of glucose and oxygen and the production of CO_2 in blood [19–21]. An easy way to locate blood collecting tubes at 2–8 °C is to insert the tubes in wet ice or position the tubes beside the ice packs.

In fact, certain components in a blood sample can be damaged, degraded, or become aggregated at 2–8 °C [22, 23]. They prefer room temperature, i.e., 20–25 °C. Therefore, suitable storage temperatures for blood collecting tubes after blood draws should be investigated for assays of plasma biomarkers associated with neurodegenerative diseases. Here, the effects of the storage temperature of blood collecting tubes on the assay results of the plasma amyloid β 1–42 peptide ($A\beta_{1-42}$) production via immunomagnetic reduction (IMR) technology are presented as an example.

K_3EDTA tubes were used to collect whole blood from cognitively unimpaired subjects and patients with Alzheimer's disease. Blood from each participant was drawn in two K_3EDTA tubes. After the blood was drawn, one tube was placed on wet ice, and the other tube was located at room temperature, i.e., in a tube rack, as illustrated in Fig. 2.2. Both tubes were centrifuged at 1500–2000 × g at room temperature (20–25 °C) for 15 min. Plasma in each tube after centrifugation was assayed for $A\beta_{1-42}$ via IMR. The results are shown in Figs. 2.3(a) and (b). In wet-ice storage, the NC concentration in plasma $A\beta_{1-42}$ was 15.89 ± 1.19 pg/ml, and the AD concentration was 16.01 ± 1.09 pg/ml. The measured levels of plasma $A\beta_{1-42}$ clearly differed non-significantly between subjects with normal cognition (NC) and patients with Alzheimer's disease (AD), when the blood collecting tubes were stored in wet ice after the blood was drawn. However, in the blood collecting tubes stored at room temperature, the measured levels of plasma $A\beta_{1-42}$ were significantly greater in the AD group (21.33 ± 3.64 pg/ml) than in the NC group (14.50 ± 1.18 pg/ml). $A\beta_{1-42}$

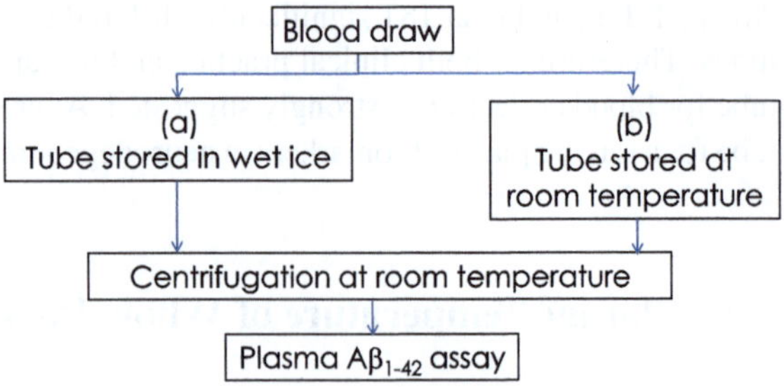

Fig. 2.2 Illustration of the flow chart for storing blood collecting tubes after blood draw, followed by room-temperature centrifugation and plasma $A\beta_{1-42}$ assay

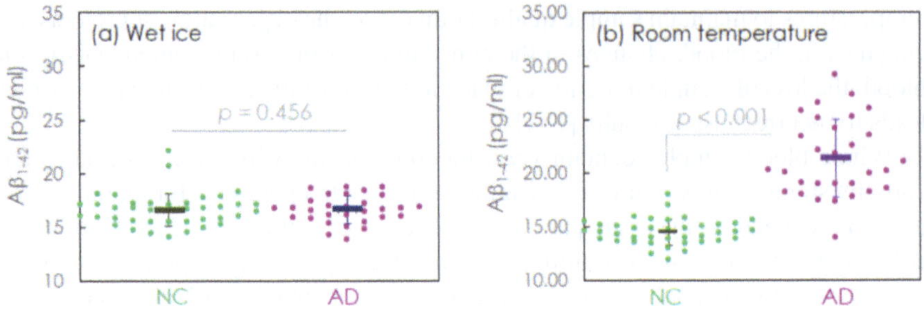

Fig. 2.3 Measured concentrations of plasma $A\beta_{1-42}$ using IMR in the case of storing blood tubes in (**a**) wet ice and (**b**) room temperature after blood draw

is a representative biomarker in Alzheimer's disease. The plasma levels of $A\beta_{1-42}$ differ significantly between AD patients and NCs. The results in Fig. 2.3 demonstrate that the storage temperature of blood collecting tubes plays a role in the accuracy of biomarker measurement.

2.4 Centrifugation Temperatures of Whole Blood

To obtain plasma samples, the blood collecting tubes were centrifuged as necessary. Although whole blood should be centrifuged at room temperature (20–25 °C) for homeostasis assays [24], centrifugation temperatures have no clear effect on the concentrations of analytes of interest in plasma. Thus, the temperature of centrifugation could be either room temperature (20–25 °C) or a low temperature (2–8 °C). For example, for performing routine coagulation tests, whole blood is frequently centrifuged at 2–8 °C [25]. It would be better to investigate the effects of the centrifuging temperature of whole blood on the measured concentrations of biomarkers associated with neurodegenerative diseases. Thus, the procedures of plasma preparation can be standardized.

Chen et al. reported the effects of centrifuging temperature on the measured concentrations of plasma α-synuclein and NfL assayed with immunomagnetic reduction (IMR) [26]. The results revealed that, as whole blood was centrifuged at room

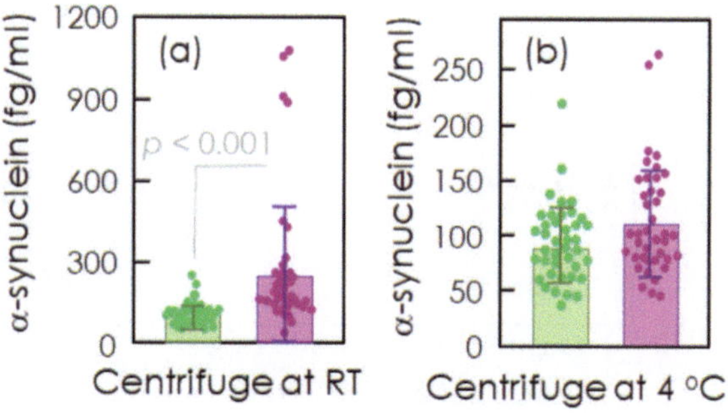

Fig. 2.4 Measured levels of plasma α-synuclein in PD and HC following centrifugation of their whole blood samples at (**a**) room temperature (RT) and (**b**) 4 °C [26]

temperature (RT), the levels of plasma α-synuclein were significantly elevated in patients with Parkinson's disease (PD) compared with healthy controls (HCs), as shown in Fig. 2.4(a). The differences in the plasma α-synuclein concentrations between normal controls and patients with Parkinson's disease were suppressed as the whole blood was centrifuged at 4 °C, as shown in Fig. 2.4(b). The differences in the plasma NfL levels between normal controls and patients with Parkinson's disease were weakened by centrifugation at 4 °C [26]. Low-temperature (2–8 °C) centrifugation may increase protein denaturation and critically lower the diagnostic potential of plasma biomarkers in diseases [27–31].

After reviewing the published literature involving the use of the IMR assay from 2012 to 2024, 19 studies used 20–25 °C for the centrifugation of whole blood [32–50], and another 4 studies used 4 °C [51–54]. In total, there were 1014 cognitively normal subjects, 533 amnestic mild cognitive impairment (aMCI) patients, and 727 AD patients. The plasma biomarkers of interest are $A\beta_{1\text{–}40}$, $A\beta_{1\text{–}42}$, or T-Tau. The discriminations in the concentrations of these biomarkers between cognitively normal subjects and patients with amnestic mild cognitive impairment (aMCI) or Alzheimer's disease (AD) are evaluated via effect size, which is expressed as

$$\text{Effect size} = \frac{\textit{Average concentration in patients}}{\textit{Average concentration in cognitively normal subjects}} \tag{2.1}$$

Once the effect size equals 1, there is no discrimination in the levels of biomarkers between cognitively normal subjects and patients. The distributions of the effect sizes of plasma $A\beta_{1\text{–}40}$, $A\beta_{1\text{–}42}$, and T-Tau at centrifugation temperatures of 20–25 and 2–8 °C reported in published studies are shown in Figs. 2.5(a)-(c). The value of 1 for the effect size is plotted with the dashed gray line.

The mean value of the effect sizes shown in Fig. 2.5(a) for plasma $A\beta_{1\text{–}40}$ centrifuged at 20–25 °C was 0.78, whereas it was 1.01 for plasma $A\beta_{1\text{–}40}$ centrifuged at 2–8 °C. For $A\beta_{1\text{–}42}$ in Fig. 2.5(b), the mean value of the effect sizes with centrifugation temperatures of 20–25 °C and 2–8 °C across studies was found to be 1.30 and 1.03, respectively. As shown in Fig. 2.5(c), the mean effect sizes of plasma T-Tau were 2.09 and 1.12 at centrifugation temperatures of 20–25 °C and 2–8 °C, respectively.

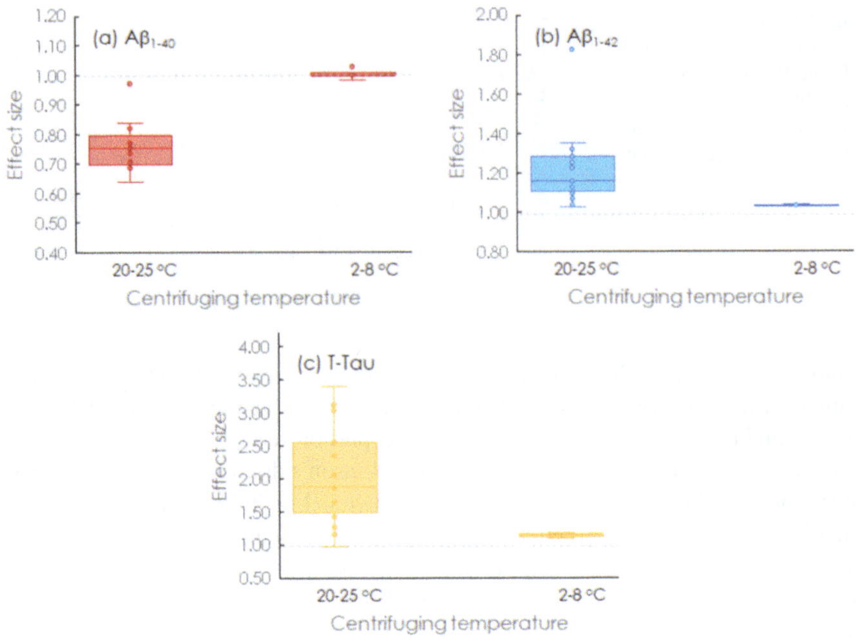

Fig. 2.5 Effect sizes of plasma (**a**) $A\beta_{1-40}$, (**b**) $A\beta_{1-42}$, and (**c**) T-Tau at centrifugation temperatures of 20–25 and 2–8 °C for blood collection tubes

The results shown in Figs. 2.5(a)–(c) clearly demonstrate that the effect sizes of plasma biomarkers are significantly greater than or less than 1 when whole blood is centrifuged at 20–25 °C, whereas they are close to 1 when the blood tubes are centrifuged at 2–8 °C; this reveals the various effects of centrifugation temperature on the discrimination of plasma biomarker levels between cognitively normal subjects and patients with aMCI and AD. In the case of IMR assay, it is preferable to centrifuge blood tubes at 20–25 °C, not 2–8 °C, to achieve obvious differences in the concentrations of plasma biomarkers between aMCI and AD patients and cognitively normal subjects.

2.5 Time Period to Centrifuge Blood Samples After Blood Draw

Prolonged contact of plasma with cells frequently manipulates the constitution of plasma in uncentrifuged blood samples. For example, red blood cells (RBCs) have 20x more potassium than plasma, and prolonged contact of plasma with RBCs will significantly change potassium levels in plasma. In addition, the hemoglobin released from hemolyzed cells contaminates the plasma due to prolonged contact [55]. Other analytes, such as glucose, phosphate, and creatinine, cannot be accurately assayed once centrifugation of blood samples is delayed by several hours

[56–58]. Boyanton et al. explained the instability and stoichiometric changes in analytes in uncentrifuged blood samples [59], namely, glucose depletion and the Na^+, K^+-ATPase pump failure; the movement of water into cells, causing hemoconcentration; and the leakage of intracellular constituents and metabolites. As a result, prolonged contact of plasma with cells commonly causes spurious test results. Plasma should be separated from cells quickly to prevent the ongoing metabolism of cellular constituents in blood collecting tubes [60, 61].

To prevent deterioration of plasma, the guidelines issued by the Clinical and Laboratory Standards Institute (CLSI) suggest the separation of plasma from cells within 2 h after blood draw [62], that is, to finish the centrifugation of blood collecting tubes within 2 h after blood draw. However, it is sometimes difficult to centrifuge blood samples within this time frame at urban hospitals or clinics. The CLSI guidelines suggest that the times to perform the centrifugation of blood samples depend on the analytes.

Owing to the stoichiometric and biochemical changes resulting from prolonged contact of plasma with cells, the proteins in plasma may be deformed, hydrolyzed, or abnormally digested [63, 64]. Thus, it is necessary to explore the effects of the time to centrifuge blood samples after blood draw on the test results of plasma biomarkers associated with neurodegenerative diseases. Chang et al. designed the study, as illustrated in Fig. 2.6, to investigate the effects [65].

Blood samples were collected from eight subjects in K_3EDTA tubes. Blood from each subject was drawn into four tubes numbered #1 to #4. The #1 tube was centrifuged at room temperature immediately (< 0.5 h) after the blood was drawn. The #2 to #4 tubes were incubated at room temperature for 3.5, 5.5, and 8.0 h before centrifugation, after which the biomarkers $A\beta_{1-40}$, $A\beta_{1-42}$, T-Tau, and α-synuclein in the plasma were assayed via immunomagnetic reduction (IMR). The measured concentrations of biomarkers in tubes #1 to #4 were denoted as ϕ_1, ϕ_2, ϕ_3, and ϕ_4, respectively. Using ϕ_1 as a reference, the recovery rates of measured concentrations in tubes #2 to #4 were calculated as follows:

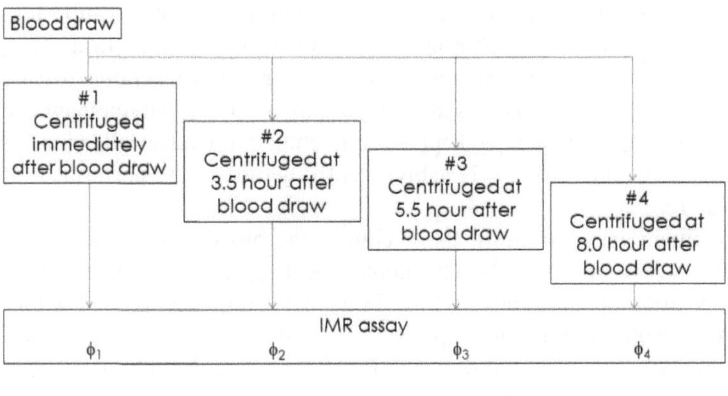

Fig. 2.6 Study design to investigate the effects of time to centrifuge blood samples after blood draw on test results using IMR [65]

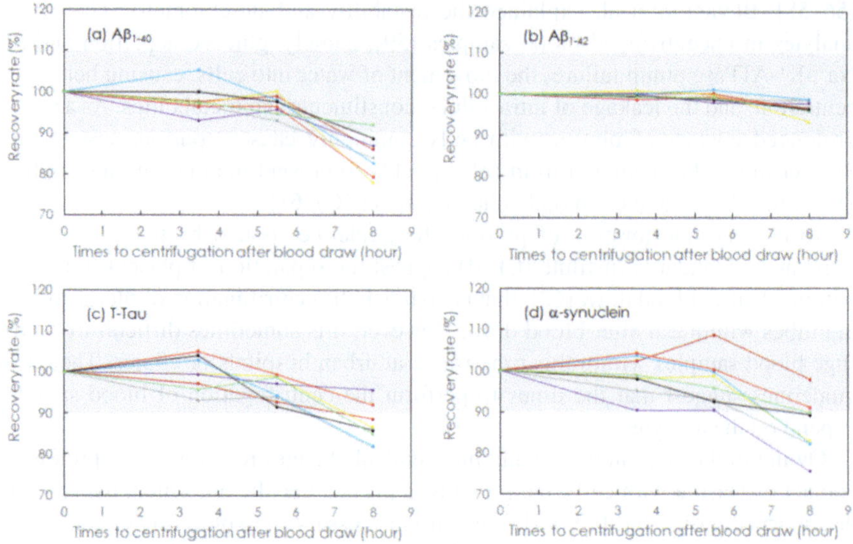

Fig. 2.7 Recovery rates at 3.5-, 5.5-, and 8.0-h centrifugation for plasma (**a**) $A\beta_{1-40}$, (**b**) $A\beta_{1-42}$, (**c**) T-Tau, and (**d**) α-synuclein of eight subjects [65]

$$\text{Recovery rate}\,(\%) = \frac{\phi_i}{\phi_1} \times 100\%, \tag{2.2}$$

where $i = 2, 3, 4$. The recovery rates at 3.5-, 5.5-, and 8.0-h centrifugations for plasma $A\beta_{1-40}$, $A\beta_{1-42}$, T-Tau, and α-synuclein from eight subjects are plotted in Figs. 2.7(a)–(d).

The acceptable range of recovery rates is from 90% to 110%, as indicated by the gray zones in Figs. 2.7(a)–(d), which implies that the measured concentrations ϕ_2, ϕ_3, or ϕ_4 do not differ significantly from ϕ_1. According to the results in Figs. 2.7(a)–(d), the recovery rates are acceptable at 3.5- and 5.5-h centrifugation for all biomarkers but are not acceptable at 8.0-h centrifugation. Thus, to avoid spurious results when plasma biomarkers associated with neurodegenerative diseases, such as $A\beta_{1-40}$, $A\beta_{1-42}$, T-Tau, and α-synuclein, are assayed via immunomagnetic reduction, blood samples, which are kept at room temperature after collection, should be centrifuged within 5.5 h. Notably, centrifugation was performed at room temperature.

Sometimes, in urban hospitals or clinics, the blood draw is performed in the morning, such as at 10:00 AM. The blood collecting tube was delivered to laboratories in the afternoon, e.g., at 1:30 PM, for centrifugation and assay. The blood samples were incubated at room temperature for 3–4 h. The time period of 5.5 h from blood draw to centrifugation, shown in Fig. 2.7, is long enough to meet the current operations in clinical practice.

2.6 Lifetime of Biomarkers in Frozen Plasma

In clinical practice, the assay is performed within 1 h or less after centrifugation. Thus, plasma is usually stored at room temperature for less than 1 h before the assay. However, for research purposes, plasma samples are frequently stored at −80 °C before being assayed. Currently, three biofluid banks have been identified for research worldwide. These biofluid banks play important roles in providing archived samples that have been stored at −80 °C for various lengths of time [66–68]. As peptides and proteins can be susceptible to degradation and denaturation during storage, investigating the stability of these biomarkers to determine the effects of storage time and conditions when interpreting results is crucial.

Chiu et al. investigated the storage stability of plasma samples at −80 °C [69]. Plasma samples were collected from cognitively normal subjects (NCs, n = 53) and patients (n = 46) with amnestic mild cognitive impairment (aMCI) or Alzheimer's disease (AD). For each plasma sample, the baseline measurements of the levels of $A\beta_{1-40}$, $A\beta_{1-42}$, and T-Tau were performed within 3 months after blood draw and are referred to as baseline concentrations. A separate aliquot from each sample was assayed to assess the stability of the measured analytes after storage for 1.1–5.4 years at −80 °C; this is referred to as a repeated concentration. To determine the stability of biomarkers in the samples, recovery rates were calculated for each sample and for each storage duration, including at baseline and after 2, 3, 4, and 5.4 years of storage. The recovery rate (%) was calculated as follows:

$$\text{Recovery rate}(\%) = \frac{\textit{Repeated concentration}}{\textit{Baseline concentration}} \times 100\%, \tag{2.3}$$

and is plotted in Figs. 2.8(a)–(c) for plasmas $A\beta_{1-40}$, $A\beta_{1-42}$, and T-Tau for various storage periods. The acceptable range (90–110%) of the recovery rate is indicated by the gray area. The recovery rates are clearly distributed in the gray area as the storage period approaches 1 year. Once the plasma samples have been stored at −80 °C for more than one year, the recovery rates are frequently distributed outside the gray area, especially for T-Tau; this implies that repeated concentrations frequently differ from the baseline concentration, as the storage period of the plasma samples is more than 1 year. The storage life of plasma $A\beta_{1-40}$, $A\beta_{1-42}$, and T-Tau at −80 °C with an immunomagnetic reduction (IMR) assay is 1 year. Notably, the storage stability results need to be considered if anybody tries to use IMR assays on archived samples from registries so that the results can be interpreted correctly depending on the context and goals of the evaluation.

Although the underlying mechanism for the instability of T-Tau in plasma stored at −80 °C is not clear, the degradation of T-Tau in frozen plasma was also shown by Zhao et al., who used a SIMOA assay [70]. According to the results reported by Zhao et al., the levels of plasma T-Tau in cognitively normal subjects (n = 229) showed a negative correlation with sample storage time at −80 °C, whereas the levels of plasma $A\beta_{1-40}$ and $A\beta_{1-42}$ did not correlate with sample storage time.

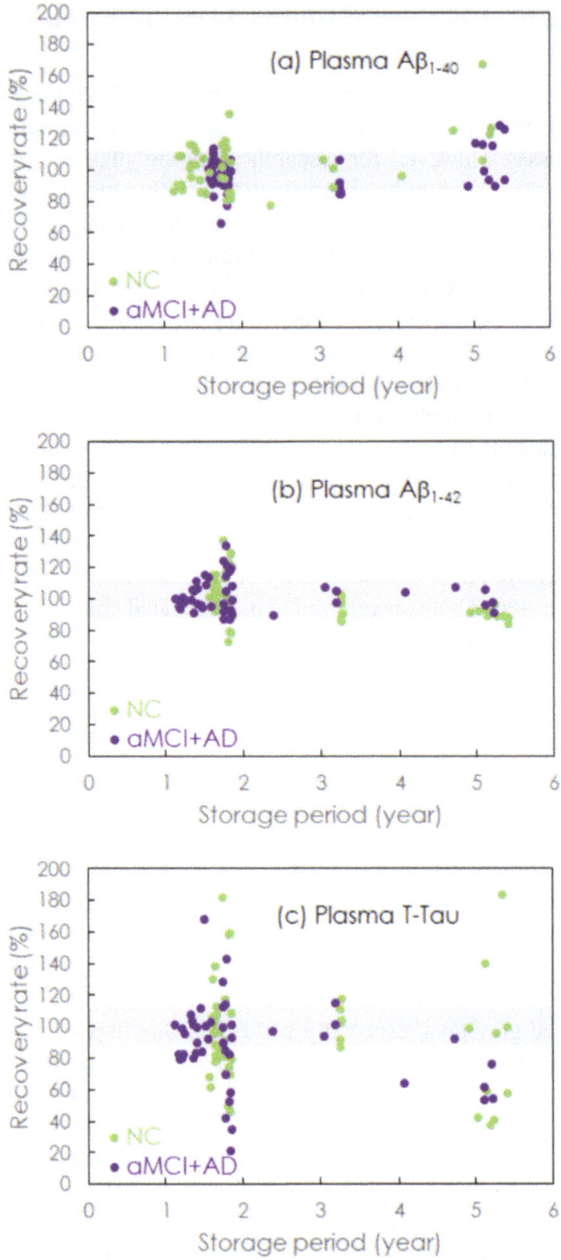

Fig. 2.8 Recovery rate versus storage period for (**a**) $A\beta_{1-40}$, (**b**) $A\beta_{1-42}$, and (**c**) T-Tau in human plasma assayed with immunomagnetic reduction. The gray area indicates recovery rates from 90 to 110% [69]

With respect to the other biomarkers in plasma, Liu et al. utilized an IMR assay and reported that the −80 °C storage stability of plasma α-synuclein and TDP-43 in middle-aged cognitively normal elderly individuals (n = 27) was no greater than 1 year [71], as shown in Figs. 2.9(a) and (b). Thus, different storage stabilities of

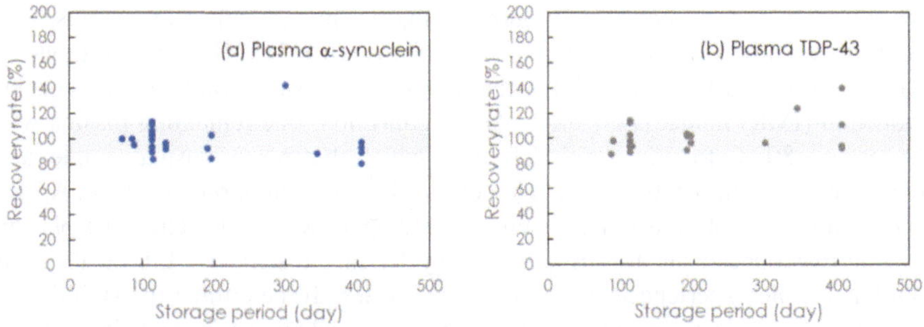

Fig. 2.9 Recovery rate versus storage period for (**a**) α-synuclein and (**b**) TDP-43 in human plasma assayed with immunomagnetic reduction. The gray area indicates recovery rates from 90 to 110%

plasma samples at −80 °C were observed with various biomarkers assayed with IMR. Researchers should pay attention to the storage time of frozen plasma samples for biomarkers of interest, such as when requesting samples from biofluid banks in retrospective studies.

2.7 Freeze/Thaw Cycles of Plasma

As mentioned in the previous section, the collected plasma samples may not be tested immediately after centrifugation, especially in research projects. The collected plasma samples were frequently stored at −80 °C until testing. Thus, the plasma samples underwent a freeze/thaw cycle from room temperature to −80 °C and then to room temperature. Sometimes, a plasma sample is subjected to repeated freeze/thaw cycles before measurement.

Some types of plasma proteins, such as fibrinogen, albumin, and immunoglobulins, are stable after repeated freeze/thaw cycles, whereas others are degraded. Lee et al. reported that five repeated freeze/thaw cycles did not cause spurious results in the assays of total protein, albumin, plasma interferon-γ, IL-8, VEGF-R2, creatinine, soluble transferrin receptor, and C-reactive protein but significantly modified the measured concentrations of MMP-7, tumor necrosis factor-α, and VEGF in plasma samples [72, 73]. Notably, the modifications in the measured concentrations could increase, such as those of MMP-7 and VEGF [72], or decrease, such as those of soluble transferrin receptor and C-reactive protein [73].

In addition to plasma protein, the freeze/thaw cycles could alter the test results of biochemical components such as vitamin D, vitamin B12, plasma glucose, high-density lipoprotein, and cholesterol [72]. However, low-density lipoprotein, ferritin, and triglyceride concentrations remained stable following freeze/thaw cycles [73].

The freeze/thaw cycles are performed with plasma samples to test biomarkers associated with neurodegenerative diseases. It is necessary to clarify the effects of the freeze/thaw cycles on the measured results of biomarkers. Liu et al. collected plasma samples from cognitively normal subjects (n = 2) and from patients with Alzheimer's disease (n = 3) or Parkinson's disease (n = 3) [74]. Concentrations of $A\beta_{1-40}$, $A\beta_{1-42}$, and T-Tau in each plasma sample were assayed via immunomagnetic reduction (IMR) immediately after plasma collection. The remaining plasma was frozen at −80 °C for more than 6 h, after which the frozen plasma sample was placed at room temperature for 1 h to defrost. The concentrations of $A\beta_{1-40}$, $A\beta_{1-42}$, and T-Tau in the plasma sample were assayed again with IMR. Thus, the plasma sample experienced the first freeze/thaw cycle. In the study by Liu et al., each plasma sample experienced five freeze/thaw cycles. To determine the stability of biomarkers in a plasma sample, recovery rates were calculated for each freeze/thaw cycle as follows:

$$\text{Recovery rate}(\%) = \frac{\textit{Measured concentration after n freeze / thaw cycles}}{\textit{Measured concentration immediately after plasma collection}} x100\%, \quad (2.4)$$

where n = 1–5. The recovery rates of individuals versus the number of freeze/thaw cycles for the $A\beta_{1-40}$, $A\beta_{1-42}$, and T-Tau assays with the IMR are plotted in Figs. 2.10(a)–(c). The acceptable range of recovery rates (90–110%) is indicated by the gray area. All the recovery rates shown in Figs. 2.10(a)–(c) are clearly distributed between 90% and 110%. The results of the IMR assay demonstrated that plasma samples could freeze/thaw for more than five cycles without any significant decrease in the concentrations of plasma $A\beta_{1-40}$, $A\beta_{1-42}$, or T-Tau.

The stabilities of plasma $A\beta_{1-40}$ (n = 14), $A\beta_{1-42}$ (n = 12), and T-Tau (n = 11) during freeze/thaw cycles were also demonstrated by Keshavan et al. via a SIMOA assay [75]. Lehmann et al. used an IP-MS assay and reported good stability for $A\beta_{1-40}$, $A\beta_{1-42}$, and APP [76]. The stability of plasma α-synuclein through three freeze/thaw cycles was revealed by Koehler et al. via a highly sensitive and specific sandwich ELISA [77].

Notably, in currently reported studies, up to five freeze/thaw cycles have been investigated. Although freeze/thaw cycles do not significantly lead to spurious results in the assay of plasma biomarkers associated with neurodegenerative diseases, the stability of the plasma biomarkers could be preserved if the number of freeze/thaw cycles is greater than 5. Attention should be given to the storage history, such as the use of plasma samples, to prevent spurious results from the assay.

2.8 Precautions

Before performing a blood draw for the assay of plasma biomarkers in neurodegenerative diseases, it is suggested to confirm the following precautions with subjects.

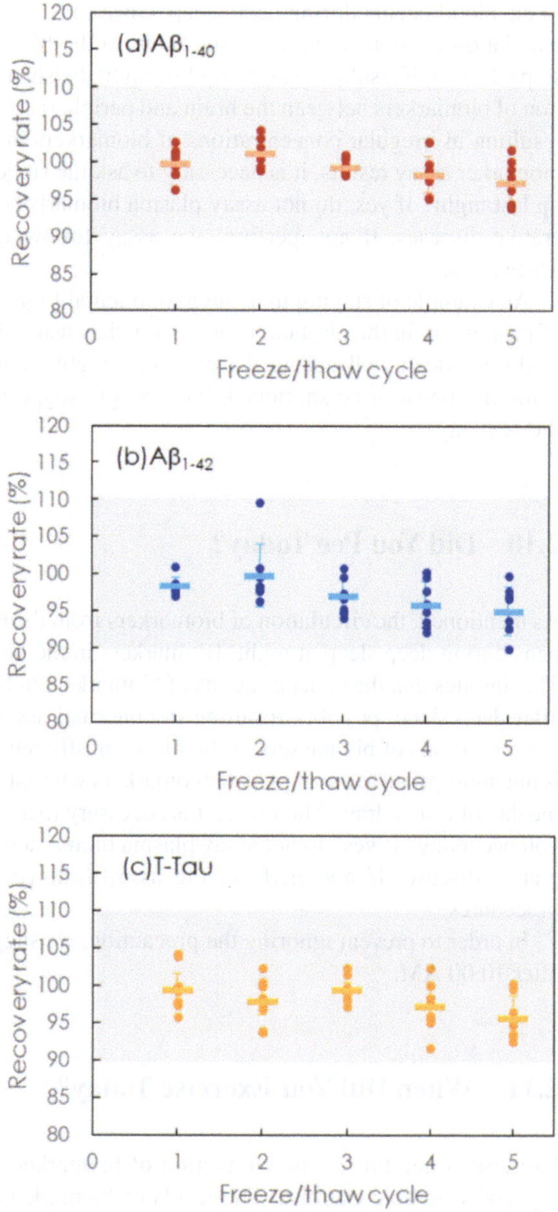

Fig. 2.10 Recovery rate versus freeze/thaw cycle for (**a**) $A\beta_{1-40}$, (**b**) $A\beta_{1-42}$, and (**c**) T-Tau in human plasma assayed with immunomagnetic reduction. The gray area indicates recovery rates from 90 to 110% [74]

2.9 Did You Stay Up Last Night?

In 2024, the group at Washington University in St. Louis found that brain waves help flush waste out of the brain to peripheral blood in humans during sleep, particularly deep sleep [78]. This implies that the circulation of biomarkers from the brain to the blood occurs during deep sleep. Once the deep sleep ends or disappears, the circulation of biomarkers from the brain to the blood becomes abnormal or even stops. Hence, if a subject stayed up last night, he/she losses deep sleep. The circulation of biomarkers between the brain and peripheral blood is not normal, potentially resulting in irregular concentrations of biomarkers in the blood. To avoid irregular biomarker assay results, it is necessary to ask the subject the question: Did you stay up last night? If yes, do not assay plasma biomarkers associated with neurodegenerative diseases. If not, perform the assay following standardized measurement procedures.

An example of staying up is intercontinental flights overnight. The sleep quality of passengers in the airplane is not as good as usual. Passengers usually lose sleep or do not sleep well during the overnight flight. Hence, the period of deep sleep could disappear or be shortened. We strongly suggest not performing the assay on the next day.

2.10 Did You Pee Today?

As mentioned, the circulation of biomarkers from the brain to the blood mainly happens during deep sleep at night. Biomarkers in the blood increase after deep sleep. This implies that the concentrations of biomarkers in blood are singular, not normal, after deep sleep, possibly resulting in abnormal assay results. After urination, the concentrations of biomarkers in the blood mostly return to normal levels. Hence, it is not appropriate to assay plasma biomarkers without finishing the first urination on the day of blood draw. Therefore, it is necessary to ask the subject the question: Did you pee today? If yes, do not assay plasma biomarkers associated with neurodegenerative disease. If not, perform the assay following standardized measurement procedures.

In order to prevent ignoring the precaution, we suggest collecting blood samples after 10:00 AM.

2.11 When Did You Exercise Today?

Exercise is helpful for the circulation of biomarkers from the brain to the blood, especially aerobic exercise. The levels of biomarkers in the blood would become abnormal after exercise. Hence, it is necessary to ask the subject the question: When did you exercise today? It is suggested to collect blood samples 1–2 h after exercise.

2.12 Does Hemolysis Occur in Tubes?

From the steps of blood drawing to centrifugation, hemolysis sometimes occurs due to the following reasons:

The needle is too thin. Red blood cells are easily squeezed and ruptured when entering and exiting a thin needle. Thus, red blood cells are highly likely to be damaged during blood drawing, resulting in hemolysis.

The amount of drawn blood is small. As the volume of blood is not enough in the collection tube, the frequency of collision of blood cells with the inner wall of the tube increases during transportation. The chance of damaging red blood cells increases, resulting in a higher likelihood of hemolysis.

It takes more than 2 min to draw blood. The use of the tourniquet for more than 2 min can easily cause red blood cells to rupture and result in hemolysis.

It is harmful to red blood cells if blood is drawn too quickly. Please pull the syringe gently, but not too slowly.

After blood draw, shake the collection tube gently. Otherwise, violent shaking of the collection tube could give a strong impact on red blood cells and damage the cells, resulting in hemolysis.

As a technician uses a needle to force blood directly into the test tube, blood cells are subjected to excessive pressure, rupture, and directly hit the test tube wall, which can easily cause hemolysis.

The blood is left for a long period of time after drawing. It is suggested to perform centrifugation as soon as possible after blood drawing.

Once hemolysis occurs, the cytoplasm of red blood cells flows into the plasma, causing a decrease in plasma protein concentration. Furthermore, many enzymes within the cytoplasm are released, which easily break down plasma proteins, causing them to become unstable. Hence, the measured concentrations of biomarkers in plasma would not be correct.

In fact, it is difficult to prevent hemolysis. According to the released levels of hemoglobin in plasma, the grades of hemolysis are classified into five levels, termed hemolysis index (HI). The relationship between the HI value and the concentration of free hemoglobin on the Beckman Coulter analyzers was classified as follows:

HI 0: hemoglobin <0.5 g/l;
HI 1: hemoglobin is 0.5–0.99 g/l;
HI 2: hemoglobin is 1–1.99 g/l;
HI 3: hemoglobin is 2–2.99 g/l;
HI 4: hemoglobin is 3–4.99 g/l.

The visual colors of the 5 classes of HI are shown in Fig. 2.11 [79]. Developers or manufactures shall inform users of the acceptable HI classes for their assay. For example, HI 0 and HI 1 are fine with the IMR assay.

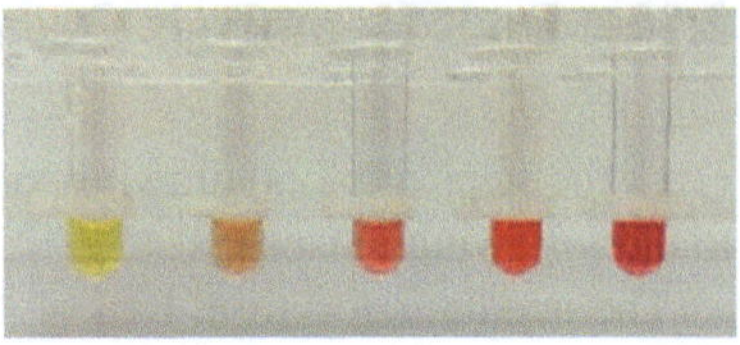

Fig. 2.11 Different grades of hemolysis in plasma (from left to right: HI 0, 1, 2, 3, and 4) [79]

References

1. Kac, P. R., Gonzalez-Ortiz, F., Simrén, J., Dewit, N., Vanmechelen, E., Zetterberg, H., Blennow, K., Ashton, N. J., & Karikari, T. K. (2022). Diagnostic value of serum versus plasma phospho-tau for Alzheimer's disease. *Alzheimer's Research & Therapy, 14*, 65-1-12.
2. Chang, C. W., Yang, S. Y., Yang, C. C., Chang, C. W., & Wu, Y. R. (2019). Plasma and serum alpha-synuclein as a biomarker of diagnosis in patients with Parkinson's disease. *Frontiers in Neurology, 10*, 1388-1-7.
3. Bibl, M., Welge, V., Esselmann, H., & Wiltfang, J. (2012). Stability of amyloid-β peptides in plasma and serum. *Electropho., 33*, 445.
4. Luque-Garcia, J. L., & Neubert, T. A. (2007). Sample preparation for serum/plasma profiling and biomarker identification by mass spectrometry. *Journal of Chromatography. A, 1153*, 259–276.
5. Kudoyarova-Zubavichene, N. M., Sergeyev, N. N., Chepurnov, A. A., & Netesov, S. V. (1999). Preparation and use of hyperimmune serum for prophylaxis and therapy of Ebola virus infections. *The Journal of Infectious Diseases, 179*, S218–S223.
6. Lavinder, J. J., Wine, Y., Giesecke, C., Ippolito, G. C., Horton, A. P., Lungu, O. I., Hoi, K., DeKosky, B. J., Murrin, E. M., Wirth, M. M., Ellington, A. D., Dörner, T., Marcotte, E. M., Boutz, D. R., & Georgiou, G. (2014). Identification and characterization of the constituent human serum antibodies elicited by vaccination. *PNAS, 111*, 2259–2264.
7. Brackexridge, C. J. (1960). The interaction of plasma fibrinogen with anticoagulants in vitro. *Clinica Chimica Acta., 5*, 552–558.
8. Mohri, M., Allahyari, L., & Sardari, K. (2007). Effects of common anticoagulants on routine plasma biochemistry of horse and comparison with serum. *Journal of Equine Veterinary Science, 27*, 313–316.
9. Palarasah, Y., Skjoedt, M. O., Vitved, L., Andersen, T. E., Skjoedt, K., & Koch, C. (2010). Sodium polyanethole sulfonate as an inhibitor of activation of complement function in blood culture systems. *Journal of Clinical Microbiology, 48*, 908–914.
10. Lima-Oliveira, G., Lippi, G., Salvagno, G. L., Montagnana, M., Picheth, G., & Guidi, G. C. (2013). Sodium citrate vacuum tubes validation: Preventing preanalytical variability in routine coagulation testing. *Blood Coagulation & Fibrinolysis, 24*, 252–255.
11. Banfi, G., Salvagno, G. L., & Lippi, G. (2007). The role of ethylenediamine tetraacetic acid (EDTA) as in vitro anticoagulant for diagnostic purposes. *Clinical Chemistry and Laboratory Medicine, 45*, 565–576.
12. Li, G., Cabanero, M., Wang, Z., Wang, H., Huang, T., Alexis, H., Eid, I., Muth, G., & Pincus, M. R. (2013). Comparison of glucose determinations on blood samples collected in three types of tubes. *Annals of Clinical and Laboratory Science, 43*, 278–284.
13. Goossens, W., Van Duppen, V., & Verwilghen, R. L. (1991). K2-or K3-EDTA: The anticoagulant of choice in routine haematology? *Clinical and Laboratory Haematology, 13*, 291–295.
14. Ashton, N. J., Suárez-Calvet, M., Karikari, T. K., Lantero-Rodriguez, J., Snellman, A., Sauer, M., Simrén, J., Minguillon, C., Fauria, K., Blennow, K., & Zetterberg, H. (2021). Effects of pre-analytical procedures on blood biomarkers for Alzheimer's pathophysiology, glial activation, and neurodegeneration. *Alzheimer's & Dementia, 13*, e12168-1-11.
15. Hughes, J. D., Macdonald, V. W., & Hess, J. R. (2007). Warm storage of whole blood for 72 hours. *Transfusion, 47*, 2050–2056.

16. Hervig, T., Kaada, S., & Seghatchian, J. (2014). Storage and handling of blood components—Perspectives. *Transfusion and Apheresis Science, 51*, 103–106.
17. Kaisar, M., van Dullemen, L. F. A., Thézénas, M. L., Zeeshan Akhtar, M., Huang, H., Rendel, S., Charles, P. D., Fischer, R., Ploeg, R. J., & Kessler, B. M. (2016). Plasma degradome affected by variable storage of human blood. *Clinical Proteomics, 13*, 26-1-11.
18. Chamchoy, K., Praoparotai, A., Pakparnich, P., Sudsumrit, S., Swangsri, T., Chamnanchanunt, S., Songdej, D., Imwong, M., & Boonyuen, U. (2021). The integrity and stability of specimens under different storage conditions for glucose-6-phosphate dehydrogenase deficiency screening using WST-8. *Acta Tropica., 217*, 105864-1-9.
19. Seymour, C. W., Carlbom, D., Cooke, C. R., Watkins, T. R., Bulger, E. M., Rea, T. D., & Baird, G. S. (2011). Temperature and time stability of whole blood lactate: Implications for feasibility of pre-hospital measurement. *BMC Research Notes, 4*, 169-1-4.
20. Turchiano, M., Nguyen, C., Fierman, A., Lifshitz, M., & Convit, A. (2013). Impact of blood sample collection and processing methods on glucose levels in community outreach studies. *Journal of Environmental and Public Health, 2013*, 256151-1-4.
21. Tobias, K. M., Serrano, L., Sun, X., & Flatland, B. (2016). The effectiveness of cooling conditions on temperature of canine EDTA whole blood samples. *PeerJ, 4*, e2732-1-9.
22. Ashworth, M., Small, B., Oldfeld, L., Evans, A., Greenhalf, W., Halloran, C., & Costello, E. (2021). The holding temperature of blood during a delay to processing can affect serum and plasma protein measurements. *Scientific Reports, 11*, 6487-1-11.
23. Hope, C. M., Huynh, D., Wong, Y. Y., Oakey, H., Perkins, G. B., Nguyen, T., et al. (2021). Optimization of blood handling and peripheral blood mononuclear cell cryopreservation of low cell number samples. *International Journal of Molecular Sciences, 22*, 9129-1-19.
24. Adcock Funk, D. M., Lippi, G., & Favaloro, E. J. (2012). Quality standards for sample processing, transportation, and storage in hemostasis testing. *Seminars in Thrombosis and Hemostasis, 38*, 576–585.
25. Lippi, G., Salvagno, G. L., Montagnana, M., Poli, G., & Guidi, G. C. (2006). Influence of centrifuge temperature on routine coagulation testing. *Clinical Chemistry, 52*, 537–538.
26. Chang, K. H., Liu, K. C., Lai, C. S., Yang, S. Y., & Chen, C. M. (2021). Assessing plasma levels of α-synuclein and neurofilament light chain by different blood preparation methods. *Frontiers in Aging Neuroscience, 13*, 759182-1-10.
27. Privalov, P. L. (1990). Cold denaturation of proteins. *Critical Reviews in Biochemistry and Molecular Biology, 25*, 281–305.
28. Ben-Naim, A. (2013). Theory of cold denaturation of proteins. *Advances in Biological Chemistry, 3*, 27790-1-11.
29. Alfano, C., Sanfelice, D., Martin, S. R., Pastore, A., & Temussi, P. A. (2017). An optimized strategy to measure protein stability highlights differences between cold and hot unfolded states. *Nature Communications, 8*, 15428-1-9.
30. Rosa, M., Roberts, C. J., & Rodrigues, M. A. (2017). Connecting high-temperature and low temperature protein stability and aggregation. *PLoS One, 12*, e0176748-1-12.
31. Bitonti, A., Puglisi, R., Meli, M., Martin, S. R., Colombo, G., Temussi, P. A., & Pastore, A. (2022). Recipes for inducing cold denaturation in an otherwise stable protein. *Journal of the American Chemical Society, 144*, 7198–7207.
32. Chiu, M. J., Yang, S. Y., Chen, T. F., Chieh, J. J., Huang, T. Z., Yip, P. K., Yang, H. C., Cheng, T. W., Chen, Y. F., Hua, M. S., & Hong, H. E. (2012). New assay for old markers-plasma beta amyloid of mild cognitive impairment and Alzheimer's disease. *Current Alzheimer Research, 9*, 1142–1148.
33. Chiu, M. J., Yang, S. Y., Horng, H. E., Yang, C. C., Chen, T. F., Chieh, J. J., Chen, H. H., Chen, T. C., Ho, C. S., Chang, S. F., Liu, H. C., Hong, C. Y., & Yang, H. C. (2013). Combined plasma biomarkers for diagnosing mild cognition impairment and Alzheimer's disease. *ACS Chemical Neuroscience, 4*, 1530–1536.
34. Tzen, K. Y., Yang, S. Y., Chen, T. F., Cheng, T. W., Horng, H. E., Wen, H. P., Huang, Y. Y., Shiue, C. Y., & Chiu, M. J. (2014). Plasma Aβ but not tau is related to brain PiB retention in early Alzheimer's disease. *ACS Chemical Neuroscience, 5*, 830–836.

35. Chiu, M. J., Chen, Y. F., Chen, T. F., Yang, S. Y., Yang, F. P. G., Tseng, T. W., Chieh, J. J., Chun, J. C., Chen, R., Tzen, K. Y., Hua, M. S., & Horng, H. E. (2014). Plasma tau as a window to the brain-negative associations with brain volume and memory function in mild cognitive impairment and early Alzheimer's disease. *Human Brain Mapping, 35*, 3132–3142.
36. Lue, L. F., Sabbagh, M. N., Chiu, M. J., Jing, N., Snyder, N. L., Schmitz, C., Guerra, A., Belden, C. M., Chen, T. F., Yang, C. C., Yang, S. Y., Walker, D. G., Chen, K., & Reiman, E. M. (2017). Plasma levels of Aβ42 and tau identified probable Alzheimer's dementia: Findings in two cohorts. *Frontiers in Aging Neuroscience, 9*, 226-1-9.
37. Yang, S. Y., Chiu, M. J., Chen, T. F., Lin, C. H., Jeng, J. S., Tang, S. C., Lee, Y. F., Yang, C. C., Liu, B. H., Chen, H. H., & Wu, C. C. (2017). Analytical performance of reagent for assaying tau protein in human plasma and feasibility study screening neurodegenerative diseases. *Scientific Reports, 7*, 9304-1-12.
38. Fan, L. Y., Tzen, K. Y., Chen, Y. F., Chen, T. F., Lai, Y. M., Yen, M. F., Huang, Y. Y., Shiue, C. Y., Yang, S. Y., & Chiu, M. J. (2018). The relation between brain amyloid deposition, cortical atrophy, and plasma biomarkers in amnesic mild cognitive impairment and Alzheimer's disease. *Frontiers in Aging Neuroscience, 10*, 175-1-10.
39. Teunissena, C. E., Chiu, M. J., Yang, C. C., Yang, S. Y., Scheltens, P., Zetterbergg, H., & Blennow, K. (2018). Plasma amyloid-β (Aβ42) correlates with cerebrospinal fluid Aβ42 in Alzheimer's disease. *Journal of Alzheimer's Disease, 62*, 1857–1863.
40. Yang, C. C., Chiu, M. J., Chen, T. F., Chang, C. L., Liu, B. H., & Yang, S. Y. (2018). Assay of plasma phosphorylated tau protein (threonine 181) and total tau protein in early-stage Alzheimer's disease. *Journal of Alzheimer's Disease, 61*, 1323–1332.
41. Morimot, S., Hatsuta, H., Minami, K., et al. (2019). The ultra sensitive biomarker for Alzheimer disease by immunomagnetic reduction (IMR) assay. In *60th Annual Meeting Japanese Society of Neurology* (p. Pj-042–6).
42. Jiao, F., Yi, F., Wang, Y., Zhang, S., Guo, Y., Du, W., Gao, Y., Ren, J., Zhang, H., Liu, L., Song, H., & Wang, L. (2020). The validation of multifactor model of plasma Aβ42 and total-tau in combination with MoCA for diagnosing probable Alzheimer disease. *Frontiers in Aging Neuroscience, 12*, 2121-1-9.
43. Chiu, M. J., Chen, T. F., Hu, C. J., Yan, S. H., Sun, Y., Liu, B. H., Chang, Y. T., Yang, C. C., & Yang, S. Y. (2020). Nanoparticle-based immunomagnetic assay of plasma biomarkers for differentiating dementia and prodromal states of Alzheimer's disease – a cross-validation study. *Nanomedicine: Nanotechnology, Biology and Medicine, 26*, 102182-1-9.
44. Lin, C. H., Chiu, S. I., Chen, T. F., Jang, J. S. R., & Chiu, M. J. (2020). Classifications of neurodegenerative disorders using a multiplex blood biomarkers-based machine learning model. *International Journal of Molecular Sciences, 21*, 6914-1-15.
45. Chang, H. T., Chiu, M. J., Chen, T. F., Hsu, Y. T., Wang, H. F., Yang, Y. C., Lien, H. T., & Hua, M. S. (2021). Boundary extension as a tool for detection of cognitive change among individuals with mild cognitive impairment: A preliminary study. *Archives of Gerontology and Geriatrics, 94*, 104329-1-7.
46. Chiu, M. J., Yang, S. Y., Chen, T. F., Lin, C. H., Yang, F. C., Chen, W. P., Zetterberg, H., & Blennow, K. (2021). Synergistic association between plasma Abeta (1-42) and pTau in Alzheimer's disease but not in Parkinson's disease or frontotemporal dementia. *ACS Chemical Neuroscience, 12*, 1376–1383.
47. Chiu, S. I., Fan, L. Y., Lin, C. H., Chen, T. F., Lim, W. S., Jang, J. S. R., & Chiu, M. J. (2022). Machine learning-based classification of subjective cognitive decline, mild cognitive impairment, and Alzheimer's dementia using neuroimage and plasma biomarkers. *ACS Chemical Neuroscience, 13*, 3263–3270.
48. Chen, P. H., Lin, S. Y., Liao, Y. Y., Hsu, W. L., & Cheng, F. Y. (2022). Associations between blood-based biomarkers of Alzheimer's disease with cognition in motoric cognitive risk syndrome: A pilot study using plasma Aβ42 and total tau. *Frontiers in Aging Neuroscience, 14*, 981632-1-10.

49. Sabbagh, M. H., Lue, L. F., Decourt, B., & Shill, H. (2022). Plasma biomarkers combined with UPDRS and MoCA for differential assessments of amnestic cognitive impairment (aMCI), Parkinson's disease (PD +PDD) and dementia with Lewy body (DLB): A pilot study. *Alzheimer's & Dementia, 18*, e059716.
50. Chao, S. P., Lin, W. C., Lu, C. H., Wu, H. C., Liu, H. C., & Hu, C. J. (2024). Changes in plasma biomarkers differentiate clinical stages of Alzheimer's disease using immunomagnetic reduction assays. *Journal of Alzheimer's Disease, 102*, 459–469.
51. Tsai, C. L., Liang, C. S., Lee, J. T., Su, M. W., Lin, C. C., Chu, H. T., Tsai, C. K., Lin, G. Y., Lin, Y. K., & Yang, F. C. (2019). Associations between plasma biomarkers and cognition in patients with Alzheimer's disease and amnestic mild cognitive impairment: A cross-sectional and longitudinal study. *Journal of Clinical Medicine, 8*, 1893-1-13.
52. Liang, C. S., Su, K. P., Tsai, C. L., Lee, J. T., Chu, C. S., Yeh, T. C., Su, M. W., Lin, G. Y., Lin, Y. K., Chu, H. T., Tsai, C. K., & Yang, F. C. (2020). The role of interleukin-33 in patients with mild cognitive impairment and Alzheimer's disease. *Alzheimer's Research & Therapy, 12*, 86-1-9.
53. Tsai, C. L., Liang, C. S., Yang, C. P., Lee, J. T., Ho, T. H., Su, M. W., Lin, G. Y., Lin, Y. K., Chu, H. T., Hsu, Y. W., & Yang, F. C. (2020). Indicators of rapid cognitive decline in amnestic mild cognitive impairment: The role of plasma biomarkers using magnetically labeled immunoassays. *Journal of Psychiatric Research, 129*, 66–72.
54. Liang, C. S., Tsai, C. L., Lin, G. Y., Lee, J. T., Lin, Y. K., Chu, C. S., Sung, Y. F., Tsai, C. K., Yeh, T. C., Chu, H. T., Su, M. W., & Yang, F. C. (2021). Better identification of cognitive decline with interleukin-2 than with amyloid and tau protein biomarkers in amnestic mild cognitive impairment. *Frontiers in Aging Neuroscience, 13*, 670115-1-10.
55. Verma, M., Dahiya, K., Malik, D., Sehgal, P. K., Devi, R., Soni, A., & Ghalaut, V. S. (2015). Effect of blood storage on complete biochemistry. *Journal of Blood Disorders & Transfusion, 6*, 6-1-4.
56. Zhang, D. J., Elswick, R. K., Miller, W. G., & Bailey, J. L. (1998). Effect of serum-clot contact time on clinical chemistry laboratory results. *Clinical Chemistry, 44*, 1325–1333.
57. Clark, S., Youngman, L. D., Palmer, A., Parish, S., Peto, R., & Collins, R. (2003). Stability of plasma analytes after delayed separation of whole blood: Implications for epidemiological studies. *International Journal of Epidemiology, 32*, 125–130.
58. Tanner, M., Kent, N., Smith, B., Fletcher, S., & Lewer, M. (2008). Stability of common biochemical analytes in serum gel tubes subjected to various storage temperatures and times pre-centrifugation. *Annals of Clinical Biochemistry, 45*, 375–379.
59. Boyanton, B. L., & Blick, K. E. (2002). Stability studies of twenty-four analytes in human plasma and serum. *Clinical Chemistry, 48*, 2242–2247.
60. Tayal, D., Gupta, M., & Goswami, B. (2017). Does prolonged storage of serum samples alter the lab results? *Indian Journal of Medical Biochemistry, 21*, 30–33.
61. Jr Boyanton, B. L., & Blick, K. E. (2002). Stability studies of twenty-four analytes in human plasma and serum. *Clinical Chemistry, 48*, 2242–2247.
62. CLSI document GP44-A4. (2010). *Procedures for the handling and processing of blood specimens* (4th ed.). Approved Guideline.
63. Baranello, R. J., Bharani, K. L., Padmaraju, V., Chopra, N., Lahiri, D. K., Greig, N. H., Pappolla, M. A., & Sambamurti, K. (2015). Amyloid-beta protein clearance and degradation (ABCD) pathways and their role in Alzheimer's disease. *Current Alzheimer Research, 12*, 32–46.
64. González, L. A., Carvalho, J. G. S., Kuinchtner, B. C., Dona, A. C., Baruselli, P. S., & D'Occhio, M. J. (2023). Plasma metabolomics reveals major changes in carbohydrate, lipid, and protein metabolism of abruptly weaned beef calves. *Scientific Reports, 13*, 8176-1-13.
65. Chang, J. F., Liu, H. C., Chen, H. H., Chen, W. P., Juang, J. L., Wang, P. N., & Yang, S. Y. (2020). Effect of times to blood processing on the stability of blood proteins associated with dementia. *Dementia and Geriatric Cognitive Disorders, 49*, 303–311.

66. Kornhuber, J., Schmidtke, K., Frolich, L., Perneczky, R., Wolf, S., Hampel, H., et al. (2009). Early and differential diagnosis of dementia and mild cognitive impairment: Design and cohort baseline characteristics of the German dementia competence network. *Dementia and Geriatric Cognitive Disorders, 27*, 404–417.
67. Mueller, S. G., Weiner, M. W., Thal, L. J., Petersen, R. C., Jack, C. R., Jagust, W., et al. (2005). Ways toward an early diagnosis in Alzheimer's disease: The Alzheimer's disease neuroimaging Initiative (ADNI). *Alzheimer's & Dementia, 1*, 55–66.
68. Visser, P. J., Verhey, F. R., Boada, M., Bullock, R., De Deyn, P. P., Frisoni, G. B., et al. (2008). Development of screening guidelines and clinical criteria for predementia Alzheimer's disease. The DESCRIPA study. *Neuroepidemiology, 30*, 254–265.
69. Chiu, M. J., Lue, L. F., Sabbagh, M. N., Chen, T. F., Chen, H. H., & Yang, S. Y. (2019). Long-term storage effects on stability of Aβ1–40, Aβ1–42, and total tau proteins in human plasma samples measured with immunomagnetic reduction assays. *Dementia and Geriatric Cognitive Disorders Extra., 9*, 77–86.
70. Zhao, L., Zhang, M., Li, Q., Wang, X., Lu, J., Han, Y., & Cai, Y. (2025). Storage time affects the level and diagnostic efficacy of plasma biomarkers for neurodegenerative diseases. *Neural Regeneration Research, 20*, 2373–2381.
71. Liu, H. C., Chen, H. H., Wang, P. N., & Yang, S. Y. (2022). One-year longitudinal stability of plasma biomarkers associated with dementia in normal controls. *Alzheimer's & Dementia, 18*, e062624.
72. Lee, J. E., Kim, S. Y., & Shin, S. Y. (2015). Effect of repeated freezing and thawing on biomarker stability in plasma and serum samples. *Osong Public Health and Research Perspectives, 6*, 357–362.
73. Abraham, R. A., Rana, G., Agrawal, P. K., Johnston, R., Sarna, A., Ramesh, S., Acharya, R., Khan, N., Porwal, A., Kurundkar, S. B., Pandey, A., Pullakhandam, R., Nair, K. M., Kumar, G. T., Sachdev, H. P. S., Kapil, U., Deb, S., de Wagt, A., Khera, A., & Ramakrishnan, L. (2021). The effects of a single freeze-thaw cycle on concentrations of nutritional, noncommunicable disease, and inflammatory biomarkers in serum samples. *Journal of Laboratory Physicians, 13*, 6–13.
74. Keshavan, A., Heslegrave, A., Zetterberg, H., & Schott, J. M. (2018). Stability of blood-based biomarkers of Alzheimer's disease over multiple freeze-thaw cycles. *Alzheimers & Dementia: Diagnosis, Assessment & Disease Monitoring, 10*, 448–451.
75. Liu, H. C., Chiu, M. J., Lin, C. H., & Yang, S. Y. (2020). Stability of plasma amyloid-β 1-40, amyloid-β 1-42, and total tau protein over repeated freeze/thaw cycles. *Dementia and Geriatric Cognitive Disorders Extra, 10*, 46–55.
76. Lehmann, S., Vialaret, J., Kindermans, J., Gabelle, A., & Hirtz, C. (2020). MALDI-TOF IP-MS quantification of plasma amyloid peptides in Alzheimer's disease. *Alzheimer's & Dementia, 16*, e047112-1-2.
77. Koehler, N. K. U., Stransky, E., Meyer, M., Gaertner, S., Shing, M., Schnaidt, M., Celej, M. S., Jovin, T. M., Leyhe, T., Laske, C., Batra, A., Buchkremer, G., Fallgatter, A. J., Wernet, D., & Richartz-Salzburger, E. (2015). Alpha-synuclein levels in blood plasma decline with healthy aging. *PLoS One, 10*, e0123444-1-16.
78. Jiang-Xie, L. F., Drieu, A., Bhasiin, K., Quintero, D., Smirnov, I., & Kipnis, J. (2024). Neuronal dynamics direct cerebrospinal fluid perfusion and brain clearance. *Nature, 627*, 157–164.
79. https://commons.wikimedia.org/wiki/File:HEMOLYSIS.jpg

Chapter 3
Preclinical Evaluations of In Vitro Diagnostic Medical Devices

Abstract The assay kits and analyzers of blood-based biomarkers for assessing/diagnosing neurodegenerative diseases are categorized as medical devices of in vitro diagnosis (IVD) medical devices. Many bench tests, including functional characterization, electrical safety, electromagnetic compatibility, biocompatibility, and sterilization validation, must be conducted to demonstrate the safety, preclinical performance, and validity of measurements. All at-bench tests are conducted in-house before launching an IVD medical device in clinical markets, even before initiating clinical trials. The evaluations of the safety of an IVD medical device to subjects or users during operation are conducted according to the ISO 14971 guidelines, IEC 61010 guidelines, IEC 61326 guidelines, EN 61000 guidelines, etc. The preclinical performance and validity of an IVD medical device must be well characterized according to the guidelines issued by the Clinical and Laboratory Standards Institute (CLSI). The tested performances include standard curve, analytical sensitivity, measuring precision, measuring accuracy, analytical specificity, and reagent stability. Although the philosophy and methods for the bench tests are clearly described in the guidelines, it is not easy for an inexperienced person to fully understand the details. In this chapter, practical examples are given to illustrate the details of bench tests for characterizing the assay kits and analyzers under developments.

Keywords In vitro diagnosis · Medical device · Safety · Preclinical performance · Validity · Guidelines

3.1 Classifications of Medical Devices

Medical devices are indispensable in clinical practice and are designed for various intended purposes. Depending on their purpose, medical devices are categorized into three groups: diagnostic devices, treatment devices, and auxiliary devices. Obviously, the use of medical devices impacts the health of subjects. The safety or

C. S.-Y. Yang, M. N. Sabbagh, *Assessing Neurodegenerative Diseases: High-Sensitivity Technologies That Assay Pathological Biomarkers in Blood*,
https://doi.org/10.1007/978-3-032-23863-4_3

risk to humans is an extremely important issue with medical devices. Hence, strict adherence to comprehensive regulations is necessary to develop medical devices.

As medical devices can have several risky impacts on subjects, there are different requirements in regulations for medical devices of various types, depending on the safety risk to the health of the subjects. This is referred to as the classification of medical devices. Notably, the classification of a medical device is not based on the complexity, innovation, or operation of the device but rather on the safety and effectiveness of the device. The classification of a medical device could vary from country to country. Here, the classifications of medical devices in the European Union and the USA are introduced.

3.1.1 European Union

For example, according to the European Union Medical Device Regulation (EU MDR), medical devices are classified into four risk classes: I, IIa, IIb, and III. The features of medical devices in various classes are briefly introduced.

3.1.1.1 Class I

The medical devices in Class I are low risk, noninvasive, and intended to be used temporarily, with durations of less than 60 min in general. Class I medical devices are very versatile, such as crutches, elastic bandages, latex gloves, masks, orthopedic shoes, tweezers, scalpels, wheelchairs (without a motor), weighing scales, patient beds, stethoscopes, and thermometers. The manufacturer or their authorized representative is required to issue the EU declaration of conformity to take full responsibility for the devices' compliance with the applicable EU law [1]. In addition, the manufacturer or their authorized representative is required to draw up the technical documentation for the devices. The technical documentation provides information on the design, manufacture, and operation of a device [1].

3.1.1.2 Class IIa

The medical devices in Class IIa are generally low- to medium-risk and are frequently referred to as devices installed within the body for approximately 60 min to 30 days. Common examples of Class IIa medical devices include dental fillings, orthodontic wires, syringes, infusion pumps, surgical suction equipment, surgical gloves, lancets, short-term corrective contact lenses, hearing aids, and diagnostic imaging devices. In addition to the EU declaration of conformity and technical documentation, the manufacturer or their authorized representative is required to issue a conformity assessment, including a clinical evaluation, conducted by a notified body to ensure that the devices meet all regulatory criteria. The detailed contents are illustrated in Annex IX and Annex XI [2, 3].

3.1.1.3 Class IIb

The medical devices in Class IIb are generally medium- to high-risk but not as high-risk as life-supporting devices and are often implanted within the body for 30 days or longer. Typical examples of Class IIb medical devices include blood bags, intensive monitoring equipment, ventilators, implantable dental devices, long-term corrective contact lenses, intraocular lenses, short-term implantable catheters, surgical lasers, incubators for babies, and high-intensity focused ultrasound equipment. Similar to Class IIb, the manufacturer or their authorized representative is required to issue the EU declaration of conformity, the technical documentation of conformity assessment conducted by a notified body. The detailed contents are illustrated in Annex IX and Annex XI [2, 3].

3.1.1.4 Class III

The medical devices in Class III are strictly high-risk and are thus subject to the highest level of regulatory scrutiny and control. Examples of Class III medical devices include implantable pacemakers, monitoring software for active implantable devices, prosthetic heart valves, diagnostic image analysis for acute treatment decisions, spinal disc replacements, breast implants, drug-eluting stents, and deep-brain stimulators. The manufacturer or their authorized representative is required to issue the EU declaration of conformity, technical documentation, and conformity assessment conducted by a notified body according to Annex IX or Annex X coupled with Annex XI [2, 3].

According to the European Database on Medical Devices (EUDAMED) of 2023, approximately 11% of approved medical devices are Class I, 45% are Class IIa, 33% are Class IIb, and 11% are Class III. Class IIa medical devices constitute the highest proportion.

3.1.2 USA

In the USA, the Food and Drug Administration (FDA) categorizes medical devices as Classes 1, 2, and 3 based on their risks and regulatory controls. The features and regulatory controls for medical devices of various classes are briefly introduced.

3.1.2.1 Class 1

The medical devices in Class 1 are noninvasive and low-risk to patients. Class 1 medical devices are subject only to General Controls. General Controls are the basic provisions (authorities) of the Medical Device Amendments to the Federal Food, Drug and Cosmetic Act, issued May 28, 1976 [4]. In accordance with the amendments, regulations related to adulteration, misbranding, device registration and listing, premarket notification, banned devices, notification (repair, replacement, or refund), records and reports, restricted devices, and good manufacturing practices are included in General Controls. General Controls are applied to all medical devices. Based on General Controls, FDA is able to regulate medical devices effectively to ensure their safety and effectiveness.

3.1.2.2 Class 2

These medical devices pose a greater risk to patients. The medical devices in Class 2 require a 501(k) premarket notification, which is a premarket submission to the FDA [5]. The submitted materials must clearly demonstrate that the medical device to be marketed is substantially equivalent to an approved and marketed device. Hence, it is necessary to perform comparison studies between a medical device to be marketed and an approved device on the market. Once 510(k) clearance is granted, the submitter can market the device immediately. The manufacturer should always be prepared for quality inspections (21 CFR 280) by the FDA after 510(k) clearance [6].

3.1.2.3 Class 3

The medical devices in Class 3 pose the highest risk to patients or are the first of their kind to pose a risk comparable to Class 2 devices. These devices usually help sustain human life or frequently present potential unreasonable risks of injury to patients. Therefore, Class 3 medical devices are subject to the highest level of regulatory control. Premarket analysis (PMA), which is the process of scientific and regulatory review, is required by the FDA for evaluating the safety and effectiveness of Class 3 medical devices [7]. Clinical trial data with a Class 3 medical device are key parts of the PMA for obtaining approval to enter the market.

3.2 Devices for Assessing Plasma Biomarkers in Neurodegenerative Diseases

The analyzers and reagent kits used to assay plasma biomarkers associated with neurodegenerative diseases are medical devices but may be categorized into different classifications in different countries. For example, an immunomagnetic reduction (IMR) analyzer is a Class I medical device in Taiwan but a Class II device in the EU, the USA, and China. The IMR reagent kits for Aβ peptides or tau protein are Class III medical devices in Taiwan but Class II devices in the EU, the USA, and China.

The assay of plasma biomarkers associated with neurodegenerative diseases is performed outside the human body, and all of the analyzers and reagent kits are referred to as medical devices for in vitro diagnosis (IVD) in the category of chemical chemistry. To market such IVD devices, the manufacturer must be certified to ISO 13485:2016 and have a quality management system (QMS) and must conduct a preclinical analysis and premarket evaluation. In this chapter, the preclinical analysis of such IVD devices is introduced. The premarket evaluation of such IVD devices will be discussed in the next chapter.

3.3 Preclinical Analysis

During the development of an IVD medical device for assaying biomarkers in human plasma, many bench tests must be conducted to demonstrate the safety, performance, and validity of measurements, which refer to the various types of bench testing, including functional characterization, electrical safety, electromagnetic compatibility, biocompatibility, and sterilization validation. The reasons why manufacturers and developers have to approve the safety, performance, and validity of measurements using an IVD medical device are briefly explained.

For the safety issue, manufacturers and developers must evaluate the harmful risk of an IVD medical device to subjects or users during operation according to the ISO 14971 guidelines. Harmful risk is defined as the possibility of loss or injury to subjects and users. Regardless of the probability of occurrence and the severity of harm, manufacturers and developers pay attention to designing safe IVD medical devices and completely proving them. In addition, notes or precautions must be clearly written in user manuals to avoid potential loss or injury.

For the performance issue, an IVD medical device must be well characterized using metrics such as standard curves, analytical sensitivity, and measurement precision. The characterizations demonstrate the ability of the IVD medical device to sensitively and precisely quantify the analyte of interest.

Before an IVD medical device is launched in clinical laboratories, bench tests are required to investigate the validity of operations, such as measuring accuracy, analytical specificity, and reagent stability, which are performed at clinical laboratories. With these results, the standard operating procedures involving the use of the IVD medical device can be correctly determined to prevent measurement errors in clinical laboratories.

All at-bench tests are conducted in-house before launching an IVD medical device to clinical markets, even before initiating clinical trials. These tests are termed preclinical analysis, preclinical evaluations, or analytical performance tests.

The Clinical and Laboratory Standards Institute (CLSI) provides global standards and guidelines for performing preclinical evaluations.

In the case of assaying plasma biomarkers associated with neurodegenerative diseases, the detection system includes an analyzer and reagent kits.

For analyzers, the preclinical evaluation mainly includes evaluations of the safety of operating IVD devices. There are several guidelines for designing and analyzing safety. For example, the IEC 61010 guidelines ensure the safety of personnel who work with or are in proximity to such IVD devices. IEC 61326 specifies requirements for immunity and emissions regarding electromagnetic compatibility (EMC) for electrical equipment, operating from a power supply or battery of less than 1000 V AC or 1500 V DC or from the circuit being measured. The EN 61000 guidelines provide more general inspections of the safety of an IVD device, including functional safety, measurement uncertainty, emission limits, immunity limits, measurement techniques, testing techniques, installation guidelines, and generic emission and immunity requirements in various environments. There are many certifying laboratories that perform evaluations for manufacturers or developers. Certifications are issued for an analyzer once the evaluation is passed.

For reagent kits, the key items of preclinical analysis, also known as analytical performance, include the following:

Standard curve, including the analytical relationship between the detected signals and the concentrations of analytes of interest, the Hook effect, and the linear range of measurements

Analytical sensitivity, including the limit of background, limit of detection, and limit of qualification

Measuring precision, including repeatability, intermediate precision, and reproducibility

Measuring accuracy, including diluted recovery, spiked recovery, and carry-over effect

Analytical specificity, including interference tests and cross-reactivity

Reagent stability, including open vials, unopened vials, and transportation stability

The manufacturers and developers of analyzers or reagent kits must follow the guidelines issued by the CLSI to characterize the preclinical analysis. The details of each item are introduced.

3.3.1 Standard Curve

In vitro diagnostic (IVD) devices require the use of a standard curve to accurately quantify the concentrations of analytes in patient samples. The standard curve is the analytical relationship between the detected signals and the concentrations of the measured analytes. Users can quantify the analyte of interest by converting the detected signal, such as the optical density or reduced magnetic signal, to the concentration of the analyte via a standard curve.

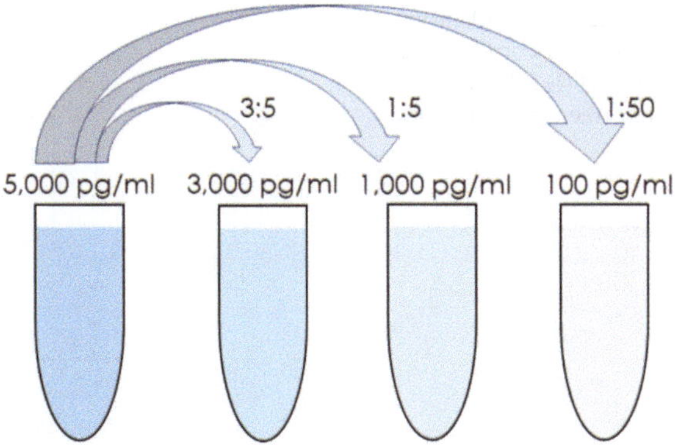

Fig. 3.1 Direct dilutions from the referenced protein solution to those of various concentrations used in establishing a standard curve. The ratio is the volume ratio of a referenced protein solution to a dilution buffer solution

The detailed methodologies used to establish a standard curve are described in Ref. 8. There are several important issues to consider when establishing a standard curve: the referenced protein solution, dilution, analytical curve, Hook effect, and linear range of measurement.

The first important issue in establishing a standard curve is to have the referenced protein solution. The referenced protein solution used for preparing samples of various concentrations to construct a standard curve should be recognized publicly. Unfortunately, assays of plasma biomarkers in neurodegenerative disease are new to clinical laboratories. Thus, no publicly recognized referenced protein solution is available. Therefore, manufacturers and developers should prepare protein solutions in-house for establishing a standard curve.

The second important issue is the dilution of the reference protein solution. The concentration of the in-house referenced protein solution is usually much greater than those of the biomarkers in human plasma. Therefore, it is necessary to dilute the referenced protein solution. Notably, the diluted concentrations must cover the concentration ranges of biomarkers in human plasma. Furthermore, at least six concentrations should be used in a standard curve.

For example, the detected concentrations of T-Tau range from 1 to 50 pg/ml in human plasma. The concentrations prepared in a standard curve could range from 0.1 to 3000 pg/ml.

The reference T-Tau concentration is 5,000 pg/ml. The solution is then diluted to concentrations of 3,000, 1,000, 100, 50, 20, 10, 1, and 0.1 pg/ml. The dilution method and dilution solutions are critical.

The stock solution should be diluted directly from 5,000 pg/ml to the target concentration, as illustrated in Fig. 3.1. The diluent could be phosphate-buffered saline (PBS) or plasma substitutes. Manufacturers and developers should discuss with their regulatory offices to select suitable dilution solutions.

The third important issue in establishing a standard curve is to formulate the analytical curve, which is a mathematical equation presenting a standard curve. The analytical curve could be logistic, linear, or other. Manufacturers and developers should fit the detected data points to various analytical curves. The most suitable

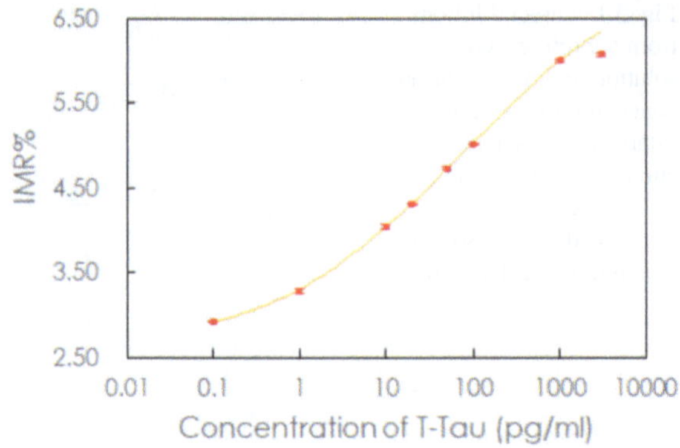

Fig. 3.2 Standard curve analytically correlating the detected signals to the biomarker concentrations

analytical curve gives the highest value of the coefficient of determination. Regardless of the type of analytical curve used, the coefficient of determination should be greater than 0.7 to reveal a good match between the detected data points and the analytical curve.

To measure the signals detected in protein solutions of various concentrations, duplicate measurements are conducted for each concentration. The average value of duplicate detected signals is used for correlation with the concentrations of the analyte. Figure 3.2 shows a typical analytical curve for T-Tau assayed with immunomagnetic reduction (IMR) technology [9]. The measured data points are well fitted to a four-parameter logistic function, as plotted with a solid line.

$$\mathrm{IMR\%} = \frac{A - B}{1 + \left(\dfrac{\phi}{\phi_o}\right)^{\gamma}} + B \tag{3.1}$$

where IMR% is the signal detected via IMR technology; A, B, ϕ_o, and γ are fitting parameters; and ϕ denotes the concentration of T-Tau. For the example in Fig. 3.2, A is 2.599, B is 7.210, ϕ_o is 76.46, and γ is 0.397. The error bar with each data point in IMR% is the standard deviation attributed to the duplicated measurements.

Because the logistic function is frequently applied to biosystems, it is worth knowing the analytical meaning of the fitting parameters A, B, ϕ_o, and γ in Eq. (3.1). In fact, the four parameters feature the logistic function.

A in Eq. (3.1) denotes the asymptotic value as the concentration of the analyte approaches zero, i.e., the low plateau in a standard curve. B denotes the saturated value as the concentration of the analyte becomes extremely high, i.e., the high plateau in a standard curve. Thus, A and B reveal the distributed range of the detected signals of an assay. Manufacturers and developers sometimes try to expand the distribution range to increase the resolution of measured concentrations, which would be helpful for increasing the discrimination between clinical groups based on the measured concentrations. ϕ_o corresponds to the concentration showing (A + B)/2 for the detected signal. Manufacturers and developers wish to position ϕ_o so that it exists in the concentration range of the analyte in real human samples. γ is the slope of a standard curve at a concentration of ϕ_o. γ is related to the binding kinetics between antibodies and analyte molecules [10].

The fourth important issue in a standard curve is to identify the analyte concentrations showing the the Hook effect. At higher concentrations, the fitted analytical curve can sometimes deviate from the data points, as shown at concentrations of approximately 3,000 pg/ml. Typically, the data points are below the fitted analytical curve. This implies that the detected signal stops or retards from increasing at higher concentrations. The most likely reason for the deviation is the so-called Hook effect. This is a result of excess captured biomarker molecules, which might block additional associations between capture antibodies and unbound biomarker molecules or block the binding sites of neighboring captured biomarker molecules to labeling antibodies. The specific concentration at which the Hook effect occurs can be tuned by adjusting the concentrations of capture antibodies, the chemical concentrations of the solution, or other factors. Manufacturers and developers should make the specific concentration much higher than the measured concentrations of the analytes in human plasma to prevent underquantification.

The fifth important issue in establishing a standard curve is to find the linear range of measurement. In Fig. 3.2, it is clear that the fitted standard curve frequently deviates from the data points at high concentrations because of the Hook effect and sometimes deviates at low concentrations because of imprecision. There exists a range in which the fitted standard curves are very close to the data points. This range is referred to as the linear range of measurement [11]. It is necessary to identify the linear range of measurement for each reagent kit. The method used to find the linear range is briefly introduced.

The detected signals at various concentrations are converted to measured concentrations via a fitted standard curve, such as that produced by Eq. (3.1). Hereafter, the concentrations calculated via the standard curve are referred to as the measured concentrations. The concentrations prepared for establishing the standard curve are referred to as spiked concentrations. The linear relationship between the measured concentrations and spiked concentrations is analyzed. Manufacturers and developers must find the concentration range in which the slope of the linear relationship is within the acceptable values, i.e., 0.9–1.1. As shown with the red dots in Fig. 3.3, the linear relationship at concentrations from 0.1 to 1,000 pg/ml is 1.01 for the slope, as indicated by the red solid line. Once the point at 3000 pg/ml is included in the analysis, the slope is 0.48, as guided by the orange dashed line. The linear range of measurement in Fig. 3.3 is from 0.1 to 1,000 pg/ml.

In addition to the criterion for the slope of the linear relationship between measured and spiked concentrations, the coefficient of variation (CV) of multiple measurements at measured concentrations is a key criterion in determining the linear range of measurement.

The coefficient of variation (CV) is defined as

$$\mathrm{CV\%} = \frac{\textit{Average value of multiple measurements}}{\textit{Standard deviation of multiple measurements}} \mathrm{x}100\% \quad (3.2)$$

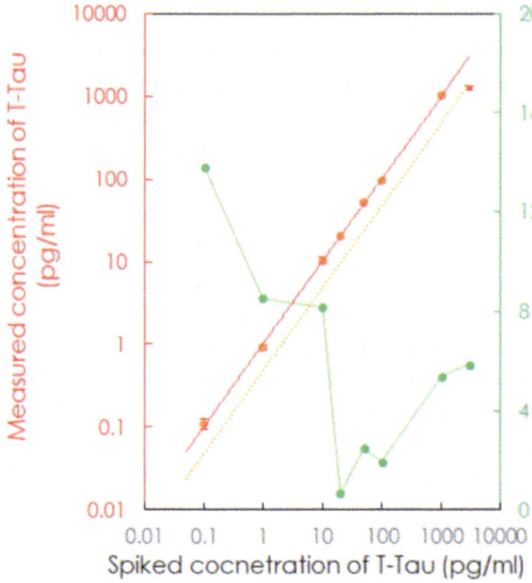

Fig. 3.3 Measured concentrations (red dots) and coefficients of variation (CVs) at various spiked concentrations

Duplicate or triplicate measurements are often performed at a given spiked concentration. The acceptable value of CV% is 3% or lower at concentrations in the ng/ml range or higher. In the case of plasma biomarkers in neurodegenerative diseases, the concentrations are in the ranges of pg/ml and fg/ml, i.e., they are extremely low. Hence, the acceptable value of the CV% could be 20% [12–14].

Manufacturers and developers must check the CV% at each spiked concentration. For the example in Fig. 3.3, the CV% values are plotted with green dots. All the CV% values are less than 20%. Hence, by combining the slope and CV criteria, the linear range of measurement for assaying T-Tau via immunomagnetic reduction is from 0.1 to 1000 pg/ml.

3.3.2 Analytical Sensitivity

The concentrations of plasma biomarkers associated with neurodegenerative diseases range from pg/ml to fg/ml, which are extremely low. The ability to assay biomarkers at such low concentrations is an important issue. It is necessary to evaluate the analytical sensitivity of an IVD medical device to ensure the ability to accurately detect ultralow concentrations of analytes in human samples. Hence, manufacturers and developers must clarify the following detection limits: limit of blank (LoB), limit of detection (LoD), and limit of qualification (LoQ) [15]. The limit of blank (LoB) should first be established, followed by finding the limit of detection (LoD) and the limit of qualification (LoQ). The global standards for these assay limits are described in CLSI EP17-A2 [1], and the conceptual interpretations of these limits are described.

If a sample does not contain a target analyte, the measured concentration of analyte might not be zero due to noise from an IVD medical device, such as that from an analyzer and unstable or nonspecific antibody–molecule association of a reagent. A nonzero concentration could cause false positives. Hence, manufacturers and developers need to clarify the background level of analyte detection for an IVD medical device to avoid causing false positives, which reflects the highest measurement result that is likely to be observed for a blank sample. A blank sample refers to a sample without the target analyte.

Limit of blank (LoB) is established starting from replicated measurements performed with a blank sample. The number of measurements should be several of tens, say 60. LoB is analyzed according to the distribution of the measured results.

The measured results follow a Gaussian distribution, and the LoB is calculated via

$$LoB = \mu_B + 1.645\sigma_B \tag{3.3}$$

where μ_B and σ_B are the average value and the standard deviation of the measured results, respectively.

For situations where values less than zero are not reported or the measured results follow a non-Gaussian distribution, the LoB is calculated as follows: the measurements are ordered according to their values, and the appropriate percentile (p) is estimated as the value of the observation with the rank value as determined below; in this case, $p = 0.95$: [16]

$$LoB = Results\ at\ position\left[0.95 \times NB + 0.5\right] \tag{3.4}$$

where NB = 60 (NB is the number of measurements) in this case. Eq. (3.4) becomes

$$LoB = Results\ at\ position\, 57.5 \tag{3.5}$$

This is a noninteger value. The distribution of the 60 test results is non-Gaussian. Linear interpolation is carried out using the 57th and 58th ranked observations according to CLSI EP17-A2.

The measured results may remain around the level of the LoB even when a sample contains extremely small amounts of the target analyte, as labeled with the brown arrow in Fig. 3.4. The measured results do not significantly deviate from the level of LoB until enough analyte exists in a sample. Confirmation of minimum detectable values to ensure sensitive detection is an important issue. The lowest amount of analyte in a sample that can be detected with probability, although perhaps not quantified as an exact value, is referred to as the low limit of detection (LLoD).

However, in a standard curve, it is difficult to determine the LLoD, which shows that the signal significantly deviates from that of LoB, as labeled with the green arrows in Fig. 3.4. CLSI EP17-A2 clearly defines the LLoD, which is expressed as [15]

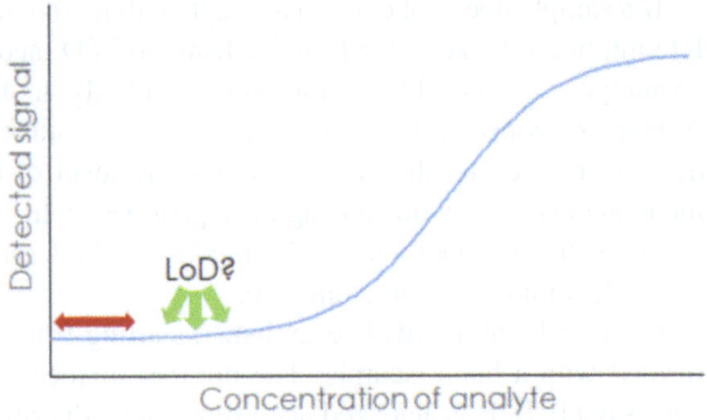

Fig. 3.4 Schematic standard curve of an assay analyte of interest

$$LLoD = LoB + 1.645\sigma_s \tag{3.6}$$

where σ_S is the standard deviation of the replicated measurements of analyte concentrations in PBS solutions at a given spiked analyte concentration, which is usually very low. The number of measurements could be the same as that in evaluating the LoB, e.g., 60.

An alternative method to evaluate the LLoD is the so-called 3-σ criterion, at which the LLoD is calculated via

$$LLoD = concentration\,showing\,the\,detected\,signal\,of\;A + 3\sigma_{s'} \tag{3.7}$$

where A is the detected signal of the low plateau in the standard curve, such as the value of A in Eq. (3.1). $\sigma_{s'}$ is the standard deviation of detected signals in duplicate or triplicate measurements at very low concentrations used to establish a standard curve. Empirically, the LLoD obtained via the 3-σ criterion is higher than that obtained via Eq. (3.6). Notably, the formally reported LLoD is that obtained via Eq. (3.6).

Limit of quantitation (LoQ) is the lowest amount of an analyte in a sample that can be quantitatively and reliably determined such that the total error meets the laboratory's requirements for accuracy. Evaluations of the total error of measurements for a device are regulated in Refs. 17 and 18. Depending on the defined goals for total error, the LoQ could be equal to the LLoD or higher than the LLoD. In most situations, the LoQ is calculated via

$$LoQ = 10SD / S \tag{3.8}$$

where SD is the standard deviation of replicated measurements of analytes at a very low concentration in the standard curve and S is the slope of the linear relationship between the measured concentration and spiked concentration, as shown in Fig. 3.3. Usually, the LoQ is greater than the LLoD. Manufacturers and developers must claim the LoQ, LLoD and LoB for their IVD medical devices.

The LoB is the highest background level in an assay with a blank sample. A blank sample refers to a sample that lacks the target analyte. However, the LLoD is often confused with the LoQ. In brief, the LLoD is taken as the lowest concentration of an analyte in a sample that can be detected, not necessarily quantified, under the stated test conditions. The LoQ is the lowest concentration of an analyte in a sample that can be determined with acceptable precision and accuracy under the stated conditions of tests. The detailed methods for determining the limit of detection and limit of quantitation of the analytical methods are reviewed in Ref. 19.

Table 3.1 lists the LoB, LLoD, and LoQ of assaying plasma biomarkers in neurodegenerative diseases using various technologies. The information is obtained from the manufacturers' data sheets or published papers [20–24].

3.3.3 Measuring Precision

Scientifically and statistically, the values of replicated measurements under fixed conditions should not be identical and should present a distribution. It is necessary to investigate the closeness of the measured values to confirm the consistency and reproducibility of an IVD medical device [25]. The precision of replicated measurements under given conditions is expressed by the imprecision of the measurements, such as the standard deviation, variance, or coefficient of variation. CLSI-EP5-A3 provides guidance for designing experiments and statistical analysis for evaluating the precision of an IVD device [26].

According to the given measurement conditions, precision can be categorized into three types, namely, repeatability, intermediate precision, and reproducibility, which are conceptually introduced separately.

The repeatibility is formerly termed "within-run precision." The measurements are performed under unchanged conditions, including the same measurement procedure, same operators, same measuring system, same operating conditions and same location, and replicate measurements on the same or similar objects over a short period of time [27, 28]. This is the lowest imprecision of measurements at a given concentration. The variation in the results of a single run is caused by intrinsic factors, such as variations in the pipetted volumes of the reagent or sample or the background noise of the analyzer. Repeatability is calculated via analysis of variance (ANOVA). Conventionally, the acceptable repeatability is less than 10%. The reported repeatability or within-run precision of assaying plasma biomarkers associated with neurodegenerative diseases is listed in Table 3.2 [29].

In real practice, an IVD medical device is located in a laboratory. The facilities, methods, and procedures of the measurements should be unchanged. However, measurements are often performed daily. Furthermore, there could be more than one medical technician operating measurements in different runs. This implies that measurements are performed on different days, in different runs, or by different operators. It is necessary to clarify the measurement imprecision under such conditions [30]. This is the so-called intermediate precision, usually termed

Table 3.1 Limit of background (LoB), low limit of detection (LLoD) and limit of quantification (LoQ) of assaying biomarkers associated with neurodegenerative diseases with various technologies

Technology	Biomarker	LoB	LLoD	LoQ
Immunomagnetic reduction (IMR)	$A\beta_{1-40}$	41 fg/ml	64 fg/ml	84 fg/ml
	$A\beta_{1-42}$	46 fg/ml	80 fg/ml	440 fg/ml
	T-tau	7 fg/ml	35 fg/ml	89 fg/ml
	pTau181	5 fg/ml	19.6 fg/ml	31 fg/ml
	pTau217	1.8 fg/ml	66 fg/ml	66 fg/ml
	α-Synuclein	0.006 fg/ml	0.21 fg/ml	0.21 fg/ml
	p-α-syn129	0.0003 fg/ml	0.072 fg/ml	1.9 fg/ml
	NfL	0 fg/ml	0.18 fg/ml	0.18 fg/ml
	BDNF	0.01 fg/ml	3.0 fg/ml	5.5 fg/ml
	GFAP	0.029 fg/ml	31.4 fg/ml	2.06 pg/ml
	TDP-43	0.53 fg/ml	0.68 fg/ml	94 fg/ml
	UCH-L1	0.002 fg/ml	3.3 fg/ml	285 fg/ml
Single molecule array (SIMOA)	$A\beta_{1-40}$	–	196 fg/ml	675 fg/ml
	$A\beta_{1-42}$	–	45 fg/ml	142 fg/ml
	T-tau	–	19 fg/ml	63 fg/ml
	pTau181	–	620 fg/ml	8000 fg/ml
	pTau231	–	621 fg/mL	1830 fg/ml
	pTau217	–	5.2 fg/mL	60 fg/ml
	NfL	–	62 fg/ml	309 fg/ml
	BDNF	–	4.2 fg/ml	68.6 fg/ml
	GFAP	–	211 fg/ml	686 fg/ml
	TDP-43	–	2480 fg/ml	8230 fg/ml
	UCH-L1	–	1050 fg/ml	3430 fg/ml
Magnetic particle-based chemiluminescent enzyme immunoassay (MP-CLEIA)	$A\beta_{1-40}$	–	440 fg/ml	440 fg/ml
	$A\beta_{1-42}$	–	270 fg/ml	430 fg/ml
	pTau181	–	50 fg/ml	470 fg/ml
	pTau217	–	30 fg/ml	30 fg/ml
High-sensitivity chemiluminescence enzymeimmunoassay (HISCL)	$A\beta_{1-40}$	570 fg/ml	1310 fg/ml	2460 fg/ml
	$A\beta_{1-42}$	70 fg/ml	130 fg/ml	160 fg/ml
Electrochemiluminescence (ECL)	$A\beta_{1-40}$	–	5410 fg/ml	20,000 fg/ml
	$A\beta_{1-42}$	–	516 fg/ml	2500 fg/ml
	$A\beta_{1-38}$	–	22,200 fg/ml	60,000 fg/ml
	T-tau	–	70 fg/ml	630 fg/ml
	pTau181	–	77 fg/ml	460 fg/ml
	pTau231	–	3100 fg/ml	15,000 fg/ml
	pTau217	–	880 fg/ml	1840 fg/ml
	NfL	–	5500 fg/ml	9600 fg/ml
	GFAP	–	8800 fg/ml	52,000 fg/ml
Immunoprecipitation mass spectrometry (IP/MS)	$A\beta_{1-40}$	–	< 41 fg/ml	1250 fg/ml
	$A\beta_{1-42}$	–	< 82 fg/ml	3130 fg/ml

Table 3.2 Coefficients of variation for within-run precision (repeatability) and within-lab precision (intermediate precision) of assaying biomarkers associated with neurodegenerative diseases with various technologies

Technology	Biomarker	Within-run precision	Within-lab precision
Immunomagnetic reduction (IMR)	$A\beta_{1-40}$	< 4%	< 6%
	$A\beta_{1-42}$	< 8%	< 9%
	T-tau	< 8%	< 8%
	pTau181	< 2%	< 10%
	pTau217	–	–
	α-Synuclein	< 3%	< 2%
	p-α-syn129	< 4%	< 3%
	NfL	< 3%	< 2%
	BDNF	< 1%	< 9%
	GFAP	< 4%	< 8%
	TDP-43	< 4%	< 3%
	UCH-L1	< 1%	< 8%
Magnetic particle-based chemiluminescent enzyme immunoassay (MP-CLEIA)	$A\beta_{1-40}$	< 3%	< 6%
	$A\beta_{1-42}$	< 4%	< 6%
	pTau181	< 3%	< 5%
	pTau217	< 7%	< 15%
High sensitivity Chemiluminescence enzyme immunoassay (HISCL)	$A\beta_{1-40}$	< 4%	< 5%
	$A\beta_{1-42}$	< 3%	< 6%
Immunoprecipitation mass spectrometry (IP/MS)	$A\beta_{1-40}$	< 5%	< 13%
	$A\beta_{1-42}$	< 8%	< 13%
Liquid chromatography tandem mass spectrometry (LC/MS/MS)	$A\beta_{1-40}$	< 5%	< 8%
	$A\beta_{1-42}$	< 10%	< 10%
	pTau217	< 8%	< 10%

"within-laboratory precision" or "within-lab precision." The within-lab precision is calculated via analysis of variance (ANOVA). Conventionally, the acceptable repeatability is less than 15%. The reported within-lab precision of assaying plasma biomarkers associated with neurodegenerative diseases is listed in Table 3.2 [29, 31–36].

As IVD medical devices are distributed in markets, they are located in several laboratories worldwide. This implies that measurements should be performed in different laboratories using various instruments. Frequently, different lots of reagents could be used in different laboratories. The medical technicians operating measurements at different laboratories are obviously different. Hence, the agreement between repeated measurements of the same sample under different reagent lots, operators, instruments, and laboratories should be investigated. Such reproducibility is termed "between-lots," "between-operators," "between-instruments," and "between-labs," respectively.

Typically, three reagent lots, three medical technicians, three instruments, and three laboratories are designated to evaluate between-lots, between-operators,

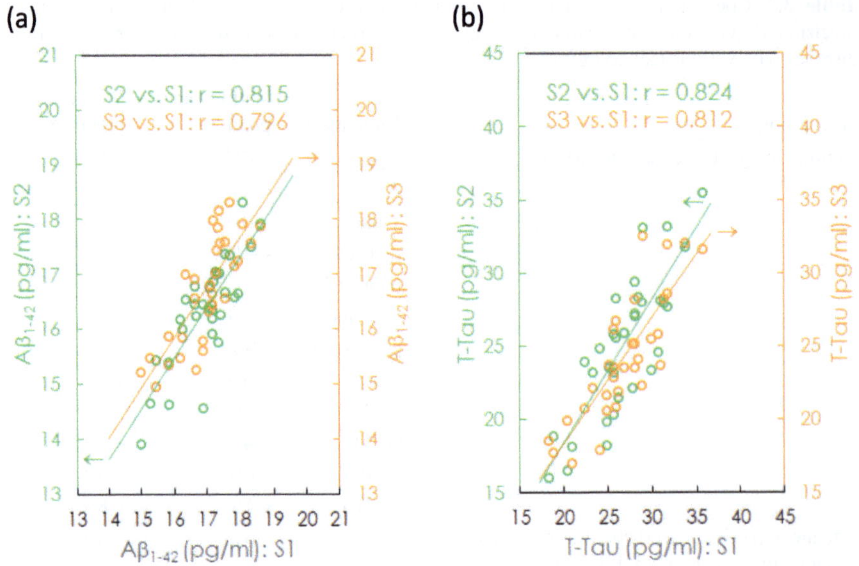

Fig. 3.5 Correlations in measured (**a**) $A\beta_{1-42}$ (**b**) T-Tau concentrations among three analyzers [37]

between-instruments, and between-labs reproducibility, respectively. For example, as reported by Lee et al. [37], three analyzers based on immunomagnetic reduction (IMR) technology were installed in three cities: Taipei (designated S1), Seoul (designated S2), and Göteborg (designated S3). Each of the 32 samples was divided into 3 aliquots. Each set of aliquots was delivered to one city where an IMR analyzer was located. The $A\beta_{1-42}$ or T-Tau concentrations measured with the analyzer S1 were used as references. The correlations of the measured $A\beta_{1-42}$ or T-Tau concentrations with analyzers S2 and S1 and S3 and S1 were calculated, as shown in Figs. 3.5(a) and (b). The coefficient of the Spearman correlation, r, was greater than 0.7, which revealed high correlations of the measured $A\beta_{1-42}$ and T-Tau concentrations between analyzers S2 and S1 and S3 and S1.

Table 3.3 gives an example of the between-lots, the between-operators, the between-instruments, and the between-labs reproducibility of assaying $A\beta_{1-40}$, $A\beta_{1-42}$, and T-Tau via immunomagnetic reduction (IMR).

3.3.4 Measuring Accuracy

The precision discussed in the previous section determines the closeness of the measured data under specific conditions. However, the closeness of the measured data to the trueness or the accepted value is unknown. The measured data could be very close to or far from the trueness. Hence, precision and accuracy have different analytical meanings. Figure 3.6 illustrates the analytical meanings of the precision and accuracy of the measured values for different measurements. The trueness is guided by the solid red line.

Table 3.3 Examples of the between-lots, between-operators, between-instruments, and between-labsreproducibility of assaying $A\beta_{1\text{-}40}$, $A\beta_{1\text{-}42}$, and T-Tau via immunomagnetic reduction (IMR)

Biomarker	Between-lots	Between-operators	Between-instruments	Between-labs
$A\beta_{1\text{-}40}$	< 5%	< 5%	< 5%	< 5%
$A\beta_{1\text{-}42}$	< 10%	< 10%	< 10%	< 5%
T-Tau	< 10%	< 10%	< 10%	< 5%

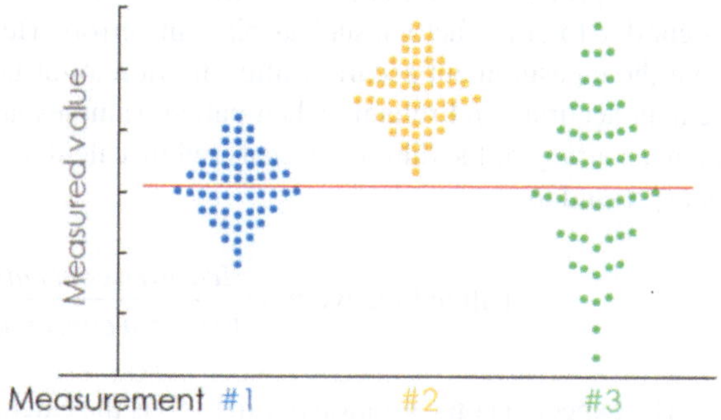

Fig. 3.6 Distributions of measured values showing different degrees of precision and accuracy

The measured values of measurements #1 and #2 are similar to the measured data. Hence, measurement #1 and measurement #2 have the same precision. However, the measured values in measurement #2 are distributed far from the trueness compared with those in measurement #1. This implies that measurement #1 is more accurate than measurement #2. In the cases of measurements #1 and #3, both are distributed around the trueness, namely, they have the same accuracy. However, the measured values in measurement #3 are distributed more widely than those in measurement #1. Hence, the precision of measurement #1 is greater than that of #3. This clearly discriminates the analytical meaning of precision from accuracy. In addition to precision, manufacturers and developers must explore the degree of agreement between the measured value and the value that is accepted either as a conventional true value or an accepted reference value. This analytical procedure is referred to as measuring accuracy, sometimes referred to as trueness. The details of the experimental designs and data analysis used to explore measurement accuracy are available in Ref. 38.

In accordance with the usual practices in laboratories, manufacturers and developers are required to clarify at least three kinds of measurement accuracies: diluted recovery, spiked recovery, and carryover effect. The three measurement accuracies are conceptually introduced to help readers understand the measurement accuracy.Plasma samples collected from different subjects could have different chemical features, such as ion suppression, which could influence the validity of the measured protein concentrations in the plasma. This is the so-called matrix effect [39]. The matrix effect could also arise from external components, such as inhibitors or stabilizers [40]. In laboratories, the most common method to suppress the matrix effect is to dilute plasma samples with suitable buffer solutions, e.g., PBS solution or a plasma substitute. The suppression of the matrix effect can significantly increase the validity of the measured protein concentrations. In addition to suppressing the matrix effect, dilution is conducted once the protein concentration in a sample exceeds the upper limit of quantification. Hence, the dilution of plasma samples is frequently conducted in laboratories. As described in Chap. 1, dilution is required before assays with several kinds of technologies, such as CLEIA (1–2-fold), ECL (1–2-fold), SIMOA (4-fold), and SMC (2-fold).Although dilution can suppress the matrix effect in plasma samples or ensure that the concentrations of target proteins are reliably quantified, inaccuracies in the measured values could occur due to other factors, such as pipetting errors. Hence, it is necessary to investigate the measurement accuracy after dilution to ensure that analyte measurements remain accurate and reliable when patient samples are diluted [41]. The measurement accuracy of the dilution is analyzed by calculating the diluted recovery, which is expressed as

$$\text{Diluted recovery} = \frac{\textit{Measured concentration}}{\textit{Expected concentration}} \times 100\% \tag{3.9}$$

The expected concentration in Eq. (3.9) is the calculated concentration after dilution of a sample. For example, the original concentration of the target protein in a plasma sample is 198 pg/ml. With 2-fold dilution, the expected concentration is 99 pg/ml. The measured concentration could be an average value of duplicate or triplicate measurements of the diluted sample. For example, the duplicated measurements of the twofold diluted plasma sample are 95.2 and 102.4 pg/ml. The measured concentration in Eq. (3.9) is 98.8 pg/ml, resulting in a diluted recovery of 99.8%.

In general, the diluted recoveries at various dilution factors, e.g., 2–100, are investigated. Figure 3.7 shows an example of the diluted recovery versus dilution factor for $A\beta_{1-40}$ (•) and $A\beta_{1-42}$ (•) assayed with immunomagnetic reduction (IMR). The original (undiluted) concentrations of $A\beta_{1-40}$ and $A\beta_{1-42}$ are 192.2 pg/ml and 97.2 pg/ml, respectively. The acceptable diluted recovery in Eq. (3.9) is within the range of 90–100%, as shadowed with a gray zone. As shown in Fig. 3.7, the diluted recovery at 100-fold dilution for $A\beta_{1-42}$ was 86%, which is outside the acceptable range for diluted recovery. Hence, the developer should clearly state that users should not dilute their plasma samples more than 100-fold for the $A\beta_{1-42}$ assay. Otherwise, the measured values would not be accurate due to dilution. Notably, the diluted concentration with a 100-fold concentration of $A\beta_{1-42}$ is expected to be 0.972 pg/ml, which is much lower than the level of $A\beta_{1-42}$ in human plasma (10–20 pg/ml). Therefore, the limitation of the 100-fold dilution for assaying $A\beta_{1-42}$ via immunomagnetic reduction does not matter in real practice. Manufacturers and developers are required to provide the suitable dilution range for their IVD medical devices.

To analyze the diluted recovery, a diluent such as PBS solution or a plasma substitute is used. The diluent is used not only for investigating diluted recovery; it is also used as a buffer solution for characterizing preclinical performance, such as standard curves, analytical sensitivity, and interference. A diluent or buffer solution is popularly used in determining the preclinical performance of an IVD device, as well as in real practice.

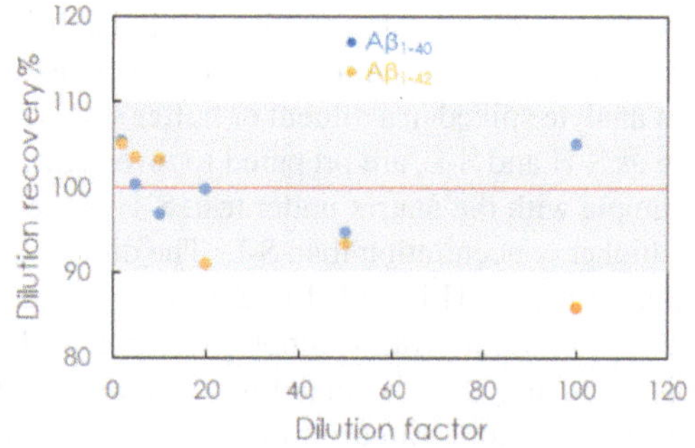

Fig. 3.7 Diluted recovery versus dilution factor in assaying $A\beta_{1-40}$ (•) and $A\beta_{1-42}$ (•) with immunomagnetic reduction (IMR)

In fact, the diluent or buffer solution is also a type of matrix. Different types of diluents or buffer solutions are composed of various ions and their concentrations and pH values, among other characteristics. For example, standard PBS solutions are usually composed of sodium chloride, potassium chloride, potassium dihydrogen phosphate, and disodium hydrogen phosphate. The standard pH of PBS solutions varies from 7 to 7.4. However, the true compositions of PBS solutions differ among providers. The contents of the PBS solution provided by Sigma-Aldrich are 137 mM NaCl, 2.7 mM KCl, 10 mM Na_2HPO_4, and 1.8 mM KH_2PO_4 at a pH of 7.4 [42]. The composition of the PBS solution provided by Qiagen is 50 mM K_3PO_4 and 150 mM NaCl at a pH of 7.2 [43]. The PBS solution provided by Beckman is 10 mM K_3PO_4 and 150 mM NaCl at a pH of 7.2 ± 0.2 [44]. The validity of assaying the target protein with an IVD device is possibly influenced by changes in the composition or concentration of the PBS solution. An inadequate matrix could cause inaccurate measurements with an IVD device.

In some countries, plasma substitutes, rather than PBS solutions, are required to be diluents or buffer solutions when performing protein assays using human plasma. The compositions of the plasma substitutes are much more diverse than those of PBS solutions. The main components are not limited to ions; protein and nonprotein molecules also play roles in plasma substitutes [45–47]. Hence, many recipes are available for plasma substitutes for specific uses.

To confirm the interference of the diluent or buffer solution in characterizing the preclinical performance and in real practice, its matrix effect should be investigated. This testing is crucial for ensuring that the assay can accurately measure the target analyte without interference from other components present in the sample.

The matrix effect is investigated by adding a known amount of analyte to a diluent or buffer solution, followed by examining the consistency between the measured and spiked concentrations in a spiked diluent or buffer solution [41]. The spiked concentration is the sum of the original concentration in the diluent or buffer solution and the added concentration to the diluent or buffer solution. The consistency between the measured concentration and the spiked concentration in a diluent or buffer solution is quantified by calculating the spiked recovery, expressed as

$$\text{Spiked recovery} = \frac{\textit{Measured concentration}}{\textit{Spiked concentration}} \times 100\% \tag{3.10}$$

The measured concentration might be the average value of duplicate or triplicate measurements. The spiked concentration is the expected value of the total amount of analyte spiked in a diluent or buffer solution.As mentioned, two samples, referred to as S-H and S-L, are prepared to investigate the matrix effect. S-L is regarded as a sample with the matrix under test. S-H is the solution with the analyte of interest at a higher concentration than S-L. The diluent or buffer solution of S-H is the same as that of S-L. S-H is added to S-L to investigate the matrix effect. S-H and S-L are assayed with a given technology to obtain the original concentrations, denoted as $\phi_{S\text{-}H}$ and $\phi_{S\text{-}L}$. S-H is spiked into S-L at various volume ratios, e.g., p%:(100-p)%. The spiked concentrations of analytes in the samples are as follows:

$$\text{Spiked concentration} = \phi_{S\text{-}H} x p\% + \phi_{S\text{-}L} x (100\text{-}p)\% \tag{3.11}$$

The concentrations of analytes in spiked samples are assayed to obtain the measured concentrations via Eq. (3.10). Thus, the spiked recoveries of spiked samples can be determined. The acceptable values of spiked recovery range from 90 to 110%. If the spiked recoveries are outside the acceptable range, the matrix effect has definitely influenced the accuracy of the measurements.An example is given to help readers understand the investigation of the spiked recovery contributed from the matrix effect. S-H and S-L samples are prepared by spiking $A\beta_{1\text{–}42}$ peptides into plasma substitutes. The measured $A\beta_{1\text{–}42}$ concentrations of S-H and S-L via immunomagnetic reduction (IMR) are 65.74 and 15.37 pg/ml, respectively. S-H is spiked with S-L at various volume ratios, as listed in Table 3.4. The spiked concentrations and measured concentrations of the spiked samples, denoted S-1, S-2, S-3, and S-4, are also listed in Table 3.4. The spiked recoveries of the mixtures are obtained via Eq. (3.10). The results in Table 3.4 show that the spiked recovery ranges from 94.7 to 104.9%, which is within the acceptable range (90–110%). This reveals that the components in the used plasma substitute do not influence the accuracy of measuring $A\beta_{1\text{–}42}$.

In laboratories, analyte measurements are usually performed sample by sample. This process generates a risk of transferring unwanted material from a measurement to the following ones, namely, contamination. The unwanted material may be from a sample, a reagent, or even the washing medium. This is the so-called carryover effect, which results in the inaccuracy of the subsequent measurements. Manufacturers and developers are required to investigate the carryover effect for their IVD devices.The details of inspecting the carryover effect are given in Refs. 48 and 49. Briefly, a sequence of samples, usually with high (H) and low (L) concentrations of the analyte to be measured, are processed. For example, sequentially triplicate measurements are performed for samples with high concentrations. The measured concentrations are H1, H2, and H3. The following process involves sequentially triplicated measurements of samples with low concentrations. The measured concentrations are L1, L2, and L3. The quantification of the carryover effect (q) is calculated as follows:

Table 3.4 Measured concentration and spiked recovery for spiked $A\beta_{1-42}$ in plasma substitute

Sample no.	Volume ratio (S-H:S-L)	Original concentration (pg/ml)	Spiked concentration (pg/ml)	Measured concentration (pg/ml)	Spiked recovery
S-L	–	15.37	–	–	–
S-1	10%:90%	–	20.41	21.29	104.3%
S-2	25%:75%	–	27.96	29.15	104.3%
S-3	50%:50%	–	40.56	38.41	94.7%
S-4	75%:25%	–	53.15	55.78	104.9%
S-H	–	65.74	–	–	–

Table 3.5 Measured concentrations of samples with high and low concentrations of $A\beta_{1-42}$

Sample no.	Expected concentration (pg/ml)	Measured concentration (pg/ml)
H	100	99.13 (H1)
	100	94.96 (H2)
	100	102.81 (H3)
L	1	0.92 (L1)
	1	1.00 (L2)
	1	0.93 (L3)

$$q = \frac{L1 - L3}{H3 - H1} \times 100\%, \tag{3.12}$$

The criterion for q is that the absolute value of q is less than 1% or 2%, which means that the carryover effect will not cause significant errors in routine analytical results. An example to illustrate the calculation of q in Eq. (3.12) is given in Table 3.5.

As shown in Table 3.5, two samples spiked with 100 and 1 pg/ml $A\beta_{1-42}$ peptide are prepared. Sequentially triplicate measurements using immunomagnetic reduction (IMR) are performed to obtain the measured concentrations H1 (= 99.13 pg/ml), H2 (= 94.96 pg/ml), and H3 (= 102.81 pg/ml) for the 100-pg/ml $A\beta_{1-42}$ sample. Another series of triplicate measurements is performed for the 1-pg/ml $A\beta_{1-42}$ sample immediately after the measurements for the 100-pg/ml $A\beta_{1-42}$ sample. The measured concentrations of the 1-pg/ml $A\beta_{1-42}$ sample are 0.92 (L1), 1.00 (L2), and 0.93 (L3) pg/ml, respectively. The q value in Eq. (3.12) is calculated to be −0.27%, whose absolute value is less than 1%. This evidence highlights the nonsignificant carryover effect of assaying $A\beta_{1-42}$ via immunomagnetic reduction (IMR).

3.3.5 *Analytical Specificity*

The common samples used in the assay of biomarkers associated with neurodegenerative diseases are human plasma samples, which contain not only target analytes but also other chemicals and biomolecules. Some colored molecules in human plasma could cause false signals in assays involving optical detections. Additionally, some intrinsic antibodies in plasma might inhibit the binding between the target analyte and capture antibodies. The false measurements caused by such caueses are referred to as interference. Moreover, some proteins similar to the analyte of interest or having high affinity to capture antibodies could bind nonspecifically with capture antibodies. The fales measurements casued by such nonspecific binding is referred to as cross-reactivity. All these nontarget materials could lead to inaccurate measurements with an IVD medical device [50]. Manufacturers and developers must investigate the possibility of the influence of other materials in plasma on the measurements. Especially, manufacturers and developers need to demonstrate the ability to assess the analyte unequivocally in the presence of other components in plasma. This ability is the so-called analytical specificity.

According to the guidelines provided in ICH Q2(R1) and CLIS-EP2-A2 [38, 51], investigations of analytical specificity are categorized into two types, which are introduced separately: interference and cross-reactivity.

Firstly, the interference effect on assay is discussed. Plasma can contain possibly interfering substances such as hemoglobin, bilirubin, or intralipid because of common diseases, such as hemolysis, jaundice, or hypertriglyceridemia [52]. Notably, hemoglobin, bilirubin, and intralipid are colorful, as shown in Fig. 3.8: red, yellow, and white. They often exist in human plasma after centrifuging a blood sample. For assay technologies involving optical detections, such as CLEIA, ECL, and SIMOA, the detected optical signals are easily interfered by the colors of hemoglobin, bilirubin, or intralipid. To reduce the interference of colorful materials in human plasma, plasma samples are frequently diluted with a transparent diluent before measurement.

Some biosubstances that exist naturally in plasma, such as uric acid, rheumatoid factor, or albumin, can also be interfering substances. Other possibly interfering substances include drugs or chemicals in medicine that are used to treat inflammatory diseases, viral and bacterial infections, cancers, and cardiovascular diseases. All these interferences may cause errors in clinical laboratory measurements. Such errors may, in some cases, represent a hazard to the patient. Manufacturers and

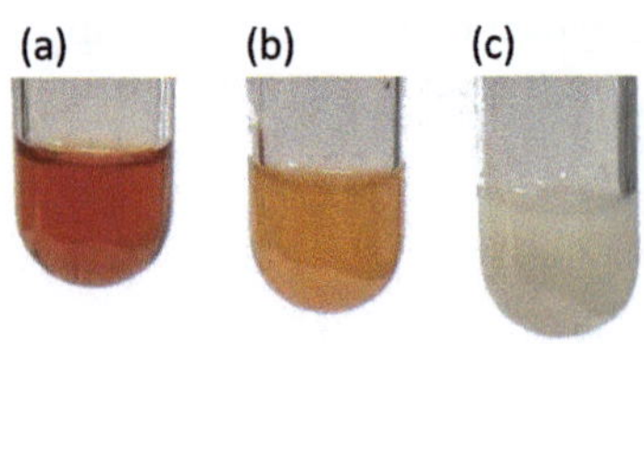

Fig. 3.8 Colorful human plasma with rich (a) hemoglobin in hemolysis, (b) bilirubin in jaundice, and (c) intralipid in hypertriglyceridemia

Table 3.6 Potentially interfering materials and their concentrations used for interference tests of the $A\beta_{1-42}$assay via IMR technology. The spiked concentration of $A\beta_{1-42}$ in each sample was 30 pg/ml. The detected $A\beta_{1-42}$concentrations of each sample are listed. The recovery percentages of the $A\beta_{1-42}$concentration for samples with interfering materials were calculated using Eq. (3.13) and are listed in the right-most column

Sample no.	Interfering material: concentration	Measured $A\beta_{1-42}$concentration (pg/ml)	Recovery%
1	None	29.22	–
2	Hemoglobin: 10,000 μg/ml	28.67	98.1%
3	Conjugated bilirubin: 600 μg/ml	27.46	94.0%
4	Intralipid: 30,000 μg/ml	29.58	101.2%
5	Uric acid: 200 μg/ml	28.78	98.5%
6	Rheumatoid factor: 500 IU/ml	27.95	95.7%
7	Albumin: 60,000 μg/ml	27.51	94.1%
8	Acetylsalicylic acid: 500 μg/ml	28.78	98.5%
9	Ascorbic acid: 300 μg/ml	28.70	98.2%
10	Ampicillin sodium: 1000 μg/ml	27.51	94.1%
11	Quetiapine fumarate: 100 ng/ml	27.91	95.5%
12	Galantamine hydrobromide: 90 ng/ml	27.64	94.6%
13	Rivastigmine hydrogen tartrate: 100 ng/ml	27.83	95.2%
14	Donepezil hydrochloride: 1000 ng/ml	28.95	99.1%
15	Memantine hydrochloride: 150 ng/ml	27.87	95.4%

developers need to clarify the potential errors of measurements with their IVD medical devices.

The quantification of interference is defined as

$$\text{Recovery\%} = \frac{\textit{Measured concentration with interfering material}}{\textit{Measured concentration without interfering material}} \times 100\% \tag{3.13}$$

The acceptable values of recovery range from 90 to 110%, in which the interfering material does not significantly interfere with the assay of the target analyte. Table 3.6 gives an example of an interference test for an $A\beta_{1-42}$ assay using immunomagnetic reduction (IMR). The concentrations of spiked interfering materials are much higher than normal levels. For example, human plasma in which hemolysis has not occurred does not contain hemoglobin. In the case of slight hemolysis, the hemoglobin concentration in human plasma is less than 1,000 μg/ml. The spiked concentration of hemoglobin in the interference test is 10,000 μg/ml, which is much higher than the concentration of 1,000 μg/ml. All the Recovery% values are within

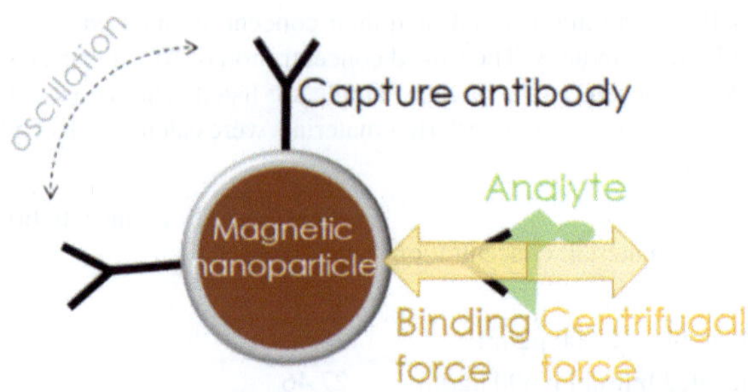

Fig. 3.9 Illustration of dominant forces acting on a bound analyte in IMR assay. The dashed line represents the oscillation of the magnetic nanoparticle during measurement

90–110%, which indicates that the $A\beta_{1-42}$ assay using IMR does not significantly interfere with these materials in human plasma. A high level of analytical specificity results.

Secondly, the cross-reactivity in investigating the analytical specificity is interpreted. In human plasma, proteins whose structures are very similar to those of the target analyte are present. For example, in the case of assaying $A\beta_{1-42}$, several structurally similar proteins, such as $A\beta_{1-38}$, $A\beta_{1-40}$, $A\beta_{1-55}$, $A\beta_{3-40}$, $A\beta_{3-42}$, and $A\beta_{pE3-42}$, exist. Such proteins potentially show nonspecific binding with antibodies against $A\beta_{1-42}$, resulting in cross-reactivity. Hence, the specificity of an antibody to a target analyte plays a role in suppressing cross-reactivity. In addition to the selection of highly specific antibodies, manufacturers and developers should demonstrate how to suppress cross-reactivity in their IVD medical devices.

The most common method to suppress cross-reactivity is the use of both capture and labeling antibodies, such as CLEIA, HISCL, ECL, ME-ECL, MDS, SIMOA, SMC, and xMAP, as listed in Table 19, Chap. 1. The proximity effect of nucleic acids linked in two capture antibodies in assays, such as PLA, SPEAR, and NULISA, is an alternative way to suppress cross-reactivity. For immunomagnetic reduction (IMR), no labeling antibody is used. Only a capture antibody is used in IMR. How can the cross-reactivity of IMR with only the capture antibody be suppressed?

IMR employs an innovative technology, the so-called spin-wash technology, to suppress cross-reactivity in measurements. As described in the introduction to IMR in Chap. 1, the reduced magnetic signals result from the association between the antibody-functionalized magnetic nanoparticles and the analyte of interest. The nanoparticle–analyte complex oscillates during measurements. This implies that the bound analyte molecule experiences a centrifugal force and a shearing force as it oscillates. Hence, the bound analyte molecule has two dominant forces: a binding force with the antibody and a centrifugal force, as illustrated in Fig. 3.9. Of course, there are minor forces due to shearing action on the analyte during oscillation and the thermal energy of water molecules, among others.

The binding force between the antibody and analyte is the key to form the nanoparticle–antibody–analyte complex, resulting in reduced magnetic signals in the IMR assay. However, the centrifugal force acting on the bound analyte disassociates the nanoparticle–antibody–analyte complex, resulting in failures in measurements. The resultant force depends primarily on the compensation between these two forces. If the binding force is stronger than the centrifugal force, reduced magnetic signals are detected. Otherwise, no signal would be observed if the centrifugal force is stronger than the binding force.

The binding force between the antibody and analyte is the total contribution from electrostatic interactions, hydrogen bonds, van der Waals forces, and hydrophobic interactions [53]. The free energy of these binding actions ranges from 4–50 KJ/mol. By comparison, the average kinetic energy of water molecules at 37 °C is 2.6 kJ/mol, which is lower than the free energy of antibody–analyte binding. Hence, the association between the antibody and analyte is stable in a water-based solution at 37 °C or room temperature.

The centrifugal force acting on the bound analyte is proportional to the square of the oscillation speed and inversely proportional to the radius of the nanoparticle. The radius of the nanoparticles used in the IMR assay is approximately 22.5 nm. The centrifugal force can be well manipulated by adjusting the oscillation frequency.

In the IMR assay, only the capture antibody is used. The capture antibody possibly binds to specific and nonspecific analytes in plasma samples, potentially resulting in cross-reactivity and false-positive results. The binding force between an antibody and its specific analyte is significantly stronger than that between an antibody and nonspecific analytes. By adjusting the oscillation frequency, e.g., to 20 kHz, to make the centrifugal force weaker than the specific binding but stronger than the nonspecific binding, the nonspecific binding is broken down, and the specific binding is still valid. The nonspecifically bound analytes are washed away with the nanoparticle oscillation. Thus, cross-reactivity is suppressed in the IMR assay with spin-wash technology.

Spin-wash technology is an intrinsically physical method for removing cross-reactivity in immunoassays and is completely different from conventional methods, such as the use of labeling antibodies or the proximity effect of nucleic acids. The discriminating power between specific and nonspecific binding is very strong. For example, the antibody is against the epitope of amino acid 1-x in both $A\beta_{1-40}$ and $A\beta_{1-42}$. However, the affinity of the antibody for $A\beta_{1-40}$ is half that for $A\beta_{1-42}$. Hence, this antibody is more useful for the assay of $A\beta_{1-42}$. Clear binding between the antibody and both $A\beta_{1-40}$ and $A\beta_{1-42}$ can be observed via Western blotting. This reveals the cross-reactivity in assaying $A\beta_{1-42}$ due to $A\beta_{1-40}$. As the antibody is utilized in IMR for assaying $A\beta_{1-42}$, spin-wash technology results in the significant suppression of the cross-reactivity with $A\beta_{1-40}$. These results indicate strong discrimination between specific and nonspecific binding.

Currently, the antibody used in the reagent for assaying $A\beta_{1-42}$ is against amino acids 35–42, i.e., the C-terminus of $A\beta_{1-42}$. An investigation of the cross-reactivity of $A\beta_{1-38}$, $A\beta_{1-40}$, $A\beta_{1-55}$, $A\beta_{3-40}$, $A\beta_{3-42}$, and $A\beta_{pE3-42}$ is performed. The quantification of cross-reactivity is performed using Eq. (3.13). Table 3.7 shows the measured recoveries for investigating the cross-reactivity of $A\beta_{1-42}$ using IMR. The measured Recovery% is within 90–110%, which clearly indicates non-cross-reactivity when $A\beta_{1-42}$ is assayed via IMR.

Table 3.7 Potentially interfering materials causing cross-reactivity in assaying Aβ_{1-42} via IMR technology. The spiked concentration of Aβ_{1-42} in each sample is 30 pg/ml. The detected Aβ_{1-42}concentrations of each sample are listed. The Recovery% values of Aβ_{1-42}concentration for samples with interfering materials are calculated using Eq. (3.13) and are listed in the right-most column

Sample no.	Interfering material: concentration	Measured Aβ_{1-42}concentration (pg/ml)	Recovery%
1	None	28.96	–
2	Aβ_{1-38}: 40 pg/ml	28.99	100.1%
3	Aβ_{1-40}: 100 pg/ml	28.69	99.1%
4	Aβ_{1-55}: 1 pg/ml	28.91	99.8%
5	Aβ_{3-40}: 2 pg/ml	28.66	99.0%
6	Aβ_{3-42}: 2 pg/ml	28.87	99.7%
7	Aβ_{pE3-42}: 1 pg/ml	29.08	100.4%

Table 3.8 Investigation of the cross-reactivity of HAHA in the production of Aβ_{1-42} via IMR technology. The spiked concentration of Aβ_{1-42} in each sample is 30 pg/ml. The detected Aβ_{1-42}concentrations of each sample are listed. The Recovery% value of Aβ_{1-42}concentration for the sample with HAHA is calculated using Eq. (3.13) and is listed in the right-most column

Sample no.	Interfering material: concentration	Measured Aβ_{1-42}concentration (pg/ml)	Recovery%
1	None	28.96	–
2	HAMA: 100 ng/ml	27.54	95.1%

In addition to structural similarities, the cross-reactivity might be attributed to antibodies induced by other biomolecules in the body. Although such antibodies are induced by nontarget analytes, they show high affinity for the analyte of interest. Hence, the target analyte is highly likely to associate with the antibodies. Specifically, the antibodies inhibit the analyte of the target and cause false-negative measurement results. The most common antibody is human anti-mouse antibody (HAMA), which was developed in humans and works directly with mice or inadvertently encounters mouse proteins or immunoglobulins by coming into contact with mouse urine or contaminated food [54, 55]. It is strongly suggested that manufacturers and developers investigate cross-reactivity of HAMA.

Table 3.8 shows the test results of the cross-reactivity of HAHA when Aβ_{1-42} is assayed via IMR technology. The Recovery% is 95.1%, which indicates nonsignificant cross-reactivity of HAMA when Aβ_{1-42} is assessed via IMR.

3.3.6 *Reagent Stability*

Reagents are definitely needed for assays of plasma biomarkers associated with neurodegenerative diseases. Reagents are commonly composed of chemicals and biomaterials such as antibodies. These compositions have shelf lives and are sometimes sensitive to environmental conditions such as temperature, light, and humidity during transportation. Manufacturers and developers are required to investigate the stability of reagents over time to maintain the consistent performance of measurements. CLIS guideline EP25 illustrates the designs and analyses for determining IVD reagent stability [56].

In real practice, there are three main tests for determining IVD reagent stability: shelf-life stability, transportation stability, and open-vial stability, which are briefly described as follows. Let's start from shelf-life stability.

After production and quality analysis, IVD reagents are stored under defined conditions by manufacturers. Manufacturers and developers must investigate the shelf life of reagents under defined conditions. This investigation is referred to as shelf-life stability testing. So that, the shelf-life stability is sometimes referred to as the storage stability.

In general, newly produced reagents are aliquoted, followed by storage at defined conditions, such as 2–8 °C in the dark. Quality analysis is performed for different aliquots on different days. The analyzed qualities among aliquots are compared to confirm the consistency in reagent qualities over time.

The shelf-life stability tests performed for the reagents used to assay $A\beta_{1-42}$, $A\beta_{1-40}$, T-Tau, and α-synuclein with immunomagnetic reduction (IMR) are used as examples here [57]. Newly produced $A\beta_{1-42}$, $A\beta_{1-40}$, T-Tau, and α-synuclein reagents are aliquoted into six tubes, referred to as #1–#6. Tube #1 is used for assaying biomarkers in a spiked sample via IMR on the first day after production. The other five tubes are stored at 2–8 °C before IMR measurements. Tube #2 is used on the 60th day, tube #3 is used on the 114th day, tube #4 is used on the 157th day, tube #5 is used on the 288th day, and tube #6 is used on the 441st day after production. The drift in the measured concentration is calculated as follows:

$$\%\,\text{drift} = \frac{\begin{array}{c}(\textit{Measured concentration on the nth day}\\ -\textit{Measured concentration on the 1st day})\end{array}}{\textit{Measured concentration in the 1st day}} \times 100\% \qquad (3.14)$$

The acceptable %drift is ±10%. Table 3.9 lists the concentrations and %drift values measured via IMR technology on different days during the storage of the reagents at 2–8 °C. The %drift is ±10% for all the measurements until the 441st day. These findings suggest that the shelf-life stability of the IMR $A\beta_{1-42}$ reagent is at least 441 days. Conventionally, shelf-life storage under defined conditions should be longer than 365 days, namely, 1 year. Notably, the shelf-life stability of IMR reagents is reduced to less than one month when they are stored at temperatures higher than 30 °C [58]. This implies that storage conditions, such as temperature, play a role in the shelf-life stability of reagents. Manufacturers and developers must clearly illustrate the conditions of storage for reagents to help users avoid reagent degradation.

Reagents are produced at manufacturers and then delivered to users or distributors. During transportation, reagents may be exposed to variations in environmental

Table 3.9 Variation in the measured $A\beta_{1\text{-}42}$, $A\beta_{1\text{-}40}$, T-Tau, and α-synuclein concentrations in a sample measured on different days during the storage of the reagent at 2–8 °C via IMR technology

Measuring day	Measured concentration (pg/ml)				%drift			
	$A\beta_{1\text{-}42}$	$A\beta_{1\text{-}40}$	T-Tau	α-synuclein	$A\beta_{1\text{-}42}$	$A\beta_{1\text{-}40}$	T-Tau	α-synuclein
1st	50.97	94.52	98.61	10.28	–	–	–	–
60th	50.75	94.90	93.98	10.35	−0.43%	0.4%	−4.7%	0.7%
114th	50.23	98.41	100.9	10.57	−1.45%	4.1%	2.3%	2.8%
157th	50.80	97.74	100.6	10.29	−0.27%	3.4%	2.0%	0.1%
288th	50.62	94.64	100.3	10.70	−0.69%	0.1%	1.7%	4.1%
441st	50.02	99.81	100.4	9.82	−1.86%	5.6%	1.8%	−4.5%

Table 3.10 Variations in the measured $A\beta_{1\text{-}42}$, $A\beta_{1\text{-}40}$, T-Tau, and α-synuclein concentrations in the spiked samples using IMR reagents before (Day 0) and after (Day 7) 7 days of storage at 2 °C

Storage period (day)	Measured concentration (pg/ml)				%drift			
	$A\beta_{1\text{-}42}$	$A\beta_{1\text{-}40}$	T-Tau	α-synuclein	$A\beta_{1\text{-}42}$	$A\beta_{1\text{-}40}$	T-Tau	α-synuclein
0	16.14	47.04	27.15	1.035	–	–	–	–
7	16.96	48.01	27.01	1.030	5.1%	2.1%	−0.5%	−0.5%

Table 3.11 Variations in the measured $A\beta_{1\text{-}42}$, $A\beta_{1\text{-}40}$, T-Tau, and α-synuclein concentrations in the spiked samples using IMR reagents before (Day 0) and after (Day 7) 7 days of 8 °C storage

Storage period (day)	Measured concentration (pg/ml)				%drift			
	$A\beta_{1\text{-}42}$	$A\beta_{1\text{-}40}$	T-Tau	α-synuclein	$A\beta_{1\text{-}42}$	$A\beta_{1\text{-}40}$	T-Tau	α-synuclein
0	16.62	49.55	27.51	1.035	–	–	–	–
7	16.40	49.01	27.53	1.009	−1.3%	−1.1	0.1%	−2.5%

conditions that could possibly degrade reagents. Hence, manufacturers and developers need to investigate the stability of reagents under simulated transport, referred to as transportation stability testing.

For example, IMR reagents are manufactured in Taiwan and then delivered to the USA, South Korea, Sweden, Kuwait, and other countries. The shipments of IMR reagents can be facilitated by professional logistic companies [59, 60]. It usually takes approximately 7 days to ship IMR reagents from Taiwan to a destination. IMR reagents must be shipped at 2–8 °C throughout the process. Hence, manufacturers and developers must demonstrate the stability of IMR reagents stored at 2 °C and 8 °C for 7 days.

Manufacturers and developers design stability tests as follows. The IMR reagents are stored in temperature monitoring boxes at 2 °C and 8 °C separately for one week, and measurements are then conducted. The measured concentrations of IMR reagents for $A\beta_{1\text{-}42}$, $A\beta_{1\text{-}40}$, T-Tau, and α-synuclein before (Day 0) and after (Day 7) 7 days of storage at 2 °C or 8 °C are listed in Table 3.10 and Table 3.11, respectively. The value of %drift calculated via Eq. (3.14) is ±10%, which demonstrates the transportation stability of the IMR reagents.

Table 3.12 Variations in the measured $A\beta_{1-42}$, $A\beta_{1-40}$, and T-Tau concentrations in human plasma using IMR reagents after being opened on different days

	Measured concentration (pg/ml)			%drift		
Storage period (week)	$A\beta_{1-42}$	$A\beta_{1-40}$	T-Tau	$A\beta_{1-42}$	$A\beta_{1-40}$	T-Tau
0	16.28	57.38	20.79	–	–	–
2	16.29	55.73	20.54	0.1%	−2.9%	−1.2%
4	16.11	56.28	20.55	−1.0%	−1.9%	−1.2%
6	16.45	54.91	21.31	1.0%	−4.3%	2.5%

Notably, the factors contributing to the transportation stability of reagents are not limited to temperature. Depending on the compositions of reagents, manufacturers and developers should list the possible factors that could compromise the transportation stability of reagents and carefully design experiments to investigate the effects of these factors on the transportation stability of their reagents.

Once users receive reagents, they might or might not be used immediately. Manufacturers and developers need to clearly illustrate how to store the reagents in users' laboratories according to the shelf-life stability. More often, users open the reagents, use a portion of them, place the rest of the reagents at room temperature for a while, and move the remaining reagents to a refrigerator for storage. The stored reagents are then be reused at a later date for measurements.

In such a reuse scenario, the reagents come into contact with air when opened. This could pollute the reagents or accelerate their degradation. Thus, manufacturers and developers need to test the open-vial stability of their reagents. Here, an example is given for investigating the open-vial stability of reagents.

Before the measurements, the reagents are incubated at room temperature for 10 min. The covers of the reagents are opened, followed by placing the reagents at room temperature for 15 min. The reagents are then directly used for measurements. The remaining reagents are then returned to 4 °C for storage. The performance of the reagents is analyzed every two weeks for six weeks. The measurements are performed via immunomagnetic reduction (IMR). These reagents are used for assaying $A\beta_{1-42}$, $A\beta_{1-40}$, T-Tau, and α-synuclein in human plasma. As shown in the results in Table 3.12, the values of %drift are ±10%, confirming the open-vial stability of the reagents with IMR measurements for 6 weeks. Manufacturers and developers should identify a suitable storage period depending on real practices. Conventionally, after being opened, the reagents should be used as soon as possible. The storage period of 6 weeks in Table 3.12 is very long in real practice. Medical technicians prefer to use reagents within a few days after vials are opened.

3.4 Analytical Quality Control

The analytical quality control can be determined by monitoring the measured results of control solutions at every run. It is impossible to have a fixed measured concentration of a control solution over measurement runs. The variations in the measured

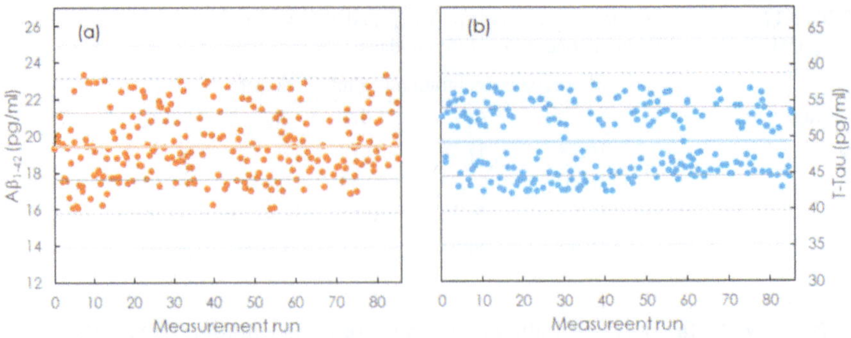

Fig. 3.10 Levey–Jennings charts of assaying (a) $A\beta_{1-42}$ and (b) T-Tau using IMR. The red/blue solid lines are the mean value of measured concentrations over runs. The gray solid lines are the standard deviation (SD) of measured concentrations from the mean value. The gray dashed lines are the 2SD from the mean value. The gray dot lines are the 3SD from the mean value

concentrations of controls solutions over measurement runs can tell the analytical qualities.

At laboratories, the measured concentration versus measurement run is plotted as the Levey–Jennings chart, which can help distinguish between normal variation and systematic error, as shown in Fig. 3.10.

In the Levey–Jennings chart, the mean value, the standard deviation (SD) from the mean value, 2SD from the mean value, and 3SD from the mean value are plotted. According to the distributions of the measured concentrations of control solutions at each measurement, the validity of this measurement or assay can be determined.

For example, in cases of high-level biomarkers such as 10 mg/dl, the criteria could be that the measured concentrations of control solutions must be distributed with the range from mean value ±1.5SD. Whereas, in case of low-level biomarkers such as 10 pg/ml, the criteria could be that the measured concentrations of control solutions must be distributed with the range from mean value ±2.5SD. Anyway, once the measured concentrations of control solutions are outside the range of mean value ±3SD, the measurement/assay fails. Repeated measurement/assay would be performed. Usually, the developer/manufacture/laboratory shall set criteria to determine the validity of measurement/assay by themselves. The criteria depend on the biomarker levels, analytical precision, etc.

As an example in Fig. 3.10(a), the mean value of the measured concentrations of control solutions for $A\beta_{1-42}$ was 19.44 pg/ml, and the standard deviation (SD) was 1.83 pg/ml. The mean value, mean value ± SD, mean value ±2SD, and mean value ±3SD are plotted with the red solid line, gray solid line, gray dashed line, and gray dot line, respectively. Since the biomarker level of $A\beta_{1-42}$ is low (pg/ml), the criteria of valid measurement/assay could be that the measured concentrations of control solutions are distributed with the range from mean value ±2.5SD. In Fig. 3.10(a), all the measured concentrations of control solutions over measurement runs are distributed within the range of mean value ±2.5SD. Hence, each measurement is valid. Similar findings are observed in the case of T-Tau measurement using IMR, as shown in Fig. 3.10(b).

References

1. https://europa.eu/youreurope/business/product-requirements/compliance/technical-documentation-conformity/index_en.htm
2. https://www.medical-device-regulation.eu/2019/08/14/annex-ix/
3. https://www.medical-device-regulation.eu/2019/08/14/annex-xi/
4. https://www.fda.gov/medical-devices/regulatory-controls/general-controls-medical-devices
5. https://www.fda.gov/medical-devices/premarket-submissions-selecting-and-preparing-correct-submission/premarket-notification-510k
6. https://www.ecfr.gov/current/title-21/chapter-I/subchapter-H/part-820
7. https://www.fda.gov/medical-devices/premarket-submissions-selecting-and-preparing-correct-submission/premarket-approval-pma
8. https://www.fda.gov/regulatory-information/search-fda-guidance-documents/bioanalytical-method-validation-guidance-industry
9. Yang, S. Y., Chiu, M. J., Chen, T. F., Lin, C. H., Jeng, J. S., Tang, S. C., Lee, Y. F., Yang, C. C., Liu, B. H., Chen, H. H., & Wu, C. C. (2017). Analytical performance of reagent for assaying tau protein in human plasma and feasibility study screening neurodegenerative diseases. *Scientific Reports, 7*, 9304-1-12.
10. Yang, S. Y., Chieh, J. J., Huang, K. W., Yang, C. C., Chen, T. C., Ho, C. S., Chang, S. F., Chen, H. H., Horng, H. E., Hong, C. Y., & Yang, H. C. (2013). Molecule-assisted nanoparticle clustering effect in immunomagnetic reduction assay. *Journal of Applied Physics, 113*, 144903-1-6.
11. CLSI (Ed.). (2020). *Evaluation of Linearity of Quantitative Measurement Procedures* (2nd ed. CLSI guideline EP06-A2.). Clinical and Laboratory Standards Institute.
12. DeSilva, B., Smith, W., Weiner, R., Kelley, M., Khan, M., Tacey, R., Hill, H., & Celniker, A. (2003). Recommendations for the bioanalytical method validation of ligand-binding assays to support pharmacokinetic assessments of macromolecules. *Pharmaceutical Research, 20*, 1885–1900.
13. Smolec, J., DeSilva, B., Smith, W., Weiner, R., Kelly, M., Lee, B., Khan, M., Tacey, R., Hill, H., Celniker, A., Shah, V., Bowsher, R., Mire-Sluis, A., Findlay, J. W., Saltarelli, M., Quarmby, V., Lansky, D., Dillard, R., Ullmann, M., Keller, S., & Karnes, H. T. (2005). Bioanalytical method validation for macromolecules in support of pharmacokinetic studies. *Pharmaceutical Research, 22*, 1425–1431.
14. Viswanathan, C. T., Bansal, S., Booth, B., DeStefano, A. J., Rose, M. J., Sailstad, J., Shah, V. P., Skelly, J. P., Swann, P. G., & Weiner, R. (2007). Quantitative bioanalytical methods validation and implementation: Best practices for chromatographic and ligand binding assays. *Pharmaceutical Research, 24*, 1962–1973.
15. CLSI (Ed.). (2020). *Evaluation of Detection Capability for Clinical Laboratory Measurement Procedures* (2nd ed. CLSI guideline EP17-A2.). Clinical and Laboratory Standards Institute.
16. Linnet, K. (2000). Nonparametric estimation of reference intervals by simple and bootstrap-based procedures. *Clinical Chemistry, 46*, 867–869.
17. CLSI. (2021). Evaluation of total analytical error for quantitative medical laboratory measurement procedures. *CLSI guideline EP21*. (2nd ed), Clinical and Laboratory Standards Institute
18. ISO. (1993). *Guide to the expression of uncertainty in measurement*. International Organization for Standardization.

19. Shrivastava, A., & Gupta, V. B. (2011). Methods for the determination of limit of detection and limit of quantitation of the analytical methods. *Chronicles of Young Scientists, 2*, 21–25.
20. Ashton, N. J., Brum, W. S., Di Molfetta, G., Benedet, A. L., Arslan, B., Jonatis, E., Langhough, R. E., Cody, K., Wilson, R., Carlsson, C. M., Vanmechelen, E., Montoliu-Gaya, L., Lantero-Rodriguez, J., Rahmouni, N., Tissot, C., Stevenson, J., Servaes, S., Therriault, J., Pascoal, T., Lleó, A., Alcolea, D., Fortea, J., Rosa-Neto, P., Johnson, S., Jeromin, A., Blennow, K., & Zetterberg, H. (2023). Diagnostic accuracy of the plasma ALZpath pTau217 immunoassay to identify Alzheimer's disease pathology. *MedRxiv*, 23292493.
21. Yamashita, K., Watanabe, S., Ishiki, K., Miura, M., Irino, Y., Kubo, T., Matsui, J., Hagino, K., Iwanaga, S., & Yoshida, T. (2021). Fully automated chemiluminescence enzyme immunoassays showing high correlation with immunoprecipitation mass spectrometry assays for β-amyloid (1–40) and (1–42) in plasma samples. *Biochemical and Biophysical Research Communications, 576*, 22–26.
22. Karikari, T., Chen, Y., Zeng, X., Olvera-Rojas, M., Sehrawat, A., Lafferty, T., Pascoal, T., Villemagne, V., Solis-Urra, P., Triviño-Ibañez, E., Gómez-Rí, M., Cohen, A., Ikonomovic, M., Esteban-Cornejo, I., Erickson, K., Lopez, O., & Yates, N. (2024). A streamlined, resource-efficient immunoprecipitation-mass spectrometry method for quantifying plasma amyloid-β biomarkers in Alzheimer's disease. *Research Square, 3*, rs-4947448.
23. Kivisäkk, P., Fatima, H., Cahoon, D. S., Otieno, B., Chacko, L., Minooei, F., Demos, C., Stengelin, M., Sigal, G., Wohlstadter, J., & Arnold, S. E. (2024). Clinical evaluation of a novel plasma pTau217 electrochemiluminescence immunoassay in Alzheimer's disease. *Science Reports, 14*, 629-1-8.
24. Kivisäkk, P., Carlyle, B. C., Sweeney, T., Trombetta, B. A., LaCasse, K., El-Mufti, L., Tuncali, I., Chibnik, L. B., Das, S., Scherzer, C. R., Johnson, K. A., Dickerson, B. C., Gomez-Isla, T., Blacker, D., Oakley, D. H., Frosch, M. P., Hyman, B. T., Aghvanyan, A., Bathala, P., Campbell, C., Sigal, G., Stengelin, M., & Arnold, S. E. (2023). Plasma biomarkers for diagnosis of Alzheimer's disease and prediction of cognitive decline in individuals with mild cognitive impairment. *Frontiers in Neurology, 14*, 1069411-1-12.
25. http://www.bipm.org/en/publications/guides/vim.html. Accessed 12 Aug 2014
26. CLSI. (2014). *Evaluation of Precision of Quantitative Measurement Procedures; Approved Guideline* (3rd ed.. CLSI document EP05-A3.). Clinical and Laboratory Standards Institute.
27. ISO. (2004). *Clinical Laboratory Medicine – in Vitro Diagnostic Medical Devices – Validation of User Quality Control Procedures by the Manufacturer. ISO 15198*. International Organization for Standardization.
28. ISO. (2007). *Clinical Laboratory Testing and in Vitro Medical Devices – Requirements for in Vitro Monitoring Systems for Self-Testing of Oral Anticoagulant Therapy. ISO 17593*. International Organization for Standardization.
29. Kirmess, K. M., Meyer, M. R., Holubasch, M. S., Knapik, S. S., Hu, Y., Jackson, E. N., Harpstrite, S. E., Verghese, P. B., West, T., Fogelman, I., Braunstein, J. B., Yarasheski, K. E., & Contois, J. H. (2021). The PrecivityAD™ test: Accurate and reliable LC-MS/MS assays for quantifying plasma amyloid beta 40 and 42 and apolipoprotein E proteotype for the assessment of brain amyloidosis. *Clinica Chimica Acta, 519*, 267–275.
30. ISO. (2006). *Statistics - Vocabulary and Symbols - Part 2: Applied Statistics. ISO 3534-2*. International Organization for Standardization.
31. Vandijck, M., Degrieck, R., Denoyette, M., Delanote, J., De Jonge, M., De Decker, B., Le Bastard, N., & Vandenbroucke, I. (2022). Analytical performance of the Lumipulse® G β-amyloid 1-40 plasma and lumipulse® G β-amyloid 1-42 plasma RUO assays. *Alzheimers Dement, 18*(Suppl. 6), e068990-1-2.
32. Vandijck, M., Dekeyser, F., Lambrechts, C., Hasumi, Y., Van Loo, M., Accou, F., Sato, H., Inaoka, D., Nojima, H., Degrieck, R., Le Bastard, N., Vanbrabant, J., Stoops, E., & Vandenbroucke, I. (2023). Preliminary performance of the Lumipulse G pTau 217 Plasmaprototype assay on plasma samples. *Alzheimers Dement, 19*(Suppl. 15), e079647-1-2.

33. Vandijck, M., D'hont, J., Dekeyser, F., Le Bastard, N., & Vandenbroucke, I. (2022). Analytical performance overview of the Lumipulse G pTau 181Plasma RUO. *Alzheimer's & Dementia, 18*(Suppl. 6), e064041-1-2.
34. Eastwood, S. M., Meyer, M. R., Kirmess, K. M., Wente-Roth, T. L., Irvin, F., Holubasch, M. S., Verghese, F. B., West, T., Braunstein, J. B., Yarasheski, K. E., & Contois, J. H. (2024). PrecivityAD2™ blood test: Analytical validation of an LC-MS/MS assay for quantifying plasma phospho-tau217 and non-phospho-tau217 peptide concentrations that are used with plasma amyloid-β42/40 in a multianalyte assay with algorithmic analysis for detecting brain amyloid pathology. *Diagnostics, 14*, 1739-1-18.
35. Yoda, R., Kaneko, N., Korenaga, A., Funatsu, S., Sakashita, N., Nishikaze, T., Ohashi, Y., Honda, M., Hioki, Y., Sekiya, S., Ogata, K., Iwamoto, S., Tsujino, K., & Tanaka, K. (2021). Analytical validation of plasma amyloid β measuring system by immunoprecipitation-mass spectrometry. *Alzheimers Dement, 17*(Suppl. 5), e051969-1-1.
36. Nakamura, A., Kaneko, N., Villemagne, V. L., Kato, T., Doecke, J., Doré, V., Fowler, C., Li, Q. X., Martins, R., Rowe, C., Tomita, T., Matsuzaki, K., Ishii, K., Ishii, K., Arahata, Y., Iwamoto, S., Ito, K., Tanaka, K., Masters, C. L., & Yanagisawa, K. (2018). High performance plasma amyloid-β biomarkers for Alzheimer's disease. *Nature, 554*, 249–254.
37. Lee, K. H., Hsu, M. H., Chen, H. H., & Yang, S. Y. (2023). Analyzer-to-analyzer variations in assaying ultralow concentrated biomarkers associated with neurodegenerative diseases using immunomagnetic reduction. *ACS Measurement Science Au, 3*, 488–495.
38. ICH Guideline (2005) Q2 (R1), Validation of Analytical Procedures: Text and Methodology. London
39. Schuhmacher, J., Zimmer, D., Tesche, F., & Pickard, V. (2002). Matrix effects during analysis of plasma samples by electrospray and atmospheric pressure chemical ionization mass spectrometry: Practical approaches to their elimination. *Rapid Communications in Mass Spectrometry, 17*, 1950–1957.
40. Voyvodic, P. L., Conejero, I., Mesmoudi, K., Renard, E., Courtet, P., Cattoni, D. I., & Bonnet, J. (2022). Evaluating and mitigating clinical samples matrix effects on TX-TL cell-free performance. *Science Reports, 12*, 13785-1-9.
41. CLSI. (2018). Establishing and verifying an extended measuring interval through specimen dilution and spiking. *CLSI guideline EP34*. (1st ed), Wayne, PA: Clinical and Laboratory Standards Institute
42. https://www.sigmaaldrich.com/TW/en/support/calculators-and-apps/1x-phosphate-buffered-saline?srsltid=AfmBOoqbsbjDfWwQayT-f4wnOesI3InjhriAAkysMiP4JsprhWgfTwe5
43. https://www.qiagen.com/us/resources/faq?id=44c54114-c4d7-4a7a-8d08-6eea922e77c0&lang=en
44. https://www.beckman.fr/reagents/coulter-flow-cytometry/qc-and-support-reagents/6603369
45. Cuming, I. H. (1964). Plasma substitutes. *The Medical Journal of Australia, 1*, 800–806.
46. Lundsgaard-Hansen, P., Collins, J. A., David-West, A. S., Lopez, C. G., Hantchef, Z. S., Lothe, F., & Von Stefens, E. (1983). Use of plasma volume substitutes and plasma in developing countries. *Bulletin of the World Health Organization, 61*, 7–22.
47. Boldt, J., & Suttner, S. (2005). Plasma substitutes. *Minerva Anestesiologica, 71*, 741–758.
48. Haeckel, R. (1991). Proposals for the description and measurement of carry-over effects in clinical chemistry (recommendations 1991). *Pure and Applied Chemistry, 63*, 301–306.
49. CLSI. (2024). Evaluation of Reagent Carryover Effects on Test Results. 1st ed. CLSI guideline EP47. Clinical and Laboratory Standards Institute.
50. Tate, J., & Ward, G. (2004). Interferences in immunoassay. *Clinical Biochemist Reviews, 25*, 105–120.
51. CLSI. (2005). *Interference Testing in Clinical Chemistry; Approved Guideline—Second Edition. CLSI document EP07-A2*. Clinical and Laboratory Standards Institute.
52. Wauthier, L., Plebani, M., & Favresse, J. (2022). Interferences in immunoassays: Review and practical algorithm. *Clinical Chemistry and Laboratory Medicine, 60*, 808–820.

53. Reverberi, R., & Reverberi, L. (2007). Factors affecting the antigen-antibody reaction. *Blood Transfusion, 5*, 227–240.
54. Klee, G. G. (2000). Human anti-mouse antibodies. *Archives of Pathology & Laboratory Medicine, 124*, 921–923.
55. Sturgeon, C. M., & Viljoen, A. (2011). Analytical error and interference in immunoassay: Minimizing risk. *Annals of Clinical Biochemistry, 48*, 418–432.
56. CLSI. (2023). Evaluation of Stability of In Vitro Medical Laboratory Test Reagents. 2nd ed. CLSI guideline EP25. Clinical and Laboratory Standards Institute.
57. Liu, H. C., Lin, C. Y., Chen, H. H., Hu, C. J., Ho, C. S., & Yang, S. Y. (2021). Using immunomagnetic reduction to assay reagent stability of biomarkers associated with Alzheimer's disease. *Journal of Neurological Disorders, 9*, 2-1-4.
58. Hong, C. Y., Chen, W. S., Jian, Z. F., Yang, S. Y., Horng, H. E., Yang, L. C., & Yang, H. C. (2007). Wash-free immunomagnetic detection for serum through magnetic susceptibility reduction. *Applied Physics Letters, 90*, 074105-1-4.
59. https://www.marken.com/
60. https://www.worldcourier.com/

Chapter 4
Validations of Plasma Biomarker Tests

Abstract Clinical validations are required for an IVD medical device to confirm their clinical performances. The clinical validation of blood-based biomarker tests is complicated because of the lack of standard/reference test. Thus, it is necessary to compare blood-based biomarker tests with the diagnoses listed in the diagnostic guidelines for neurodegenerative diseases, such as brain tissue staining, neuroimaging, biomarker tests in cerebrospinal fluid (CSF), clinical diagnosis, etc., depending on the types of neurodegenerative diseases. In this chapter, clinical evidence revealing the correlations between blood-based biomarker test and the standard diagnosis is reviewed. For example, in the case of Alzheimer's disease, blood-based biomarker tests are compared with amyloid neuropathology in postmortem brain, amyloid positron emission tomography (PET), tau PET, CSF biomarkers, magnetic resonance imaging (MRI), neuropsychological tests, and behavioral observations. In the case of Parkinson's disease and dementia with Lewy bodies, the correlations between blood-based biomarker tests and postmortem Lewy-type synucleinopathy, TRODAT scans, MRI, CSF biomarkers, behavioral observations, and neuropsychological tests are considered. In the case of frontotemporal dementia, the clinical validity of blood-based biomarker tests is evidenced in comparison with postmortem TDP neuropathology, neuropsychological tests, and behavior observations. The contents can help readers bridge blood-based biomarkers to the standard diagnosis.

Keywords Clinical validation · Blood biomarkers · Tissue staining · Neuroimaging · CSF biomarkers · Neuropsychological tests

4.1 Philosophy of Validating Plasma Biomarker Assays

The most common validations of an IVD medical device follow FDA 510(k) regulations, in which the clinical performance of a to-be-approved IVD medical device is compared with that of an approved IVD medical device with the same clinical indications. For example, before a new ELISA kit for assaying plasma biomarkers of colorectal cancer, such as carcinoembryonic antigen (CEA), enters the

C. S.-Y. Yang, M. N. Sabbagh, *Assessing Neurodegenerative Diseases: High-Sensitivity Technologies That Assay Pathological Biomarkers in Blood*,
https://doi.org/10.1007/978-3-032-23863-4_4

clinical market, clinical validation is needed to confirm the clinical validity of this assay. In the clinical validation, developers usually collect several hundred plasma samples, which are simultaneously assayed with the new ELISA kit and an approved ELISA kit. Once the correlation between the CEA concentrations measured using the two ELISA kits is significantly high, the clinical validity of the new ELISA kit can be confirmed.

At present, very few IVD medical devices have been approved for the assessment of neurodegenerative diseases by assaying plasma biomarkers except for genetic markers which usually use PAXgene or buffy coat. This finding implies that IVD medical devices for assessing neurodegenerative diseases by assaying plasma biomarkers are not very popular or perhaps not very well developed since there have not been many approved by the FDA to date. The clinical performance of IVD devices for assaying plasma biomarkers associated with neurodegenerative diseases should be compared with that of the methods listed in the guidelines for diagnosing neurodegenerative diseases. This is a stepped approach though. The FDA required imaging correlation to autopsy for the amyloid positron emission tomography (PET) agents and will again require them for the tau PET agents. Therefore, the amyloid PET becomes the new gold standard for IVD of plasma markers.

For example, the clinical diagnosis of Alzheimer's disease (AD) includes neuropsychological tests; magnetic resonance imaging to exclude other pathologies; assays of biomarkers in cerebrospinal fluid (CSF), including $A\beta_{1-42}$, T-Tau, and pTau; or amyloid positron emission tomography (PET).

A comparison of the clinical performance of plasma biomarker assays with that of all clinical diagnostic methods for AD is difficult. Sometimes, the correlation between the clinical performance of plasma biomarker assays and certain clinical diagnostic methods is investigated. Hence, the clinical indications of plasma biomarker assays are not the diagnosis of AD but rather the agreement between plasma biomarker assays and neuropsychological tests, magnetic resonance imaging, CSF biomarker assays, or amyloid PET.

For example, IMR kits, which were approved by the Taiwan FDA in 2020 to assay plasma biomarkers, have been validated clinically by investigating the agreement between plasma $A\beta_{1-42}$xT-Tau (or $A\beta_{1-40}/A\beta_{1-42}$) levels and neuropsychological tests with magnetic resonance imaging. Independent validation using IMR kits was conducted to determine the agreement between plasma $A\beta_{1-40}/A\beta_{1-42}$ levels and amyloid PET. MP-CLEIA kits approved by the US FDA in 2025 have been validated by investigating the agreement between plasma pTau217/$A\beta_{1-42}$ levels and amyloid PET. However, two levels of confidence have been established for blood-based biomarkers (BBMs) including confirmatory (>90% sensitivity and specificity) or screening [1].

Therefore, a plasma biomarker assay is regarded as an assistant diagnosis (or assessment), not a confirmatory diagnosis unless the clinical sensitivity and specificity are greater than 90%, of neurodegenerative diseases.

In this section, we wish to provide a comprehensive understanding of the clinical performance of plasma biomarker assays for readers. Comparisons between the results of the plasma biomarker assay and all of the clinical diagnostic methods are illustrated in this chapter. The comparisons are described for various types of neurodegenerative diseases separately. The neurodegenerative diseases of interest here include Alzheimer's disease (AD), Parkinson's disease (PD), dementia with Lewy bodies (DLB), and frontotemporal dementia (FTD).

A survey of publicly available documents revealed that not all assay technologies introduced in Chap. 1 have been completely clinically validated. We aimed to collect more information to show the agreement between plasma biomarkers and standard diagnostic methods used in the guidelines for various types of neurodegenerative diseases.

4.2 Clinical Validation of Assays for Alzheimer's Disease

Amyloidosis has been proposed to explain the development of Alzheimer's disease (AD) [2]. Amyloid-β peptides are 36–43 amino acids long and result from cleavage of the amyloid precursor protein (APP) by the β and γ secretases in the membrane of neurons. Monomers of amyloid-β are soluble in the cytoplasm of neurons. The abnormal aggregation of monomers of misfolded amyloid-β (Aβ) molecules to form oligomers, protofibrils, fibrils, and plaques in the brain plays a role in AD [3]. Amyloid neuropathological evaluations are the gold standard for diagnosing AD.

In addition to amyloid neuropathology, neurofibrillary tangles (NFTs) are a neuropathological hallmark of AD [4]. The main component of NFTs is tau proteins, which have six isoforms. The formation of NFTs is predominantly due to the phosphorylation of tau proteins. Hence, the observation of NFTs in the brain provides a direct inspection of the neuropathology of AD. Conventionally, NFTs are also termed tau tangles. Biologically, the tau protein serves as an anchor to stabilize the microtubules of axons. Once the tau protein is phosphorylated, the anchoring function is disrupted, resulting in damage to axons and the death of neurons. Neurodegeneration subsequently occurs.

Since performing antemortem brain tissue staining for investigating amyloid or tau neuropathology is impossible, amyloid or tau positron emission tomography (PET) for identifying amyloid plaques or tau tangles in the antemortem brain is proposed as a proxy assessment of the neuropathological changes associated with AD [5, 6].

Positron emission tomography (PET) is a type of medical imaging performed with the aid of a tracer [7, 8]. PET tracers are created by attaching a radioactive isotope to a biologically active molecule, such as a sugar or a drug. Common PET radioisotopes include fluorine-18 (^{18}F), carbon-11 (^{11}C), oxygen-15 (^{15}O), and nitrogen-13 (^{13}N). The tracer can specifically bind to target biomolecules in the human body. These tracers are injected into the body and accumulate in specific tissues, where the radioactive decay of the isotope is detected by the PET scanner to create images of metabolic and biochemical activity.

In addition, the tracer is usually functionalized with an isotope for positron emission under excitation. Thus, PET imaging can visualize the distributions of target biomolecules in the human body. Using PET with the aid of suitable tracers (^{18}F and

^{11}C), the distributions of amyloid plaques and tau tangles in the brain can be visualized. The three approved amyloid specific tracers are florbetapir, florbetaben, and flutemetamol, and they all specifically identify the mature senile plaque in vivo. The approve F18 tau PET tracer is flortaucipir. ^{11}C tracers have limited clinical applicability because of their short half-life (~20 min).

Many clinical results have shown the spread of amyloid plaques and tau tangles in the brain of patients with AD. Since the approval of tracers by the US FDA in since 2012, amyloid PET became a desirable technology to deploy but was limited in utilization because of the national coverage determination requirement (NCD) and non-reimbursement. The NCD requirement was lifted in 2023. In 2020, a tracer for tau PET (flortaucipir) was approved by the US FDA, which provided an opportunity to use tau PET to identify the formation of tau tangles in the brain of patients with AD or other neurodegenerative diseases related to tauopathy, such as progressive supranuclear palsy (PSP), corticobasal degeneration (CBD), and frontotemporal dementia.

The assays of Aβ or tau biomarkers in cerebrospinal fluid (CSF) has also been developed to evaluate and measure concentrations of amyloid or tau neuropathology. In the hypothetical model of dynamic biomarkers of the Alzheimer's pathological cascade proposed in 2010, CSF $A\beta_{1-42}$ and T-Tau levels were suggested as relevant biomarkers [9]. The abnormal changes in CSF $A\beta_{1-42}$ levels represented amyloid neuropathology, and the increases in CSF T-Tau levels represented neuronal injury and dysfunction. This model was approved clinically 2 years later. pTau181 was added later and is now part of CSF testing for AD.

In 2018, new criteria were proposed for a biological definition of AD which incorporated biomarkers such as CSF, namely, the A-T-N criteria, where A denotes aggregated Aβ or the associated pathological state, T denotes aggregated tau angles or the associated pathological state, and N denotes neurodegeneration or neuronal injury, as shown in Fig. 4.1 [10, 11]. The CSF biomarkers in A could be $A\beta_{1-42}$ or $A\beta_{1-42}/A\beta_{1-40}$. The CSF biomarker in T could be phosphorylated tau, pTau181, with other forms of pTau currently being investigated. The CSF biomarker in N could be T-Tau.

Remarkably, the Alzheimer's continuum is identified only when A is positive. This confirms the indispensability of $A\beta_{1-42}$ or $A\beta_{1-42}/A\beta_{1-40}$ levels in diagnosing AD. In the latest guidelines issued in 2024, CSF $A\beta_{1-42}$ levels were suggested in A [12]. More types of tau proteins in CSF were included in T. The neurofilament light chain (NfL) level in CSF has been suggested as a biomarker in N, along with pTau181, pTau217, and pTau231, as well as nonphosphorylated mid-region tau fragments. There is a move to expand A-T-N to A-T-X (N), with consideration of biomarkers for inflammation (GFAP, YKL-40, and TREM2) and synucleinopathy and vascular changes. At present, definite biomarkers in CSF are listed in the guidelines issued by the National Institute of Aging and Alzheimer's Association to assist in the diagnosis of AD.

The formation of NFTs suggests damage to neural cells in the brain, which leads to brain atrophy. Hence, brain atrophy as a proxy marker of neurodegeneration is a typical hallmark of advancing AD. Magnetic resonance imaging (MRI) is a

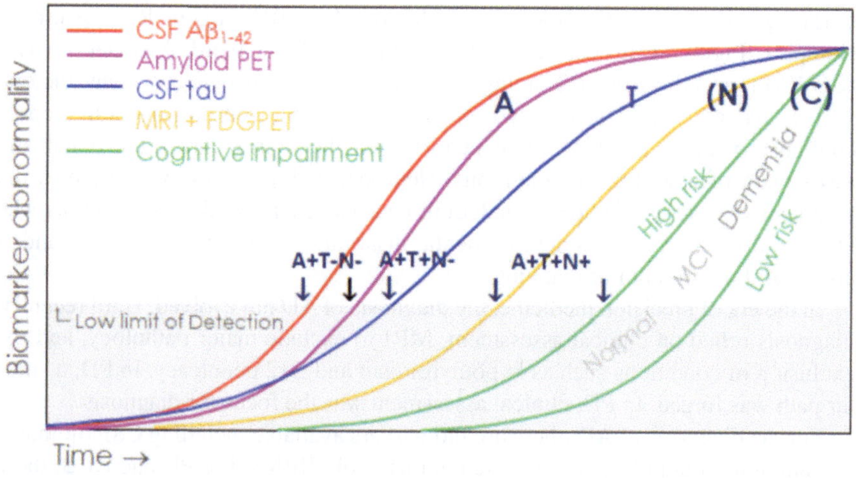

Fig. 4.1 Dynamic biomarkers in A-T-N criteria of Alzheimer's pathology [11]

powerful tool for observing structural/volumetric changes in the brain. As a result, MRI is listed as a standard examination in the evaluation of AD to exclude other pathology but now to include regional volumetry, quantification of white matter rarefaction, and quantification of microhemorrhages. MRI examination to exclude patients who have brain ischemia, mass, or hydrocephalus is customary during the evaluation of AD dementia. Patients with vascular cognitive impairment/dementia could have similar clinical symptoms as AD patients. Without excluding patients with dementia combined with brain ischemia or injuries, false-positive diagnosis for AD can occur, but there are situations where vascular dementia and AD can co-occur.

As the brain of patients progresses due to AD dementia, brain function abnormally degrades, resulting in disorders of cognition and behavior, which are terminal findings. The most typical symptom of AD is sporadic memory impairment.

Prior to the advent of biomarkers such as PET, CSF, or even BBMs, we relied on the clinical phenotypic description and ancillary tests such as bedside cognitive evaluations, global evaluations, and neuropsychological evaluations. Bedside cognitive evaluations include the Mini-Mental State Examination (MMSE), Montreal Cognitive Assessment (MoCA), the Mini-Cog, and the Saint Louis University Mental Status (SLUMS) Examination. For global assessments, the Clinical Dementia Rating (CDR) scale has become the standard outcome measure in clinical trials. For functional assessment, activities of daily living are assessed using, e.g., the Alzheimer's Disease Cooperative Study/Activities of Daily Living (ADCS-ADL) scale and the Disability Assessment for Dementia (DAD) score. Neuropsychological testing can be very detailed for clinical assessment by brain regions including attention, concentration, visual memory, verbal memory, fluency (naming and categorical), and visuospatial abilities. Tests could include Trails A, Trails B, the Wisconsin Card Sort Test, Rey Auditory Verbal Learning Test (R-AVLT), the Hopkins Verbal Learning Test (HVLT), the California Verbal

Learning Test (CVLT), the Brief Visual Memory Test (BVMT), the Rey–Osterrieth Complex Figure Test (ROCFT), the Controlled Oral Word Association Test (COWAT), and the Boston Naming Test (BNT); scores depend on many factors, such as age, education, gender, and social level [13–16]. Neurologists or neuropsychologists need to combine these factors with the scores during the diagnostic assessment of neurological conditions. Neuropsychological testing is used in the diagnosis of several conditions, including neurodegenerative diseases such as AD, DLB, and FTD, as well as developmental disabilities, epilepsy, and stroke, and is also useful in brain tumor mapping.

In the era of precision medicine, the diagnosis of AD has evolved. Until recently, diagnosis relied on clinical assessment, MRI to exclude major pathology, and the exclusion of conditions such as hypothyroidism and B12 deficiency. In PD, a similar path was forged. In PD, clinical assessment was the focus for diagnosis.

For the diagnosis of AD, there are biomarkers available including CSF biomarkers, amyloid or tau PET, and, as noted in this book, BBMs. For PD and DLB, there are biomarkers available including dopamine scan, SynOne skin biopsy assessment of α-synuclein, and seed amplification assays or RT-QuIC tests of CSF for α-synuclein. For FTD/ALS, there are now genetic panels to detect C9orf72, FUS, progranulin, and TDP-43.

To demonstrate validity, tests have had to be comparable in sensitivity and specificity to brain tissue staining. Developers of plasma biomarker assays must show the correlations between the biomarker measurements and these examinations to validate the clinical efficacy of plasma biomarkers, which is an incredible challenge. With great contributions worldwide in the past 20 years, many results using the assay technologies described in Chap. 1 demonstrate the correlations between plasma biomarkers and these examinations. These efforts not only create innovations in the diagnosis of AD but also facilitate convenient and precise assessments of AD. Importantly, successful developments in assays of plasma biomarkers have enabled assessments of AD at home or in communities at low cost and low risk, which would be very helpful in identifying prodromal or preclinical AD and would be very significant in preventing the onset of AD in high-risk individuals.

In this chapter, we describe the key results used to determine the clinical validity of plasma biomarker assays for AD via the correlations between plasma biomarkers and brain tissue staining, PET, CSF biomarkers, MRI, or neuropsychological tests. According to the results, a paradox occurs among assay technologies. A discussion of this paradox is provided here.

4.2.1 Plasma Biomarkers Versus Brain Tissue Staining

The direct and convincing evidence for AD is brain tissue staining of senile plaques with a core component including amyloid and neurofibrillary tangles with a core component of phosphorylated tau. The formation of amyloid plaques and tau tangles in the brains of AD patients have been observed, described, and quantitated

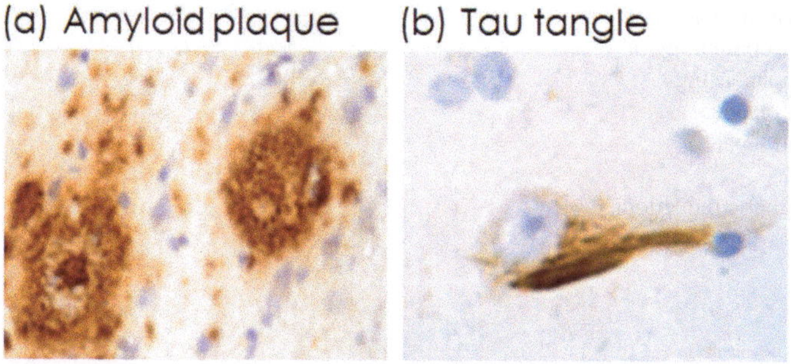

Fig. 4.2 Brain tissue staining of (**a**) amyloid plaques and (**b**) tau tangles [17]

since the time of Alois Alzheimer, MD, himself. There are several staging scales used to quantify the neuropathology of AD, including Braak staging, scores issued by the National Institute on Aging/Reagan Institute (NIA–Reagan), and the Consortium to Establish a Registry for Alzheimer's Disease (CERAD). Typical images of amyloid plaques and tau tangles in the brain are shown in Fig. 4.2a, b, respectively [17].

Biopsies of the brain are not used in real practice. Hence, comparisons between brain biopsy results and plasma biomarkers are rare in the real world. However, validations of antemortem plasma biomarkers with postmortem neuropathology are feasible.

The precedent for comparison of comparing in vivo diagnostics to postmortem samples in the neurodegenerative space comes from FDA mandate for radiological imaging tracers of neuropathology to correlate directly with neuropathological staining of cerebral tissue from the very same brain regions. This was the standard of proof that was set forth for florbetapir, flutemetamol, florbetaben, and flortaucipir [18–20]. Other agents—PI-2620 for the detection of tangles and NAV4694 for the detection of plaque—are in development [21]. All tracers that are FDA approved are F18 compounds that can be used on commercial PET scanners.

The analysis of samples from brain banks, which could provide paired antemortem plasma samples and brain tissue from autopsies for staining, would be a better approach to accelerate validation. More than 150 brain banks exist globally, such as in Phoenix, Melbourne, Mount Sinai, Munich, Singapore, Taipei, Hefei, and Tokyo. Investigators should pay attention to the time interval between the antemortem collection of plasma samples and the harvesting of postmortem brain tissue. The time interval should not be too long and should usually be shorter than 1.5 years. Otherwise, the biological conditions of the plasma will not be consistent with the pathological conditions of the brain. In addition, investigators should pay attention to the preparation and storage of human plasma samples to confirm the quality of the plasma. Ideally investigators should avoid enrolling patients with dementia diagnosed with mixed types of neurodegenerative diseases. However, most neurodegenerative diseases have mixed pathology.

A typical example of verifying the correlation between antemortem plasma biomarkers and postmortem neuropathology is illustrated [22]. Antemortem plasma

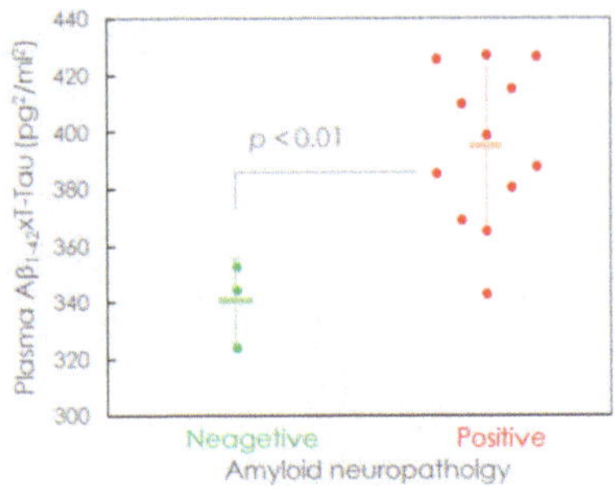

Fig. 4.3 Discrimination of amyloid neuropathology in autopsy samples using antemortem plasma Aβ_{1-42}xT-Tau levels assayed with immunomagnetic reduction (IMR)

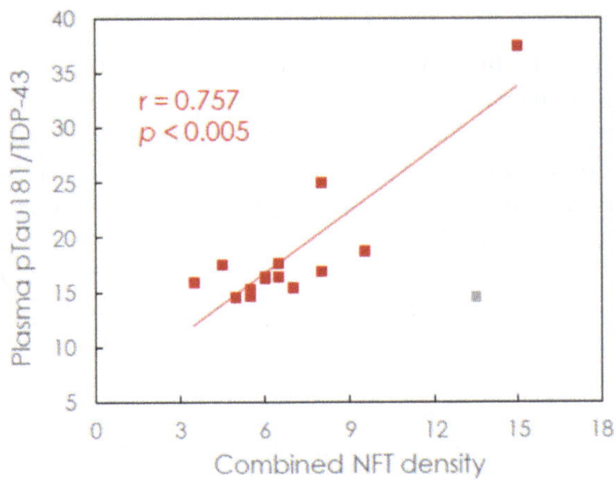

Fig. 4.4 Correlation between the combined NFT density in autopsy tissues with antemortem plasma pTau181/TDP-43 levels assayed with immunomagnetic reduction. The gray dot is an outlier

samples were collected within 1.5 years before death. Plasma Aβ_{1-42}, T-Tau, pTau181, and TDP-43 levels were assayed using immunomagnetic reduction (IMR). Amyloid and tau neuropathology was examined after death. The brain regions used for the analysis of amyloid and tau neuropathology included the frontal, temporal, and parietal lobes plus the hippocampal and entorhinal/transentorhinal areas.

As shown in Fig. 4.3, the plasma Aβ_{1-42}xT-Tau levels were significantly higher ($p < 0.01$) in patients positive for amyloid neuropathology than in patients negative for amyloid neuropathology. Although the number of subjects in Fig. 4.3 is limited, the discrimination of positive/negative amyloid neuropathology in terms of plasma Aβ_{1-42}xT-Tau levels is promising. The cutoff value for plasma Aβ_{1-42}xT-Tau levels for discriminating amyloid neuropathology is approximately 360–380 (pg/ml)^2.

Figure 4.4 shows the correlation between the antemortem plasma pTau181/TDP-43 levels and combined NFT density in the postmortem brain tissue [22]. A high correlation was detected (Pearson's correlation coefficient $r = 0.757$, $p < 0.005$). These findings support the powerful ability of plasma biomarkers to predict tau neuropathology in the brain.

For both amyloid and tau neuropathology, combined but not individual plasma biomarkers are able to predict the status of neuropathology. This finding implies that the formation of amyloid plaques and tau tangles in the brain is not dominated by an individual protein but by complicated mechanisms involving several proteins. More work should be conducted to clarify the details in the future.

4.2.2 *Plasma Biomarkers Versus Amyloid or Tau PET*

With the successful development of ^{18}F- and ^{11}C-isotope tracers, which have specific affinities for amyloid plaques or tau tangles in the brain, imaging of amyloid plaques and tau tangles with the aid of positron emission tomography (PET) in the antemortem brain tissue has become feasible. Once more amyloid plaques or tau tangles form in the brain, higher values of the standardized uptake value ratio (SUVr) are obtained using amyloid PET or tau PET. The quantification scale for SUVr for amyloid is now called the Centiloid scale. The quantification scale for SUVr for tau is now called the centaur scale. The density of brain amyloid-β neuritic plaque accumulation, a marker of Alzheimer's disease (AD), can be visualized and quantified using amyloid PET. Amyloid PET data can be obtained using different tracers and methodologies; therefore, comparison across studies can be difficult. The introduction of Centiloid scale allowed a common scale to be created for amyloid PET quantitative data, enabling comparisons using a common scale. Since then, Centiloid values have been used increasingly in research and clinical trials for multiple purposes, being tested and validated with a variety of clinical, biomarker, and pathological standards of truth. In clinical trials, Centiloid values have been used for patient selection and confirmation of the presence of AD pathology, with outcomes of therapeutic studies often quantifying amyloid burden using the Centiloid (CL) scale [23].

Typical images of amyloid PET scans are shown in Fig. 4.5 [24]. The red regions correspond to the locations of the amyloid plaques. The original approval for amyloid PET tracers included the stipulation of assessment using the gray scale and reports of a binary outcome: positive or negative. In research, quantification was always used and presently is slowly making its way into clinical practice. The image in Fig. 4.5a does not have an obvious red region, whereas the image in Fig. 4.5b shows many clear red regions. The image in Fig. 4.5a shows negative amyloid neuropathology, and the image in Fig. 4.5b shows positive amyloid neuropathology in the brain. Similar images have also been obtained for tau PET. Remarkably, the determination of positive/negative amyloid or tau pathology for AD by PET images depends not only on the total area of red regions but also on the distributions of red regions in the brain. Therefore, determining positive/negative amyloid or tau pathology using PET images to diagnose AD is complicated.

A common approach is to determine positive/negative amyloid or tau pathology by a visual ranking. Such determinations might be prone to interpretation. But it is a marker of where amyloid pathology is emerging, with the range between 10 and 30 CL representing evolving pathology. Values below 10 CL generally indicate the absence of amyloid pathology, while values above 30 CL typically signify

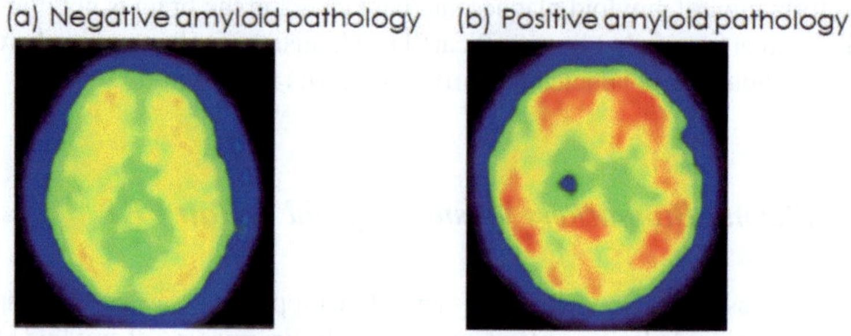

Fig. 4.5 PET images of (**a**) negative and (**b**) positive amyloid pathology [24]

established pathology. Thresholds between 20 and 26 CL are frequently used in studies to define a "positive" result, but the exact cutoff can vary [25]. Many artificial-intelligent algorithms have been applied to achieve scientific examinations of AD using amyloid or tau PET and to perform an objective analysis of PET images and quantify amyloid and tau respectively [26].

Undeniably, amyloid PET or tau PET prevents the high risks associated with brain biopsy and staining to examine the neuropathology in AD and might be a surrogate given the strong clinical–pathological correlation. Hence, PET examinations involving the injection of amyloid or tau tracers can provide evidence of target pathology when diagnosing AD. Developers of plasma biomarker assays must verify the correlations between plasma biomarkers and amyloid or tau PET.

Many results have been reported to reveal the prediction of amyloid pathology, which is determined by amyloid PET, using plasma biomarkers. Figure 4.6 shows the values of the areas under the receiver operating characteristic curves (AUCs) to discriminate amyloid pathology using plasma biomarkers assayed with magnetic particle-based chemiluminescent enzyme immunoassay (MP-CLEIA), single molecule array (SIMOA), immunomagnetic reduction (IMR), electrochemiluminescence (ECL), and immunoprecipitation mass spectrometry (IP/MS) technologies [24, 27–42]. Some plasma biomarkers, such as $A\beta_{1-42}/A\beta_{1-40}$, $A\beta_{1-42}$xT-Tau and $A\beta_{1-42}$ + APOE genotype with IMR, and pTau217 with MP-CLEIA or SIMOA, have AUC values greater than 0.9 in predicting amyloid pathology with PET. These results indicate the high consistency and strong predictive value and high likelihood ratio between plasma biomarkers and amyloid PET images.

The possibility of using plasma biomarkers for prescreening tau PET is also desirable. Figure 4.7 shows an example of the correlation between plasma pTau181 levels and the SUVr of tau PET. The assay of plasma pTau181 levels was performed using immunomagnetic reduction (IMR). The SUVr shown in Fig. 4.7 was analyzed in the temporal lobe. Pearson's correlation coefficient, r, is 0.656 ($p < 0.0001$), revealing a significantly positive correlation. In fact, significant positive correlations were also observed for the frontal lobe, parietal lobe, occipital lobe, amygdala, caudate, putamen, gray matter, white matter, etc. This evidence indicates that the density of tau tangles can be evaluated by assaying plasma pTau181 levels. Hence, the plasma levels of biomarkers are correlated with the NFT density in the brain.

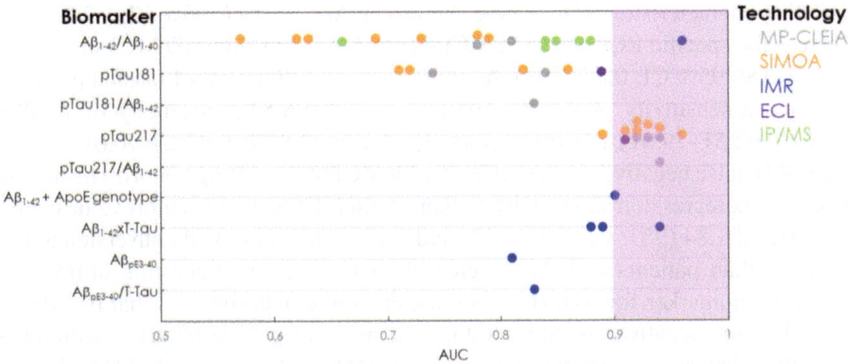

Fig. 4.6 Reported areas under the receiving operating characteristic curves (AUCs) to discriminate amyloid pathology using plasma biomarkers assayed with various technologies

Fig. 4.7 Correlation between plasma pTau181 levels and the SUVr of tau PET in the temporal lobe

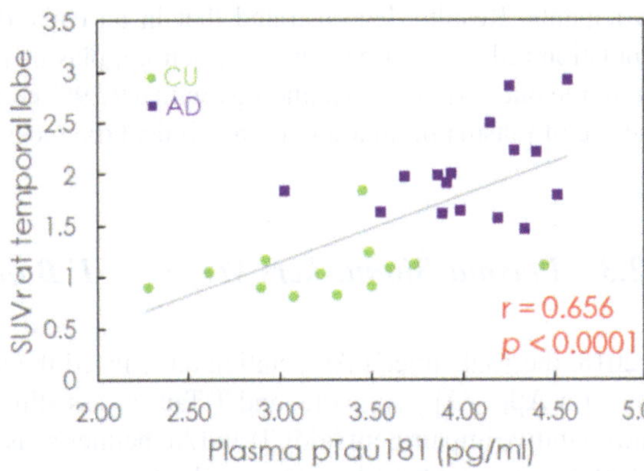

Another important feature shown in Fig. 4.7 is that patients with Alzheimer's disease (AD) (▪) have higher plasma pTau181 levels than cognitively unimpaired (CU) patients (▪). Plasma pTau181 level has shown promise for differentiating patients with AD from CU patients but with a lot of overlap.

In a study of patients drawn from three cohorts from real-world clinics, referred to as the modelling cohort (n = 991, 59.7% female), the testing cohort (n = 642, 56.2% female), and the validation cohort (n = 441, 55.1% female), patients with subjective cognitive decline (SCD), mild cognitive impairment (MCI), AD dementia, and other dementias were included [43]. Paired cerebrospinal fluid (CSF) and plasma samples were used to measure AT(N) CSF biomarkers and plasma pTau181. Quantification of CSF biomarkers was performed with magnetic particle-based chemiluminescent enzyme immunoassay (MP-CLEIA) or a standard enzyme-linked immunosorbent assay. Plasma pTau181 was measured with MP-CLEIA. The study found that CSF and plasma pTau181 showed correlation in all phenotypes except in SCD and other dementias. Age significantly influenced the biomarker's performance. The general amyloid PET-positive versus amyloid PET-negative receiver operating characteristic (ROC) curve showed an AUC = 0.77 (95% CI: 0.74–0.80), whereas the specific ROC curve of MCI due to AD versus non-AD MCI showed an AUC = 0.89 95% CI: 0.85–0.93). A cutoff value of 1.30 pg/ml of plasma pTau181 exhibited a sensitivity of 93.57% (95% CI: 88.72–96.52), specificity of 72.38% (95% CI: 62.51–79.01), positive predictive value of 77.85% (95% CI: 70.61–83.54), and 8.30% false negatives in the subjects with MCI of the testing cohort. The hit rate (HR) of Cox regression showed that patients with MCI up to this cutoff value exhibited a HR of 1.84 (95% CI: 1.05–3.22), indicating a higher risk of conversion to AD dementia than patients with MCI below the cutoff value, suggesting utility as a screening biomarker for AD. However, the emerging data suggests that pTau181's strength is the negative predictive value. The assay of plasma pTau181 with magnetically enhanced electrochemiluminescence (ME-ECL) is the only FDA-cleared, blood-based Alzheimer's test indicated as an aid in the initial assessment for Alzheimer's and other causes of cognitive decline in the primary-care setting. The performance was evaluated in a multicenter, non-interventional clinical study in 312 participants. Results demonstrated that in an early disease-stage, low-prevalence population reflective of primary-care setting, plasma pTau181 assay with ME-ECL could rule out Alzheimer's pathology with a 97.9% negative predictive value (NPV). The use of plasma biomarkers in the clinical diagnosis is discussed in Chap. 5.

4.2.3 Plasma Biomarkers Versus CSF Biomarkers

In 2018, the Alzheimer's Association announced the utility of the CSF biomarkers $A\beta_{1-42}$ (or $A\beta_{1-42}/A\beta_{1-40}$), pTau, and T-Tau for assisting in the diagnosis of amnestic mild cognitive impairment (aMCI) and Alzheimer's disease (AD) [10]. Examinations of CSF biomarkers have become standard examinations for aMCI and AD patients. The reported concentrations of CSF biomarkers associated with AD are several hundreds of pg/ml or higher. Hence, the common assay technologies for these CSF biomarkers are enzyme-linked immunosorbent assay (ELISA) and chemiluminescence enzyme immunoassay (CLEIA). Several reagent kits based on ELISA or CLEIA are approved for clinical use in the USA, Europe, and other countries. However, side effects such as headache, back pain, bleeding, infection, tingling or numbness, and brain compression or herniation appear hours to up to a day after lumbar puncture. Unfortunately, examinations of CSF biomarkers for AD, PD, or other neurodegenerative diseases are not popular in routine practice. CSF testing is commonly used to identify paraneoplastic, autoimmune, or infectious causes. Except in Europe, CSF biomarker assays for AD detection are rarely executed in clinical practice. The desire to use plasma biomarkers instead of CSF biomarkers for assisting in the diagnosis of aMCI and AD remains strong. Hence, the correlations between biomarkers in plasma and CSF must be determined.

Table 4.1 Reported correlations of AD biomarkers between plasma and CSF

Biomarker	Technology for assaying plasma biomarkers	Correlation
$A\beta_{1-42}$	ELISA	$r = 0.364$ ($p < 0.005$)
$A\beta_{1-42}$	IMR	$r = -0.335$ ($p < 0.005$)
$A\beta_{1-42}$	MP-CLEIA	$r = 0.54$ ($p < 0.0001$)
pTau181	MP-CLEIA	$r = 0.47$ ($p < 0.001$)
pTau181	SIMOA	$r = 0.649$ ($p < 0.01$)
NfL	SIMOA	$r = 0.86$ ($p < 0.01$)
$A\beta_{1-42}$	SIMOA	$r = 0.274$ ($p < 0.001$)
$A\beta_{1-40}$	SIMOA	$r = 0.136$ ($p < 0.001$)

Significant correlations between plasma and CSF biomarker levels have been reported since 2016 [44, 45]. Several groups independently used different assay technologies to explore the levels of AD biomarkers in plasma and CSF, as listed in Table 1 [44–48]. Most of the results revealed positive correlations between plasma and CSF AD biomarkers. However, in the case of $A\beta_{1-42}$, the IMR shows a negative correlation, but ELISA, MP-CLEIA, and SIMOA show positive correlations. Although a paradox exists in the detection of $A\beta_{1-42}$ among assay technologies, the significant correlations in AD biomarkers between plasma and CSF are clear.

With the reported evidence in Table 4.1, the correlations in biomarkers between plasma and CSF were confirmed.

Another important finding supported by the results in Table 4.1 is that the biomarkers assayed in peripheral fluid (blood) originate from the central nervous system (CNS). Before the results in Table 4.1 were reported, researchers suspected that the biomarkers assayed in plasma might originate from platelets, red blood cells, or organs such as the liver. However, the biomarkers assayed in CSF originate from the CNS. Thus, the concentrations of these biomarkers in plasma would not be correlated with those in CSF. The results in Table 4.1 dispel this myth. The assay of these biomarkers in plasma is clinically equivalent to their analysis in CSF.

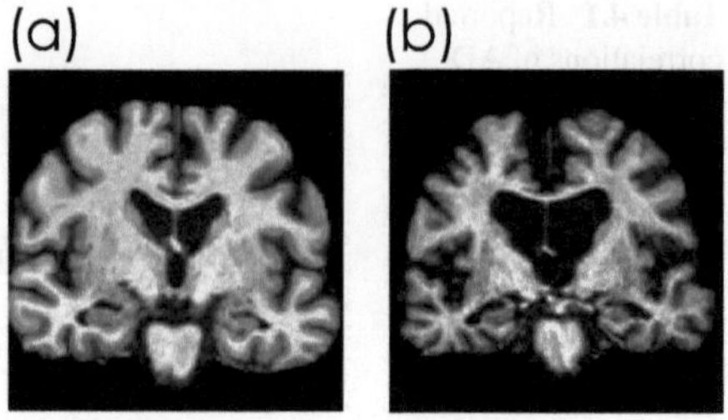

Fig. 4.8 Coronal views of brain MR images of (**a**) a subject without dementia and (**b**) an AD patient

4.2.4 Plasma Biomarkers Versus Magnetic Resonance Imaging

Atrophy of the brain, especially atrophy of the hippocampus, is a common symptom of AD. Magnetic resonance (MR) imaging, which is able to construct the shape of the whole brain and calculate the brain volume, provides a regular examination of brain atrophy in clinical practice. The MR imaging methods include 1.5-, 3-, or 7-Tesla T1- or T2-weighted imaging, although 7T is quite rarely available and mostly used for research. The higher the magnetic field strength, the higher the spatial resolution.

Typical coronal MR images of healthy and AD brains are shown in Fig. 4.8a, b, respectively [49]. Clearly, many interstices are present in the AD brain. Smaller brain volume observed in AD patients using MR imaging has been shown to be a proxy marker of neurodegeneration (ATN).

The atrophy that reflects neurodegeneration of the AD brain is caused mainly by the death of neurons and other cells. As a neuron dies and lyses, its axon is broken, resulting in the expression of tau protein in the brain and CSF. The abundant death of neurons leads to the overexpression of tau protein in the brain. A clear increase in the concentration of tau protein in human plasma is also expected. A multicenter study in Taiwan revealed a negative correlation between plasma T-Tau levels and brain volume [50, 51].

The enrolled subjects included people who were cognitively unimpaired and patients with amnestic mild cognitive impairment or early-stage Alzheimer's disease. Plasma T-Tau levels were assayed with immunomagnetic reduction (IMR). The brain volume of each enrolled subject was analyzed using 1.5-T MR imaging. Notably, the enrolled subjects were never diagnosed with brain ischemia, brain injury, or depression. The atrophy of the brain in the enrolled subjects was almost completely contributed by AD. The plasma T-Tau concentration versus left and right hippocampal volumes are shown in Fig. 4.9a, b, respectively. The plasma T-Tau levels clearly increased with decreasing hippocampal volume but did not linearly increase. The Spearman correlation coefficients in Fig. 4.9a, b are −0.477 ($p < 0.0001$) and − 0.431 ($p < 0.0001$), respectively.

In addition to the hippocampus, further analysis revealed a negative correlation between the plasma T-Tau level and volume in regions such as the thalamus and gyrus in this study, as listed in Table 4.2 [50], suggesting that plasma T-Tau is a

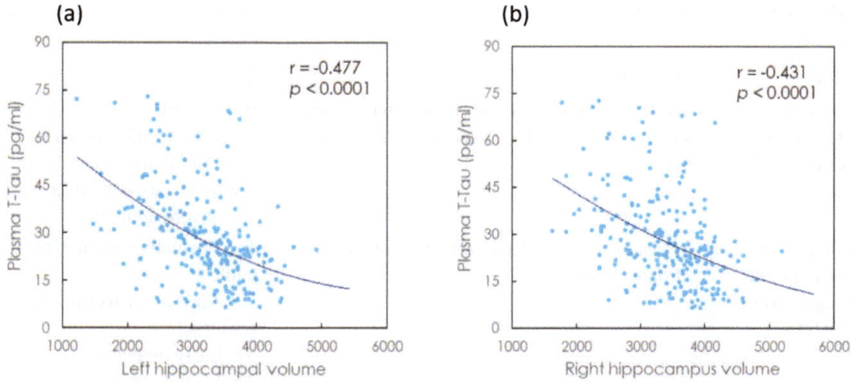

Fig. 4.9 Plasma T-Tau concentration versus the (**a**) left and (**b**) right hippocampal volumes of subjects who are cognitively unimpaired and patients with aMCI or early-stage AD

Table 4.2 Correlation between the plasma T-Tau concentration and the regional volume of the brain in cognitively unimpaired subjects and patients with amnestic mild cognitive impairment or early-stage Alzheimer's disease

Region	Spearman correlation coefficient
Left hippocampus	−0.477 ($p < 0.0001$)
Right hippocampus	−0.431 ($p < 0.0001$)
Thalamus	−0.397 ($p < 0.01$)
Superior frontal gyrus	−0.432 ($p < 0.01$)
Middle frontal gyrus	−0.411 ($p < 0.01$)
Interior gyrus	−0.350 ($p < 0.01$)
Superior temporal gyrus	−0.337 ($p < 0.01$)
Middle temporal gyrus	−0.445 ($p < 0.01$)
Supramarginal gyrus	−0.401 ($p < 0.01$)

marker of neurodegeneration as well. Other studies involving the use of other technologies for assaying plasma T-Tau levels also revealed an increase in plasma T-Tau concentrations with decreasing brain volume in aMCI patients or AD patients [52, 53]. Thus, plasma T-Tau levels may serve as a proxy marker for neurodegeneration as measure by brain atrophy in AD patients. On the other hand, in aMCI and AD patients, plasma T-Tau levels are promising for the routine monitoring of brain atrophy, which would be a very convenient and low-cost examination for monitoring progression.

4.2.5 *Plasma Biomarkers Versus Neuropsychological Tests*

Several types of neuropsychological tests, including tests of many domains of cognition and behavior, have been developed. The most common tests used in clinical practice are listed in Table 4.3. The details of diagnosing AD via neuropsychological tests are described in the guidelines issued by the National Institute on

Table 4.3 Common neuropsychological tests for diagnosing Alzheimer's disease in clinical practice

Test	Domain	Score
Clinical Dementia Rating (CDR)	Memory, orientation, judgment, and problem-solving; community affairs; home and hobbies; and personal care	0 = Normal 0.5 = Very mild dementia 1 = Mild dementia 2 = Moderate dementia 3 = Severe dementia
Functional Assessment Staging Tool (FAST)	Personal ability to function and perform tasks of daily living	1 = Normally functioning adult 2 = Normally functioning senior adult 3 = Early dementia 4 = Mild dementia 5 = Mid-stage dementia 6 = Moderately severe dementia 7 = Severe dementia
Cognitive Abilities Screening Instrument (CASI)	Attention, concentration, orientation, short-term memory, long-term memory, language abilities, visual construction, list-generating fluency, abstraction, and judgment	26–30 = Normal cognition 18–25 = Mild cognitive impairment 10–17 = Moderate cognitive impairment < 10 = Severe cognitive impairment
Mini-Mental State Examination (MMSE)	Orientation, attention and calculation, recall, language, and registration	26–30 = Normal 20–26 = Mild dementia 10–20 = Moderate dementia 0–10 = Severe dementia
Montreal Cognitive Assessment (MoCA)	Ability to process and understand visual information about where objects are, ability to manage cognitive processes (executive functions), language, short-term memoryrecall, attention, concentration, working memory, awareness of time and place	26–30 = Normal cognition 0–25 = Cognitive impairment
Activities of daily life (ADL)	Personal hygiene, dressing, eating, maintaining continence, transferring or mobility	100 = Complete independence 91–99 = Mild dependence 61–90 = Moderate dependence 21–60 = Severe dependence 0–20 = Complete dependence
Instrumental activities of daily life (IADL)	Basic communication, transportation, meal preparation, shopping, homework, doing laundry, managing medications, managing personal finances	3 of basic communication, transportation, meal preparation, homework, and doing laundry are dependent = mild dependence

(continued)

Table 4.3 (continued)

Test	Domain	Score
Wechsler Memory Scale-3rd Edition (WMS-III)	Tests of different types of memory, including verbal, auditory, visual, short-term, and working memory. The details include information and orientation, logical memory, faces, verbal paired associates, family pictures, word lists, visual reproduction, letter–number sequencing, spatial span, mental control, digit span, etc.	Complicated statistical analysis

Aging–Alzheimer's Association (NIA–AA), the International Working Group (IWG), and other associations.

According to the comprehensive scores of neuropsychological tests, a subject is categorized into various stages: cognitively unimpaired (CU), amnestic mild impairment (aMCI), and AD dementia (AD). Thus, the correlations between plasma biomarker levels and disease stages should be verified. Furthermore, the correlations of plasma biomarker levels with scores for various domains have been explored.

Remarkably, several medical conditions, such as heart failure, recent myocardial infarction (in the past 6 months), malignancy (in the past 2 years), poorly controlled diabetes (HbA1c level > 8.5), end-stage renal disease, brain ischemia, and a significant history of depression (Geriatric Depression Scale score > 8), influence the scores. Careful evaluations of aMCI or AD in subjects with these medical conditions should be performed. In addition, the scores of neuropsychological tests should be and are adjusted for education, age, and other factors, and the scores are always corrected for these factors using normative data. For example, the MMSE scores for AD are 0–26 for subjects with 12 years of education, 0–24 for subjects with 6 years of education, and 0–13 for subjects with less than 6 years of education.

In addition, the diagnosis of aMCI or AD via neuropsychological tests depends on the clinical experience of psychologists. Not all practitioners rendering these tests are board-certified neuropsychologists, and the quality of reports can sometimes make interpretation difficult. Collaboration with senior neuropsychologists is preferred for plasma biomarker validation. In clinical trials of plasma biomarker validation, the exclusion of subjects who are suspected of having aMCI or AD is strongly recommended. Instead, patients with clear symptoms of aMCI or AD are preferred. Although such tasks involving two or more neuropsychologists are difficult, the diagnostic accuracy of aMCI or AD can be high.

Subjects with comorbidities may have abnormal levels of plasma biomarkers [54]. Specifically, chronic kidney disease (CKD) has been shown to affect the interpretation of biomarkers such as neurofilament light chain (NfL). CKD is associated with higher levels of NfL, a biomarker for axonal injury, indicating that reduced kidney function may impact neurodegeneration. Studies show that as kidney function declines, i.e., eGFR decreases, plasma NfL levels tend to rise, suggesting that impaired renal function can influence blood levels of this neurodegeneration marker

[55]. A better approach would be to exclude these subjects from clinical trials or studies to obtain true results for plasma biomarkers in CU, aMCI, and AD patients.

Many studies have described the significant discrimination among CU and aMCI using various plasma biomarkers. For example, Chiu et al. reported a significant decrease in plasma $A\beta_{1-40}$ levels but significant increases in plasma $A\beta_{1-42}$ and T-Tau levels in aMCI and AD patients compared with those in CU individuals using immunomagnetic reduction (IMR) technology in 2013 [56]. Álvarez-Sánchez et al. showed decreases in plasma $A\beta_{1-42}$ levels in patients with aMCI and AD compared with CU individuals using SIMOA [57]. Mielke et al., Fossati et al., Jiao et al., and other groups used SIMOA and observed elevated plasma T-Tau levels in AD patients [58–61]. Many groups have utilized various assay technologies to detect increased levels of pTau181, pTau217, NfL, and other biomarkers in aMCI or AD patients compared with those in CU individuals [62–65]. These results indicate the abnormal levels of plasma biomarkers in aMCI or AD patients and the feasibility of using plasma biomarkers for differentiating aMCI patients and AD patients from CU individuals. The differentiation accuracy can be greater than 90%.

The correlations of plasma biomarker levels with neuropsychological test scores have been reported by individual groups. The consistency in the reported levels of plasma biomarkers among groups is important for validating the clinical repeatability of plasma biomarker assays. Chao et al. reviewed the published plasma $A\beta_{1-40}$, $A\beta_{1-42}$, and T-Tau levels measured in CU individuals using IMR and SIMOA [66]. The coefficient of variation among the 30 studies was 8.1% for assaying plasma $A\beta_{1-40}$ levels, 4.1% for assaying plasma $A\beta_{1-42}$ levels, and 19.4% for assaying plasma T-Tau levels with IMR. Using SIMOA, the coefficient of variation among the 13 studies was 18.4% for assaying plasma $A\beta_{1-40}$ levels, 61.5% for assaying plasma $A\beta_{1-42}$ levels, and 45.7% for assaying plasma T-Tau levels. Owing to the limited number of studies using MP-CLEIA, the coefficient of variation between the three studies was 4.1% for assaying plasma $A\beta_{1-40}$ and 3.5% for assaying plasma $A\beta_{1-42}$ levels in CU individuals [67–69].

In addition to individual biomarkers, combinations of biomarkers are recommended to enhance the discrimination of CU individuals and aMCI/AD patients. In terms of assay technologies, the combinations include $A\beta_{1-42}/A\beta_{1-40}$, $A\beta_{1-42}$/T-Tau, $A\beta_{1-42}$xT-Tau, pTau181x$A\beta_{1-42}$, and pTau217/$A\beta_{1-42}$. The clinical accuracy of differentiating aMCI/AD from CU can increase by 10–15% when combined biomarkers are used compared with individual biomarkers.

A better approach would be to identify biomarkers that are not only able to discriminate CU from aMCI/AD but also aMCI from AD. Thus, biomarkers that are able to assess subjects at the CU, aMCI, or AD stages are needed. Biomarkers for differentiating severity are helpful for determining disease management by medical doctors and are useful for monitoring the effects of treatments/interventions on subjects. For example, Chiu et al. finished a prior study and reported that the plasma $A\beta_{1-42}$xT-Tau levels assayed with IMR can differentiate CU individuals ($n = 66$) from aMCI patients ($n = 22$) ($p < 0.0001$) and aMCI patients from early-stage AD patients ($n = 45$) ($p < 0.005$) [56]. Further study demonstrates the discrimination from CU subjects ($n = 774$) from aMCI patients ($n = 81$) ($p < 0.0001$) and aMCI patients from AD patients (n = 238) ($p < 0.0001$), as shown in Fig. 4.10a. These findings strongly imply that the plasma $A\beta_{1-42}$xT-Tau level is a stage discrimination biomarker.

In fact, the plasma T-Tau level, not the $A\beta_{1-42}$ level, depends on the disease severity, as illustrated in Fig. 4.10b, c. Furthermore, plasma T-Tau levels correlate with scores on specific tests, such as the MMSE score. As shown in Fig. 4.11, plasma T-Tau levels negatively correlated with MMSE scores ($r = -0.625$, $p < 0.0001$) in CU (•), aMCI (•), and AD (•) patients. Hence, the plasma T-Tau level is a promising biomarker for monitoring the progression of AD or monitoring the effects of interventions/treatments. For example, the results of the longitudinal study shown in Fig. 4.12 show a negative correlation ($r = -0.619$, $p < 0.0001$) between the annual percentage change in plasma T-Tau levels and the annual percentage change in MMSE scores. The plasma T-Tau concentration was assayed using IMR, as shown in Fig. 4.12. As the MMSE score decreases, i.e., as cognition becomes worse, the plasma T-Tau level increases.

The plasma pTau181 level also depended on the disease severity, as shown in Fig. 4.13 (CU: $n = 94$, aMCI: $n = 100$, AD: $n = 138$). Once the scores of the neuropsychological tests worsen, the plasma pTau181 concentrations increase. Plasma pTau181 levels can indicate the degree of cognitive impairment in AD patients. Figure 4.14a, b show the typical negative correlation between plasma pTau181 levels and MoCA scores ($r = -0.412$, $p < 0.005$), plasma pTau181 levels and MMSE scores ($r = -0.362$, $p < 0.005$). The correlations between plasma biomarkers and neuropsychological test results are evidenced with results shown in Figs. 4.10, 4.11, 4.12, 4.13, and 4.14.

In summary, the clinical validations of plasma biomarkers are shown, with clear correlations between biomarker levels and neuropathology, amyloid or tau PET, MR imaging, CSF biomarker levels, and neuropsychological test scores.

4.2.6 *Paradox in Detecting Plasma $A\beta_{1-42}$ Levels Among Technologies*

Notably, the plasma $A\beta_{1-42}$ levels assayed with IMR were higher in aMCI/AD patients than in CU patients, as shown in Fig. 4.10b. However, the reported results for plasma $A\beta_{1-42}$ levels assayed with SIMOA, MP-CLEIA, ME-ECL, or other technologies revealed decreases in aMCI/AD patients compared with CU individuals [46, 68, 70, 71]. The opposite changes in plasma $A\beta_{1-42}$ levels in aMCI/AD patients were obtained using different assay technologies. Li et al. suggested that the opposite changes in plasma $A\beta_{1-42}$ levels in aMCI/AD patients are possibly because different assays detect different conformations of plasma $A\beta_{1-42}$ [72].

Technologies such as SIMOA, MP-CLEIA, and ME-ECL are sandwich assays, as described in Chap. 1, and use capture and labeling antibodies. Therefore, two different epitopes of an $A\beta_{1-42}$ molecule should be available simultaneously for the assay, as illustrated in Fig. 4.15a. IMR technology is a single-antibody assay, as described in Chap. 1. Only the capture antibody is used, without a labeling antibody.

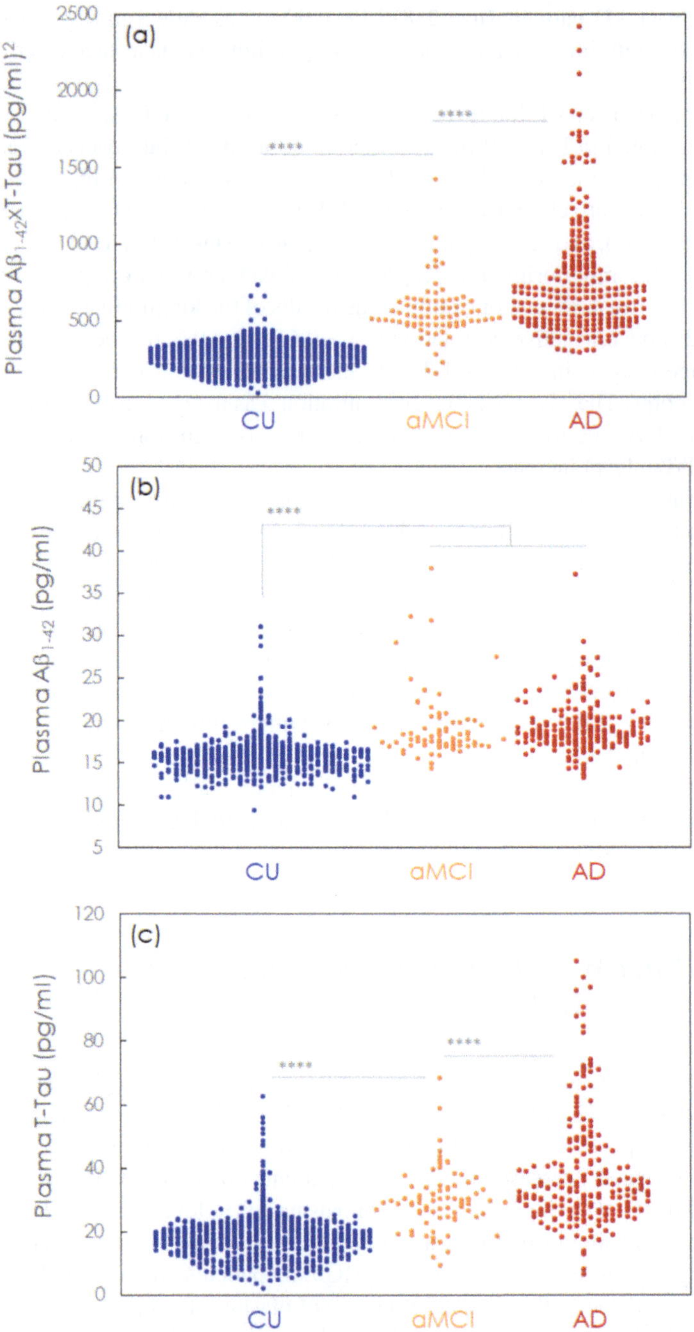

Fig. 4.10 Plasma (**a**) $A\beta_{1-42}$xT-Tau, (**b**) $A\beta_{1-42}$, and (**c**) T-Tau levels in CU, aMCI, and AD patients. The plasma $A\beta_{1-42}$ and T-Tau levels were assayed using immunomagnetic reduction (IMR). ****: $p < 0.0001$

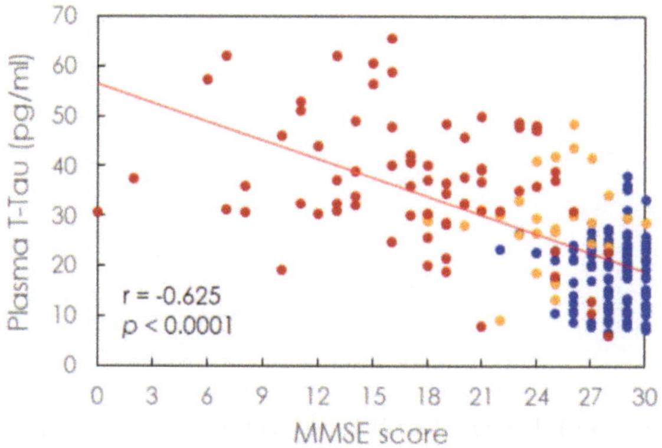

Fig. 4.11 Plasma T-Tau level versus MMSE score. The plasma T-Tau level was assayed using immunomagnetic reduction (IMR)

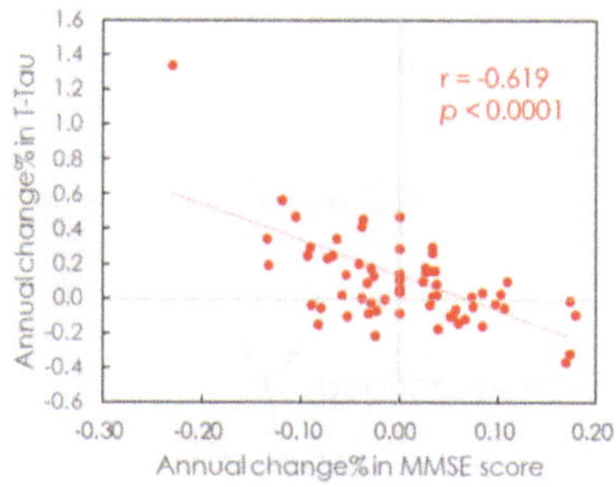

Fig. 4.12 Annual percentage change in the plasma T-Tau level versus annual percentage change in the MMSE score. The plasma T-Tau level was assayed using immunomagnetic reduction (IMR)

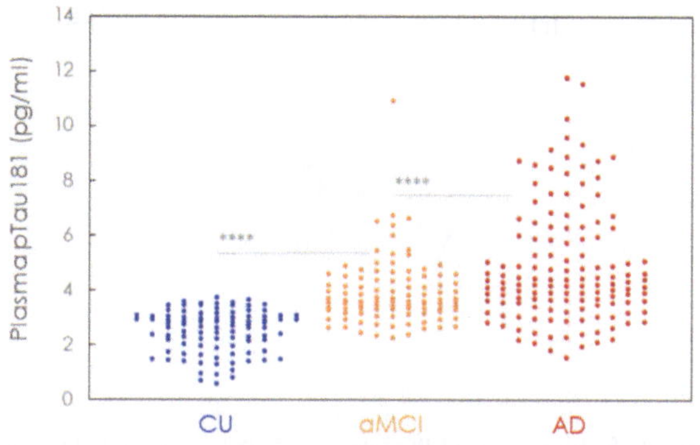

Fig. 4.13 Plasma pTau181 levels in CU, aMCI, and AD patients. The plasma pTau181 level was assayed using immunomagnetic reduction (IMR). **** $p < 0.0001$

Thus, only one epitope of an $A\beta_{1-42}$ molecule needs to be exposed to be assayed, as illustrated in Fig. 4.15b. Monomers of $A\beta_{1-42}$ can aggregate due to protein misfolding to form soluble oligomers, protofibrils, and then insoluble fibrils and plaques deposited in the brain. As illustrated in Fig. 4.15a, b, sandwich and single-antibody technologies can detect monomers of $A\beta_{1-42}$ because the epitopes are exposed and bind to capture and labeling antibodies. However, for oligomers and protofibrils of $A\beta_{1-42}$, sandwich and single-antibody technologies might have different detection abilities.

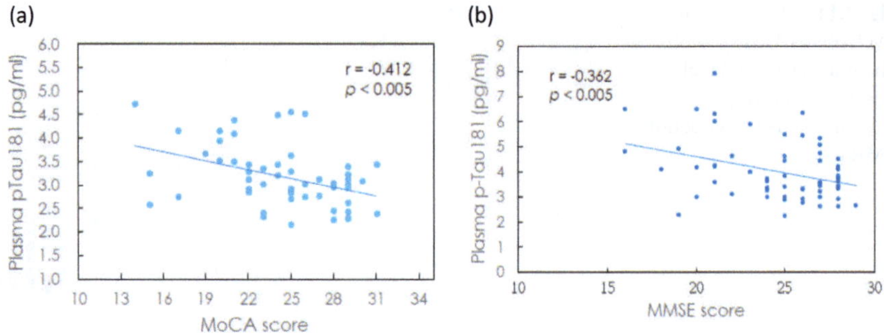

Fig. 4.14 Plasma pTau181 level versus (**a**) MoCA and (**b**) MMSE scores. The plasma pTau181 level was assayed using immunomagnetic reduction (IMR)

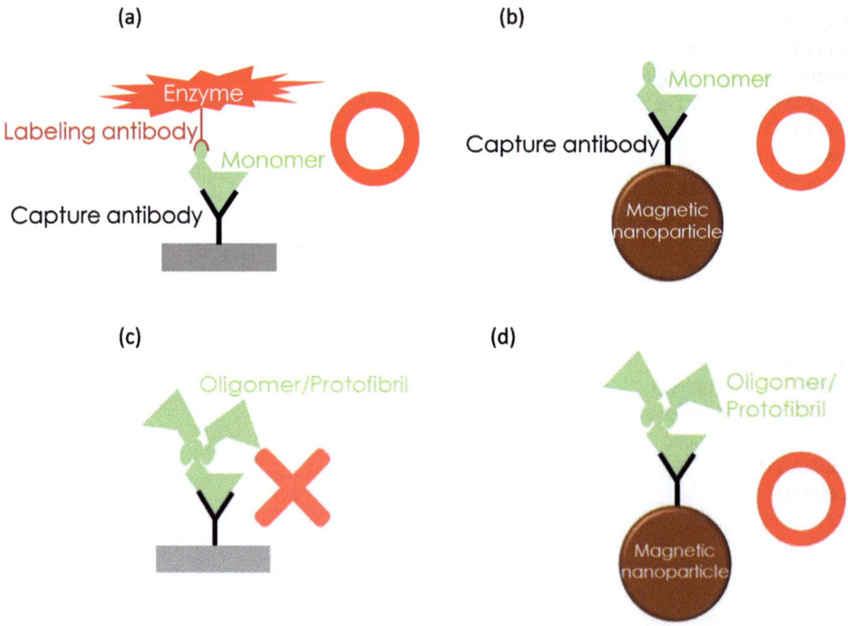

Fig. 4.15 Illustration of the interactions between antibodies and monomers, oligomers, or protofibrils of $A\beta_{1-42}$ using (**a**, **c**) sandwich technology and (**b**, **d**) single-antibody technology

As $A\beta_{1-42}$ oligomers or protofibrils are formed, the epitopes of $A\beta_{1-42}$ molecules are easily blocked, resulting in the inhibition of the interaction between $A\beta_{1-42}$ oligomers and the labeling antibody, as illustrated in Fig. 4.15c. Hence, sandwich technologies are potentially unable to detect oligomers and protofibrils of $A\beta_{1-42}$. However, the IMR assay requires only one epitope of $A\beta_{1-42}$. The epitope is highly likely to be exposed in oligomers and protofibrils. Thus, the interaction between the oligomers and protofibrils of $A\beta_{1-42}$ and the capture antibody on a magnetic

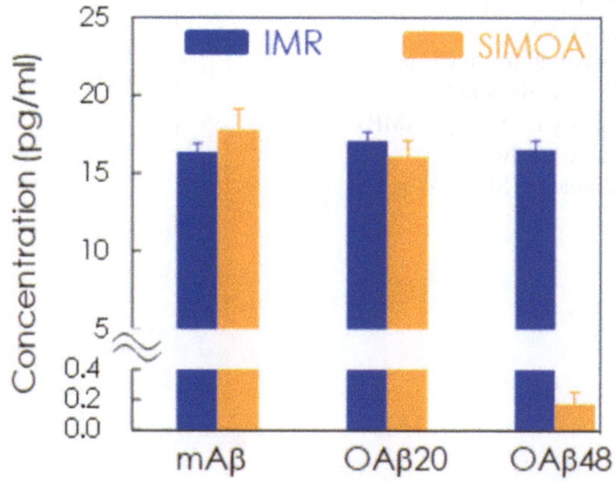

Fig. 4.16 $A\beta_{1-42}$ concentrations in monomer-rich (mAβ), oligomer-rich (OAβ20), and more oligomer-rich (OAβ48) solutions measured using IMR (single-antibody technology) and SIMOA (sandwich technology)

nanoparticle in the IMR assay is active, as shown in Fig. 4.15d. Theoretically, monomers of $A\beta_{1-42}$ dominate the measurements of sandwich technologies, whereas all forms of soluble $A\beta_{1-42}$, including monomers, oligomers, and protofibrils, can be measured using single-antibody technology.

The hypothesis shown in Fig. 4.15 was confirmed by Li et al. [72]. In Li's study, three solutions (denoted as mAβ, OAβ20, and OAβ48) with the same total concentrations of $A\beta_{1-42}$ were prepared. According to the results of the Western blot analysis, the mAβ solution is monomer-rich, OAβ20 is more oligomer-rich, and OAβ48 is the most oligomer-rich. The measured $A\beta_{1-42}$ concentrations assayed with SIMOA (sandwich technology) and IMR (single-antibody technology) are shown in Fig. 4.16.

In the SIMOA results, clear concentrations were detected for mAβ and OAβ20 solutions, but almost nondetectable concentrations were obtained for the OAβ48 solution. This result implies that SIMOA (sandwich technology) rarely detects oligomer-rich $A\beta_{1-42}$ samples, but detects monomer-rich $A\beta_{1-42}$. According to the IMR results, clear concentrations were detected for the mAβ, OAβ20, and OAβ48 solutions. These findings support the ability of IMR (single-antibody technology) to detect monomers and oligomers of $A\beta_{1-42}$. Therefore, different assay technologies detect different conformations of $A\beta_{1-42}$ molecules.

The head-to-head comparison in detected plasma $A\beta_{1-42}$ using the single-antibody technology (IMR) and the sandwich technology (SIMOA) in CU (•) and AD (•) was conducted. The results are shown in Fig. 4.17. The measured plasma $A\beta_{1-42}$ levels are higher in AD than CU in the case of IMR, whereas the measured plasma $A\beta_{1-42}$ levels do not significantly differ between AD and CU. Furthermore, there is no correlation in the measured plasma $A\beta_{1-42}$ levels between IMR and SIMOA. The results in Fig. 4.17 evidence that different conformations of $A\beta_{1-42}$ in human plasma are detected using the single-antibody technology (IMR) and the sandwich technology (SIMOA).

The results in Fig. 4.10a, showing the elevation in plasma $A\beta_{1-42}$ in aMCI/AD patients compared with that in CU individuals using IMR (single-antibody

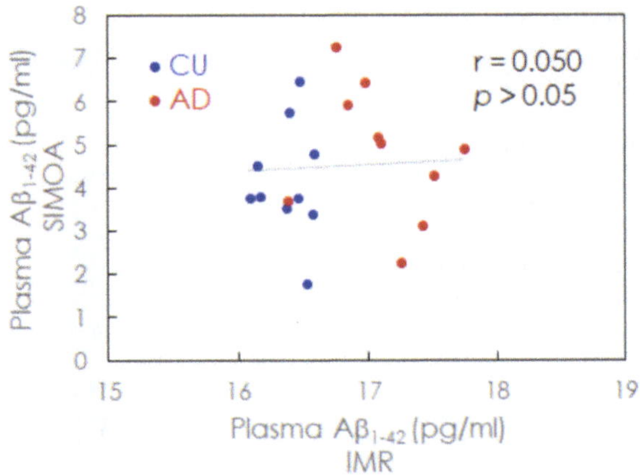

Fig. 4.17 Measured plasma $A\beta_{1-42}$ concentrations in CU and AD using the single-antibody technology (IMR) and the sandwich technology (SIMOA)

technology), demonstrate increases in the effective amount of $A\beta_{1-42}$ molecules, which might be due to oligomers/protofibrils or increased total amounts of $A\beta_{1-42}$ in aMCI/AD patients. By considering the results assayed with SIMOA or MP-CLEIA (sandwich technology), plasma $A\beta_{1-42}$ levels decrease in AD patients, revealing decreases in the amounts of $A\beta_{1-42}$ monomers. A reasonable assumption is that oligomers or protofibrils of $A\beta_{1-42}$ become dominant in aMCI/AD patients. These results imply that the conformational profiles of $A\beta_{1-42}$ in human plasma evolve from monomer dominance to oligomer/protofibril dominance as a CU subject develops aMCI/AD.

The increases in the levels of oligomers or protofibrils of $A\beta_{1-42}$ play a role in the development of aMCI/AD. In vitro studies have shown that oligomers and protofibrils of $A\beta_{1-42}$ are highly toxic to neurons, whereas monomers of $A\beta_{1-42}$ are not [73, 74]. This finding also explains why pharmaceutical companies are developing antibody drugs against oligomers or protofibrils of Aβ in the brain to delay disease progression or prevent the development of AD. Therefore, assays of oligomers and protofibrils of $A\beta_{1-42}$ instead of monomers of $A\beta_{1-42}$ are preferred to predict aMCI/AD or cognitive impairments.

Since IMR detects oligomers and protofibrils of $A\beta_{1-42}$ in plasma, the higher measured levels of $A\beta_{1-42}$ in plasma reveal greater possibilities of positive amyloid neuropathology, as shown in Fig. 4.3. Moreover, the higher measured $A\beta_{1-42}$ levels in plasma with IMR indicate a greater possibility of developing cognitive impairment in the population at risk. Yang et al. reviewed articles and concluded that the plasma levels of $A\beta_{1-42}$, not T-Tau, assayed with IMR are the most effective at predicting cognitive decline in patients with Down syndrome, stroke, and aMCI [75]. In addition, Hsieh et al. used IMR to quantify plasma biomarkers and found that higher plasma $A\beta_{1-42}$ levels significantly discriminated participants with subjective cognitive decline from healthy controls [76].

The roles of oligomers and protofibrils of pathological proteins in neurodegenerative diseases are not limited to the case of $A\beta_{1-42}$ in AD, but also include α-synuclein in PD. As mentioned later, the pathological evidence of PD is Lewy bodies, which are formed from soluble monomers of α-synuclein into oligomers, protofibrils, and insoluble Lewy bodies. Oligomers and protofibrils of α-synuclein are central to the development of PD. The abilities to assay oligomers and protofibrils of α-synuclein in plasma are key issues in assessing PD via blood tests. IMR technology and RT-QuIC technology are particularly useful for the assay of oligomers and protofibrils of α-synuclein in plasma. The seed amplification assay or RT-QuIC has been shown to reliably detect α-synuclein in CSF but has not been correlated with plasma α-synuclein to date.

Not only aggregated forms but also truncated forms of pathological biomarkers play roles in the development of AD. Hence, the detections of truncated forms of biomarkers are crucial to correlate the measured results with the clinical diagnosis of AD. We should pay attention to investigating the feasibility of detecting truncated forms of biomarkers using the assay technologies illustrated in Chap. 1.

4.2.7 Detections of Truncated and Aggregated Tau Protein in Human Plasma

Although amyloid-β peptides are not easily truncated, the tau protein which consists of 400–500 amino acids can be truncated. Zilka et al. performed an animal test to reveal that truncated tau protein isoforms from sporadic AD were sufficient to drive neurofibrillary degeneration in vivo [77]. In fact, tau protein is truncated at multiple sites by various proteases in the brain of AD patients. Gu et al. found that the first 150 or 230 amino acids enhanced tau's site-specific phosphorylation, self-aggregation, and binding to hyperphosphorylated and oligomeric tau protein isolated from AD brain tissue, but deletion of the first 50 amino acids did not produce a significant effect [78]. Deletion of the last 50 amino acids was found to modulate Tau's site-specific phosphorylation, promote its self-aggregation, and enable its capture and aggregation seeding by AD oligomeric tau protein, whereas deletion of the last 20 amino acids had no such effects. Liu et al. showed that both soluble truncated tau species and NFTs could cause cognitive dysfunction in the brain [79]. Hence, soluble truncated tau protein plays a role in the early stages of AD

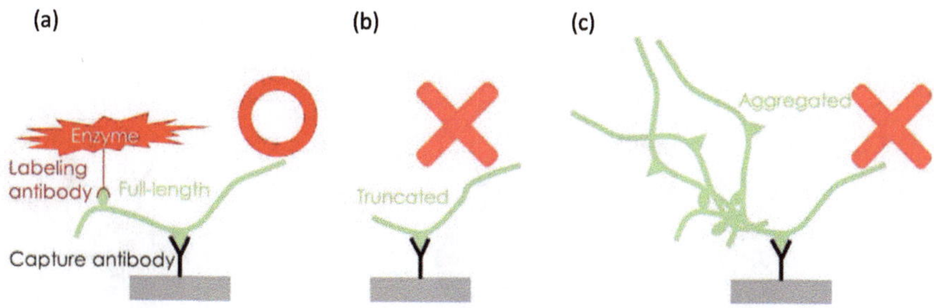

Fig. 4.18 Illustration of the interactions between antibodies and (**a**) full-length and (**b**) truncated forms of tau protein using sandwich technology

development and other tauopathy-specific neurodegenerative diseases (e.g., PSP and CBD). In addition, the aggregations of tau protein also play a role in developing Alzheimer's disease and tauopathies. This emphasizes the demands of detecting soluble truncated and aggregated tau protein isoforms in human plasma.

With the use of the sandwich technologies such as SIMOA, MP-CLEIA, ECL, ELISA etc., the two sites of the epitopes of tau protein binding to the capture antibody and the labeling antibody should be close to each other, as illustrated in Fig. 4.18a. Otherwise, once tau protein is truncated or aggregated, the two epitopes could be in two truncated segments and be separated. Although the capture antibody can catch its segment, the labeling antibody cannot find its segment because the segment is removed at wash steps or blocked, as illustrated in Fig. 4.18b.

In order to prevent the separation of the two epitopes as tau protein is truncated, the sites of the two epitopes should be very close. Thus, the two epitopes could be in the same segment as tau protein is truncated. The epitope with the labeling antibody will not be removed at wash steps. However, the two epitopes cannot be very close because the epitope for the labeling antibody may be blocked when the tau protein is caught by the capture antibody. Hence, it is truly difficult to have suitable capture and labeling antibodies for assaying truncated tau protein with the use of sandwich technologies. Unfortunately, sandwich technologies hardly detect truncated tau protein.

As tau protein aggregates, one of the two epitopes can be easily blocked. The labeling or capture antibody is not able to bind aggregated tau protein, resulting the loss of measuring aggregated tau protein. Therefore, full-length tau protein dominates the measured results when using sandwich technologies.

Immunomagnetic reduction (IMR), one of the assay technologies introduced in Chap. 1, only uses the capture antibody. Regardless of full-length or truncated segments, the site of the epitope for the capture antibody is always present, as illustrated in Fig. 4.19a, b. Hence, both the full-length and truncated tau proteins can be detected.

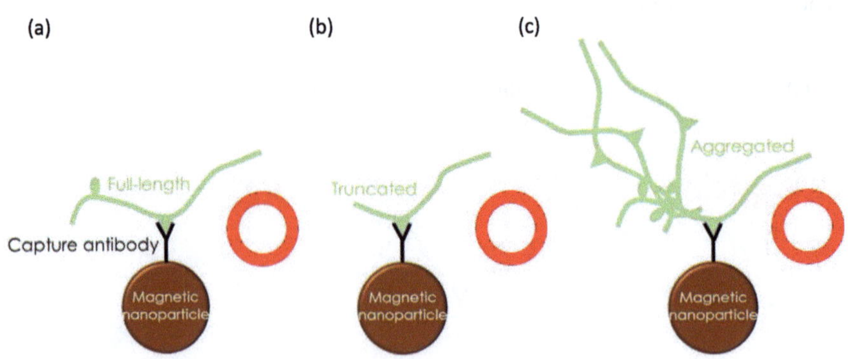

Fig. 4.19 Illustration of the interactions between antibodies and (**a**) full-length and (**b**) truncated forms of tau protein using single-antibody technology

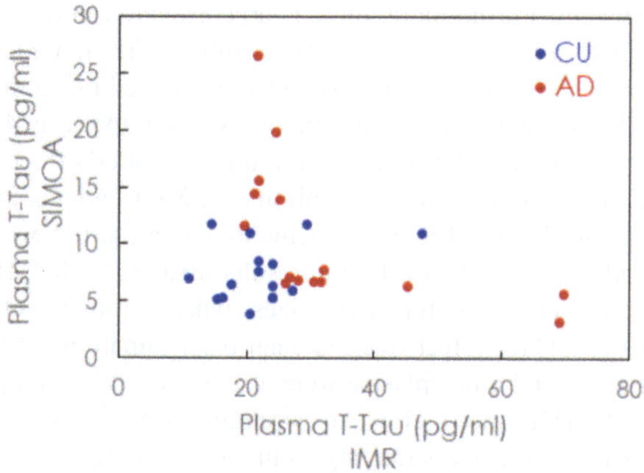

Fig. 4.20 Measured plasma T-Tau concentrations in CU and AD using the single-antibody technology (IMR) and the sandwich technology (SIMOA)

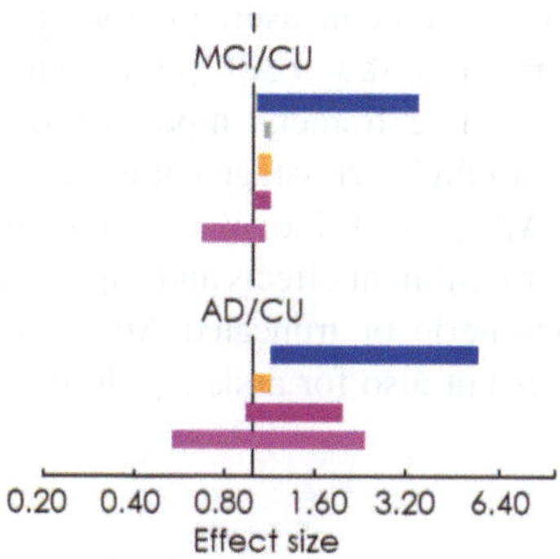

Technology	Study numbers	
	MCI/CU	AD/CU
IMR	11	15
IP/MS	1	-
ECL	2	3
SIMOA	6	11
ELISA	5	3

Fig. 4.21 Effect sizes of tau protein concentrations in human plasma between MCI patients and CU subjects (MCI/CU), AD patients, and CU subjects (AD/CU) using various assay technologies such as IMR, IP/MS, ECL, SIMOA, and ELISA, reviewed by Zabala-Findlay et al. in 2023 [80]

In fact, the antibody used in IMR is against amino acids 210–229, which is the common nonphosphorylated mid-region of six isoforms of tau protein. The mid-region fragments of tau protein are recommended as the plasma biomarkers presenting tau pathology in the revised criteria for diagnosing and staging AD issued in 2024 [12].

Since truncated tau protein is more related to the early stages of AD development, its contribution to measured results should be more significant in early-stage AD patients than in CU subjects. This implies that the discrimination in terms of measured tau protein concentrations between early-stage patients and CU subjects should be more obvious as truncated tau protein is detected.

A head-to-head comparison of detected plasma T-Tau using single-antibody technology (IMR) and sandwich technology (SIMOA) was conducted in CU (•) and AD (•) subjects. The results are shown in Fig. 4.20. The measured plasma T-Tau levels are higher in AD than in CU in the case of IMR, whereas the measured plasma T-Tau levels do not significantly differ between AD and CU using

SIMOA. Furthermore, there is no correlation in the measured plasma T-Tau levels between IMR and SIMOA. The results in Fig. 4.20 indicate that different conformations of T-Tau in human plasma are detected using the single-antibody technology (IMR) and the sandwich technology (SIMOA) (Fig. 4.21).

Zabala-Findlay et al. reviewed the reported effect sizes of measured concentrations of tau protein in the plasma of MCI patients and CU subjects, as shown in Fig. 4.21 [80]. The assay technologies in the review include IMR, ECL, SIMOA, and ELISA. Obviously, the results assayed with IMR show the definitely higher values (1.1–3.3) for effect sizes, whereas other technologies show lower values (0.7–1.4) for effect sizes. Greater discriminations of measured tau protein concentrations in human plasma were also evidenced in AD patients compared to CU subjects. IMR shows effect sizes ranging from 1.3 to 5.8, whereas other technologies exhibit effect sizes ranging from 0.6 to 2.4. Therefore, different assay technologies might assay different forms of $A\beta_{1-42}$ and T-Tau in human plasma. Developers should clearly reveal the forms of biomarkers to be measured in their assay kits.

Since the oligomeric or truncated forms of T-Tau or $A\beta_{1-42}$ are highly related to the development of Alzheimer's disease, it is extremely important to assay these disease-derived $A\beta_{1-42}$ and T-Tau in order to find early-stage patients. With drugs in development targeting disease-derived $A\beta_{1-42}$ or T-Tau, it is important to detect changes in their concentrations to monitor treatment effects and support drug development. Thus, the ability to assay oligomeric or truncated $A\beta_{1-42}$ and T-Tau is advantageous not only for early diagnosis but also for assisting drug development and monitoring treatment effects.

4.2.8 Correlations Among Oligomerized $A\beta_{1-42}$, pTau181, and T-Tau in AD

In AD pathogenesis, the toxicity of oligomers and protofibrils of $A\beta_{1-42}$ triggers the hyperphosphorylation of tau protein with media IGFBP-3 or GSK-3β [81–84]. Notably, such pathogenesis occurs only in AD, not in PD or FTD. This suggests that increases in oligomers and protofibrils of $A\beta_{1-42}$ levels could lead to increased pTau levels in AD. As discussed, the IMR assay mainly measures $A\beta_{1-42}$ oligomers and protofibrils in plasma. By using IMR to assay plasma $A\beta_{1-42}$ and pTau, a clear relationship can be observed in AD, but not in PD and FTD.

Chiu et al. demonstrated a positive linear correlation between plasma $A\beta_{1-42}$ and pTau181 levels assayed with IMR in AD ($r = 0.699$, $p < 0.001$), as shown in Fig. 4.22a [63]. In contrast, PD and FTD patients did not show a significant correlation between plasma $A\beta_{1-42}$ and pTau181 levels (PD: $r = 0.208$, $p > 0.05$; FTD: $r = 0.052$, $p > 0.05$).

Hyperphosphorylation of tau protein damages the stability of microtubules in neurons and causes the death of neurons. Tau protein is overexpressed due to the death of neurons. Hence, in AD, T-Tau levels should elevate with increasing pTau181

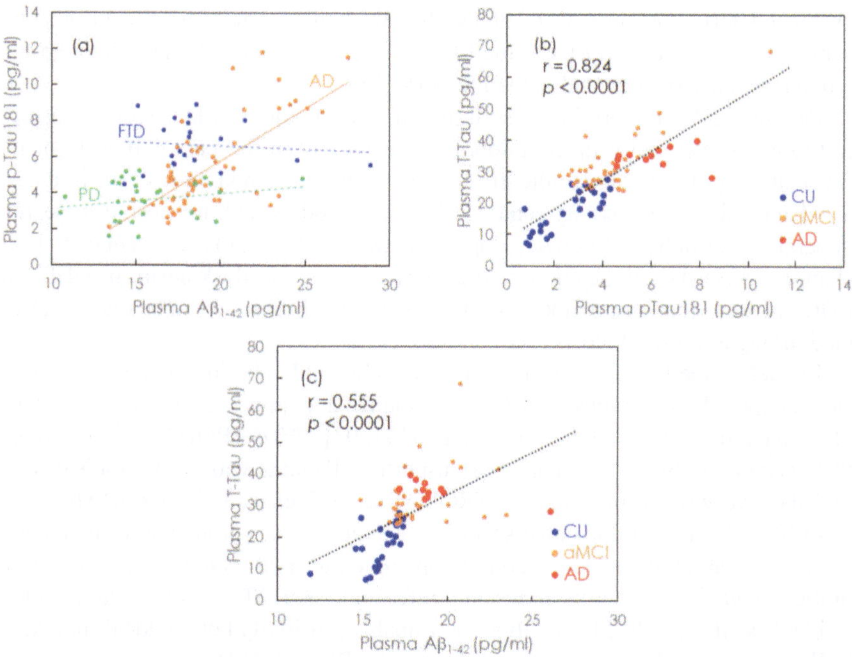

Fig. 4.22 Plasma (**a**) $A\beta_{1-42}$ and pTau181 in AD, PD, and FTD; (**b**) pTau181 and T-Tau; and (**c**) $A\beta_{1-42}$ and T-Tau in CU, aMCI, and AD. Plasma biomarkers were assayed with IMR technology

levels. Figure 4.22b shows the significantly positive correlation between plasma T-Tau and pTau181 levels in CU/aMCI/AD ($r = 0.824$, $p < 0.0001$). Consequently, AD patients show a significant positive correlation between plasma T-Tau and $A\beta_{1-42}$ ($r = 0.555$, $p < 0.0001$). The pathogenesis of AD involving $A\beta_{1-42}$, pTau, and T-Tau is clearly evidenced with results in Fig. 4.22a–c.

4.3 Clinical Validation of Assays for Parkinson's Disease and Dementia with Lewy Bodies

Parkinson's disease (PD) is one of the most common neurodegenerative disorders, affecting more than 1% of the global population over the age of 65, with an incidence rate that is likely to double by 2030 [85]. The typical symptoms of PD are motor disorders, such as bradykinesia, cogwheel rigidity, resting tremor, and postural instability [86]. Additionally, nonmotor symptoms such as rapid eye movement (REM) sleep behavior disorders, constipation, depression, and olfactory loss can precede motor symptoms by decades [87]. Until recently, the diagnosis of PD is primarily based on the clinical presentations. Confounding PD diagnosis and management is the development of PD-related dementia when patients have PD for prolonged periods of time. Dementia with Lewy bodies is an overlap syndrome that some consider essentially the same disease as PD dementia (PDD), although this remains extensively debated in the literature [88].

Pathologically, the original description of Lewy bodies in PD is their deposition in the brain, especially in the midbrain, specifically in the pars compacta of the substantia nigra [89]. Dopaminergic neurons in the substantia nigra of the midbrain are damaged by the toxicity rendered by Lewy bodies and are unable to secrete dopamine normally. Moreover, dopamine transporter levels are reduced. Both changes cause the dysfunction of the basal nucleus and dyskinesia. In DLB and PDD, the Lewy bodies are not sequestered to the midbrain and can be found distributed throughout the neocortex [90].

In nuclear medicine, clinicians use ^{99m}Tc-TRODAT-1 brain single photon emission computed tomography (SPECT) to visualize dopamine transporters at the ends of presynaptic neurons in the basal ganglia [91, 92]. ^{99m}Tc-TRODAT-1 is an isotope that specifically binds to dopamine transporters. Patients with Parkinson's disease will show very weak intensities on TRODAT scans. Hence, ^{99m}Tc-TRODAT-1 brain SPECT is also called dopamine scanning, which provides clinicians with valuable information about the degeneration of dopamine transporters in patients and assists in the diagnosis of PD. It is commonly referred to as DAT scanning and is widely available as it is a SPECT scan. It has reasonable sensitivity but not ideal specificity [93]. It is approved as an ancillary diagnostic in PD and DLB.

A biological analysis revealed that the main fundamental component of Lewy bodies is α-synuclein [94]. α-Synuclein monomers aggregate to form oligomers due to phosphorylation, nitrosylation, ubiquitination, or sumoylation. Oligomers further aggregate to form fibrils and finally Lewy bodies [95–97]. Hence, α-synuclein plays a role in PD. Many studies have reported abnormal levels of α-synuclein in the CSF of patients with PD [94]. Currently, α-synuclein is recognized as a biomarker of PD. One CSF assay with seed amplification assay or RT-QuIC has shown excellent sensitivity and specificity but does not yet have FDA clearance as a diagnostic [98].

In addition to motor impairment, cognitive impairment or dementia often occurs in patients with PD [99, 100]. Neuropsychological tests are sometimes ordered for patients with PD to examine their cognitive profile, since PD cognitive impairment tends to be subcortical rather than the cortical profile seen in AD. Depending on the severity of the cognitive impairment, patients with PD can be categorized as having Parkinson's disease with normal cognition (PD-NC), mild cognitive impairment (PD-MCI), or dementia (PDD).

In the clinic, another neurodegenerative disease—dementia with Lewy bodies (DLB)—is characterized by motor disorders and cognitive impairment [101]. Hence, patients with DLB can have similar clinical symptoms to patients with PDD.

In addition, the name DLB clearly shows that Lewy bodies are the pathological hallmark, found throughout the neocortex and brainstem, with or without comorbid AD pathology. Therefore, α-synuclein could also serve as a biomarker for DLB, as it does in PD. Therefore, DLB patients have the same brain pathology and biological biomarkers as PDD patients. Both DLB and PDD are termed Lewy body diseases. Obviously, differentiating DLB from PDD according to motor disorders, cognitive impairment, brain pathology, and abnormal α-synuclein levels in CSF can sometimes be difficult. This difficulty sometimes leads to the misdiagnosis of PDD or DLB.

In fact, DLB should be diagnosed when cognitive impairment (or dementia) occurs before or concurrently with motor disorders. PDD should be diagnosed when cognitive impairment (or dementia) occurs after a motor disorder. However, patients and their family members are not always aware of the onset of cognitive impairment or motor disorders. Clinicians cannot obtain determinant information from patients and their families during clinical visits. Therefore, clinicians are usually unable to differentiate between DLB and PDD.

Some clinicians believe they are the same conditions. Others feel they overlap but are separate. The clinical features certainly overlap, but the chronicity does not.

In clinical practice, the 1-year rule between the onset of dementia and motor disorders is recommended to differentiate DLB from PDD [102, 103]. If the time period between the onset of dementia and motor disorders is less than 1 year, the condition is considered as DLB. However, if the time period between the onset of dementia and motor disorders is greater than 1 year, the condition is considered as PDD. In DLB, the conventional wisdom is that the motor and cognitive symptoms are separated by less than 1 year, but some consider it less than 3 years. In PDD, the cognitive symptoms come many years (>10) after the motor decline.

In addition, significant information about rapid eye movement (REM) sleep behavior disorder (SBD), DAT scanning, and iodine-123-metaiodobenzylguanidine (MIBG) myocardial scintigraphy have been incorporated into the diagnosis of DLB [104, 105]. Iodine-123-metaiodobenzylguanidine (MIBG) myocardial scintigraphy has been considered in the diagnosis of DLB. In autopsy studies, diffuse Lewy body pathology in the limbic and cortical areas is helpful for distinguishing PDD from DLB. The presence of concomitant AD pathology can be seen in both conditions. Furthermore, potential biomarkers that are helpful for discriminating between DLB and PDD are expected to be useful in clinical practice in the future.

Both PD and DLB are clinically characterized based on motor disorders, dopamine scans, Lewy body pathology, and abnormal α-synuclein levels in the CSF. Hence, the clinical validation of plasma biomarkers for assessing PD or DLB should be performed by correlating plasma biomarker levels with all the clinical characteristics. The reported results of the clinical validations are described.

4.3.1 Plasma Biomarkers Versus Brain Tissue Staining

A neuropathological hallmark of PD or DLB is the formation of neuronal protein aggregates termed Lewy bodies (LBs). Postmortem studies have suggested that the gradual appearance of LBs correlates with disease progression in PD patients [106]. Lewy bodies appear as dark brown areas in stained tissue. Lewy bodies are composed mainly of α-synuclein. Typical α-synuclein staining in the substantia nigra is shown in Fig. 4.23 [107].

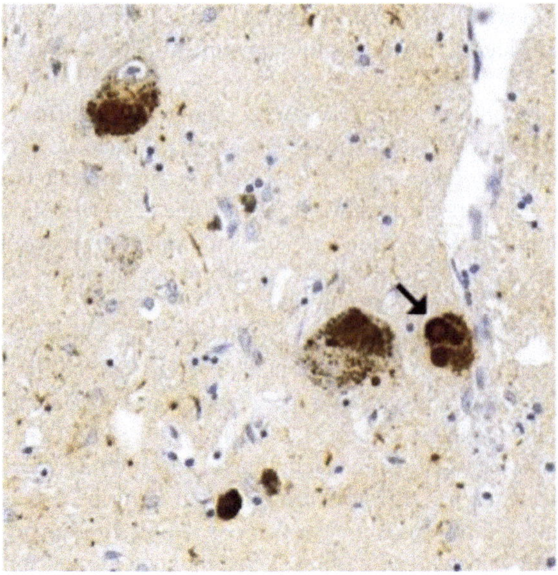

Fig. 4.23 Immunohistochemical staining for α-synuclein in the substantia nigra of a PD patient [107]

Fig. 4.24 Relationship between the antemortem plasma total α-synuclein level and postmortem Lewy-type synucleinopathy (LTS) total brain load [22]

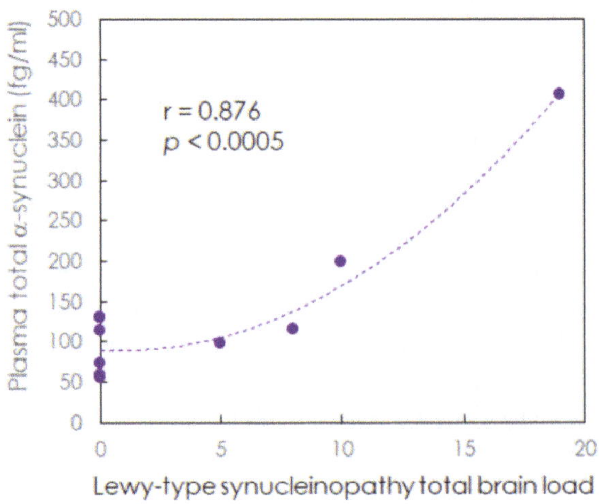

Mastrangelo et al. used RT-QuIC to assay postmortem α-synuclein levels in CSF and analyzed the Lewy body density in the brain [108]. The Lewy body density was analyzed in ten standard brain regions, including the olfactory bulb and tract, anterior medulla, anterior and mid-pons, substantia nigra, mid-amygdala, transentorhinal area, anterior cingulate gyrus, middle temporal gyrus, middle frontal gyrus, and inferior parietal lobule. A positive correlation ($r = 0.62$, $p < 0.0001$) was detected between CSF α-synuclein levels and Lewy body density in the brain. This result implies that CSF α-synuclein levels predict the Lewy body density in the brain.

In addition to CSF, Yang et al. explored the correlation between the postmortem Lewy-type synucleinopathy (LTS) total brain load and antemortem plasma total α-synuclein level [22]. The plasma samples were collected within 1.5 years before death. The LTS total brain load is the sum of LTS in the olfactory bulb, dorsal motor nucleus of the vagus nerve (in the medulla), locus coeruleus (in the pons), substantia nigra, amygdala, transentorhinal area, cingulate gyrus, middle frontal gyrus, middle temporal gyrus, and inferior parietal lobule. The total α-synuclein level in plasma was assayed using IMR. The relationship between postmortem Lewy-type synucleinopathy (LTS) total brain load and antemortem plasma total α-synuclein level is plotted in Fig. 4.24.

Although the number of enrolled subjects used to obtain the results shown in Fig. 4.24 was limited, a strong correlation ($r = 0.876$, $p < 0.0005$) between the antemortem plasma total α-synuclein level and postmortem LTS total brain load was obtained. Hence, the plasma total α-synuclein level can be used to predict the LTS total brain load.

4.3.2 Plasma Biomarkers Versus Dopamine Scanning

Studies of the relationships between plasma biomarkers and dopamine scans are rare. We found a paper published in 2020 in which Chen et al. investigated the correlations between dopamine scans and plasma biomarkers, including $A\beta_{1-40}$, $A\beta_{1-42}$, T-Tau, and α-synuclein [109]. The plasma biomarkers were assayed using IMR. The score for ^{99m}Tc-TRODAT-1 uptake in the striatum was analyzed by visual inspection. Chen et al. used a 5-point scoring system:

0 = Normal striatal uptake
1 = Normal caudate uptake but decreased putamen uptake in the tail portion
2 = Normal caudate uptake but no putamen uptake
3 = Decreased caudate uptake with no putamen uptake
4 = Total loss or near total loss of striatal uptake
The scores of the left and right striata were analyzed separately.

The results revealed a significant correlation between the uptake of ^{99m}Tc-TRODAT-1 via imaging and plasma $A\beta_{1-40}$ levels in patients with PD ($n = 60$), but not $A\beta_{1-42}$, T-Tau, or α-synuclein levels. These findings suggest that $A\beta_{1-40}$ plays a role in the dysfunction of dopaminergic neurons in PD patients. More studies are needed to clarify the detailed mechanisms involved.

4.3.3 Plasma Biomarkers Versus Magnetic Resonance Imaging

α-synuclein is recognized as a potential biomarker for assessing PD or DLB, and the correlation between plasma α-synuclein levels and atrophy of the brain has been investigated. The volume of the brain is analyzed using volumetric magnetic resonance (MR) imaging. The most popular method for assaying plasma α-synuclein levels is IMR.

In 2020, Chen et al. reported a weak negative correlation between plasma α-synuclein levels and limbic cortex thickness ($r = -0.18$, $p < 0.05$), but not the frontal cortex thickness, temporal cortex thickness, parietal cortex thickness, occipital cortex thickness, insular cortex thickness, or the thickness of the hippocampus and caudate in PD patients [110].

In 2021, Chen et al. demonstrated that the plasma α-synuclein level did not correlate with atrophy of the gyri, putamen, or hippocampus in PD patients [111]. Three years later, Huang et al. revealed the lack of a correlation between plasma α-synuclein levels and the surface atrophy of the gyrus [112]. These results suggest that plasma α-synuclein levels do not strongly correlate with atrophy of the brain in PD patients.

Instead of α-synuclein levels, Chen et al. reported in 2020 that plasma NfL levels are negatively correlated with the frontal cortex thickness ($r = -0.25$, $p < 0.005$), temporal cortex thickness (-0.25, $p < 0.005$), parietal cortex thickness ($r = -0.26$, $p < 0.005$), insular cortex thickness ($r = -0.21$, $p < 0.05$), and the hippocampal thickness ($r = -0.21$, $p < 0.05$) in PD patients [110]. In 2021, Chen et al. reported that plasma NfL levels increase as gyrus atrophy occurs in PD patients ($r = -0.462$, $p < 0.05$) [111].

Other studies have revealed roles for tau protein in brain atrophy in PD patients. For example, Lin et al. in 2021 and Huang et al. in 2024 reported increases in plasma T-Tau levels as gray matter atrophy ($r = -0.35$, $p < 0.01$) and gyrus atrophy ($r = -0.587$, $p < 0.0005$) occurred [112, 113]. Yu et al. reported negative correlations between plasma pTau181 levels and the bilateral cerebellar cortex ($r = -0.32$, $p < 0.005$), the right caudate nucleus ($r = -0.342$, $p < 0.005$), and the right accumbens area ($r = -0.321$, $p < 0.005$) [114]. Notably, Yu et al. also showed that plasma $A\beta_{1-42}$ levels are negatively correlated with the bilateral cerebellar cortex ($r = -0.36$, $p < 0.005$), the bilateral caudate nucleus ($r = -0.41$, $p < 0.001$), and the right accumbens area ($r = -0.301$, $p < 0.01$) in PD patients. These results indicate that plasma NfL, T-Tau, pTau181, or $A\beta_{1-42}$ levels, but not α-synuclein levels, predict brain atrophy in PD patients [114].

The atrophy of the brain results from cellular damage or death, primarily neuronal death, and serves as a marker of neurodegeneration. Hence, the structure-related proteins of neurons, such as NfL and Tau, are overexpressed in the brain, which is attributed to the increases in the plasma NfL, T-Tau, or pTau levels as brain atrophy occurs. Therefore, measurements of changes in the plasma NfL, T-Tau, or pTau181 levels are promising for follow-up investigations of brain atrophy in PD or DLB patients. Such measurements would be very powerful alternatives in clinical practice because of the low availability of MR imaging.

Notably, the finding that plasma $A\beta_{1-42}$ levels negatively correlate with brain atrophy in PD patients reported by Yu et al. is noteworthy [114]. Yu et al. suggested that $A\beta_{1-42}$ deposits in these brain areas might be key factors contributing to neuronal damage. $A\beta_{1-42}$ could play a role in PD patients. More studies are needed to clarify the contributions of $A\beta_{1-42}$ to PD or DLB.

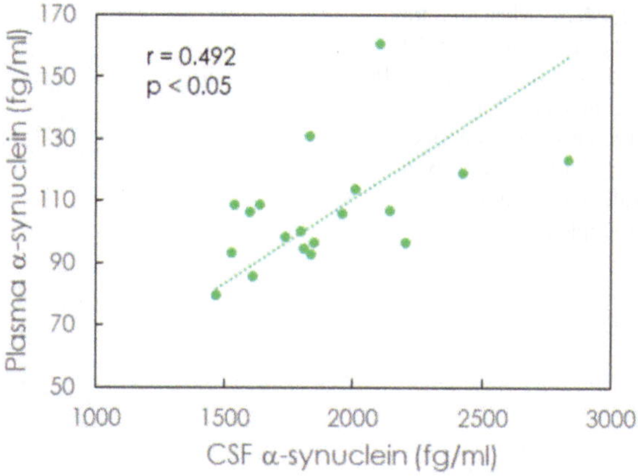

Fig. 4.25 Relationship between the plasma total α-synuclein level and CSF total α-synuclein level assayed using IMR

4.3.4 *Plasma Biomarkers Versus CSF Biomarkers*

To verify that the measured α-synuclein levels in plasma originate from the central nervous system (CNS), the correlation of α-synuclein levels measured in plasma and CSF should be determined. Ishii et al. used ELISA to measure α-synuclein levels in plasma and CSF from PD patients ($n = 30$) and normal controls ($n = 103$) [115]. A significant positive correlation of α-synuclein levels between plasma and CSF was detected ($p < 0.01$) in PD patients, whereas no significant correlation was detected in the control group ($p > 0.50$).

Malaty et al. used IMR to assay total α-synuclein levels in paired plasma and CSF samples from PD patients and normal controls ($n = 20$). A moderately positive correlation ($r = 0.492$, $p < 0.05$) was detected, as shown in Fig. 4.25.

These results indicate that the measured α-synuclein is mostly from the CNS. Researchers have proposed that the α-synuclein measured in plasma is derived from brain α-synuclein leakage across the blood–brain -barrier, blood–CSF circulation, and the glymphatic system.

4.3.5 *Plasma Biomarkers Versus Clinical Diagnosis*

The clinical diagnosis of PD is based on the United Kingdom Parkinson's Disease Society Brain Bank clinical diagnostic criteria [116]. Subjects must have symptoms of bradykinesia and at least one of the following symptoms: muscular rigidity; resting tremor (4–6 Hz); or postural instability unrelated to primary visual, cerebellar, vestibular, or proprioceptive dysfunction. In addition, three or more of the following symptoms must be present: unilateral onset, presence of resting tremor, progressive disorder, persistent asymmetry mainly affecting the side of onset, an excellent response to levodopa, severe levodopa-induced chorea, levodopa response for over 5 years, and a clinical course of over 10 years. PD patients with MoCA scores

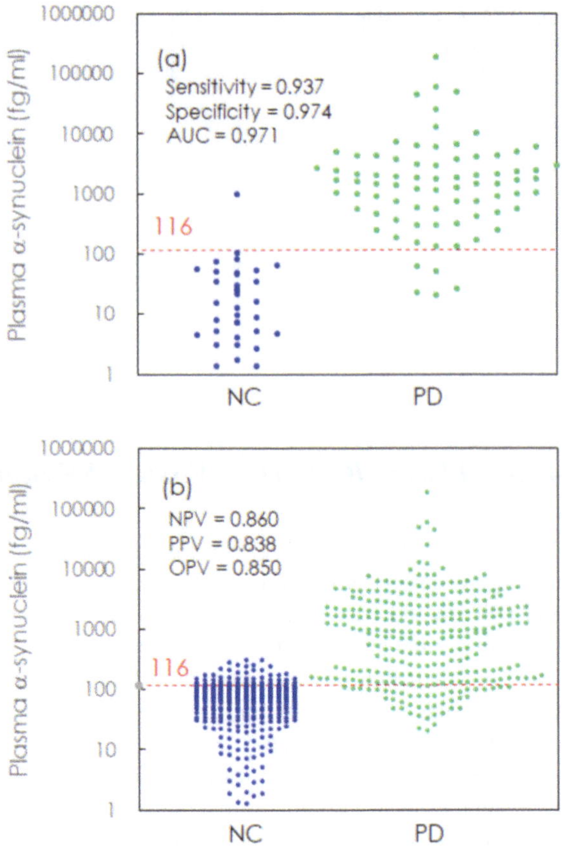

Fig. 4.26 Measured total α-synuclein levels in NC and PD patients using IMR technology in (**a**) a reference cohort at one hospital [117] and (**b**) a validation cohort at multiple centers

higher than 26 are diagnosed with PD with normal cognition (PD-NC), those with scores less than 21 are diagnosed with PDD, and those with scores between 21 and 26 are diagnosed with PD-MCI.

Subjects with the following physical or psychiatric conditions are excluded: subjects with a significant history of depression, repeated strokes with stepwise progression, repeated head injury, use of antipsychotic or dopamine-depleting drugs, definite encephalitis and/or oculogyric crises without drug treatment, a negative response to large doses of levodopa (if malabsorption is excluded), strictly unilateral features after 3 years, other neurological features (supranuclear gaze palsy, cerebellar signs, early severe autonomic involvement, Babinski sign, or early severe dementia with disturbances of language, memory, or praxis), exposure to a known neurotoxin, or the presence of cerebral tumor or communicating hydrocephalus on neuroimaging. Hence, the enrolled subjects are diagnosed as normal controls (NCs), PD-NC, PD-MCI, or PDD. The plasma biomarkers of the enrolled subjects were assayed with various technologies. The differences in measured plasma biomarker levels among NC, PD-NC, PD-MCI, and PDD patients were analyzed to explore the ability of plasma biomarkers to predict PD-NC, PD-MCI, and PDD.

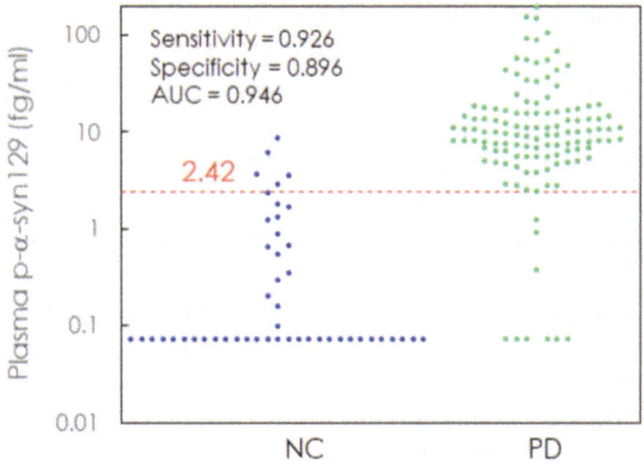

Fig. 4.27 Plasma p-α-syn129 levels measured in NCs and PD patients using IMR technology [118]

A typical method for distinguishing PD patients from NCs using plasma α-synuclein was reported by Lin et al. in 2017, as shown in Fig. 4.26a [117]. The NCs ($n = 35$) had plasma α-synuclein levels lower than 100 fg/ml, while PD patients ($n = 79$) had plasma α-synuclein levels higher than 100 fg/ml, resulting in a clear difference in the plasma α-synuclein levels between NCs and PD patients. Through an analysis of the receiver operating characteristic curve, the cutoff value for plasma α-synuclein levels to discriminate PD patients from NCs was 116 fg/ml, as plotted with the dashed line in Fig. 4.26a. The clinical sensitivity was 0.937, the clinical specificity was 0.947, and the area under the curve (AUC) was 0.971.

The study in Fig. 4.26a is a one-site cohort, termed the reference cohort. Other studies at multiple centers, termed the validation cohort, were executed to validate the cutoff value, i.e., 116 fg/ml. The measured plasma α-synuclein levels in the NC ($n = 371$) and PD ($n = 284$) groups in the validation cohort are plotted in Fig. 4.26b. When 116 fg/ml was used as the cutoff value to discriminate PD patients from NCs, the positive percentage agreement (PPA) was 0.860, the negative percentage agreement (NPA) was 0.838, and the overall percentage agreement (OPA) was 0.850.

In addition to α-synuclein, phosphorylated α-synuclein (Serine 129) (p-α-syn129) is an alternative biomarker for differentiating patients with PD from NCs. Lin et al. used IMR technology to assay p-α-syn129 in human plasma from NCs ($n = 48$) and PD patients ($n = 122$) [118]. The results are shown in Fig. 4.27. Notably, PD patients had higher plasma p-α-syn129 levels than did NC patients ($p < 0.0001$). The ROC curve analysis revealed that the cutoff value for discriminating PD patients from NCs was 2.42 fg/ml, with values of 0.926, 0.896, and 0.946 for clinical sensitivity, specificity, and AUC, respectively. These results in Figs. 4.26 and 4.27 show the good agreement between plasma biomarker assessments using total α-synuclein or its phosphorylated form and the clinical diagnosis of PD.

The technology used in Figs. 4.26 and 4.27 was IMR. Other technologies, such as ELISA, SIMOA, and immunoprecipitation RT-QuIC (IP/RT-QuIC), used for assaying plasma α-synuclein levels to differentiate PD patients from NCs are listed in Table 4.4. Although the cutoff values for plasma α-synuclein levels for discriminating PD patients from NCs differ among assay technologies, significant

Table 4.4 Reported clinical sensitivity, specificity, and AUC for assessing PD using plasma biomarkers assayed with various technologies

Biomarkers	Assay	Cutoff	Sensitivity	Specificity	AUC	*p* value
α-synuclein	IMR [117]	116 fg/ml	0.937	0.974	0.971	< 0.0001
α-synuclein	ELISA [119]	87.65 ng/ml	0.76	0.62	0.74	< 0.05
α-synuclein	SIMOA [120]	–	–	–	–	< 0.05
α-synuclein	IP/RT-QuIC [121]	Forming rate: 662.4	0.946	0.921	0.96	< 0.0001
α-synuclein	ECL [122]	–	–	–	–	< 0.05
p-α-syn129	IMR [118]	2.42 fg/ml	0.926	0.896	0.946	< 0.0001
T-Tau	IMR [123]	20.41 pg/ml	0.623	0.626	0.705	< 0.0001
T-Tau	IMR [113]	–	–	–	–	< 0.001
pTau181	SIMOA [120]	–	–	–	–	< 0.05
pTau181	IMR [114]	–	–	–	–	< 0.05
$A\beta_{1-40}$	IMR [111]	–	–	–	–	< 0.01
$A\beta_{1-42}$	IMR [123]	16.17 pg/ml	0.585	0.578	0.614	< 0.0001

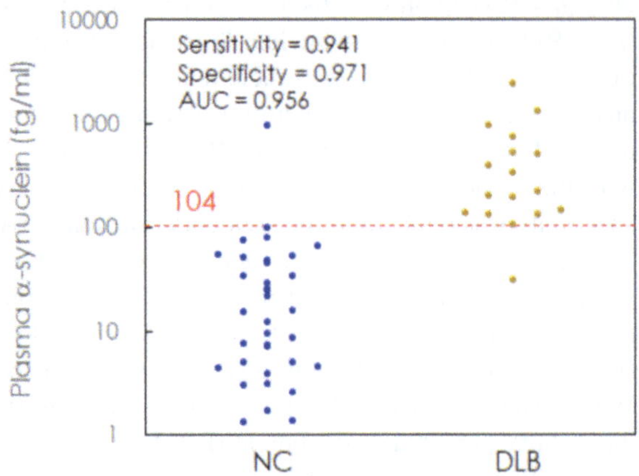

Fig. 4.28 Plasma total α-synuclein levels measured in NCs and DLB patients using IMR technology [124]

differences in plasma α-synuclein levels between PD patients and NCs have been obtained. Therefore, α-synuclein levels in human plasma are a promising biomarker for predicting PD.

In Table 4.4, IMR and IP/RT-QuIC assays show high discrimination between PD and NCs by measuring plasma α-synuclein, compared to other technologies. A possibly key reason is that IMR and IP/RT-QuIC mainly detect oligomers and protofibrils of α-synuclein in plasma. Other technologies that are based on sandwich immunoassays hardly detect oligomers and protofibrils of α-synuclein in plasma, which are mostly monomers of α-synuclein. As we know, oligomers and protofibrils of pathological protein are critical to the development of neurodegenerative diseases. Assay technologies that can detect oligomers and protofibrils of pathological proteins can show high accuracy in finding patients with neurodegenerative diseases. Detailed illustrations of the detected protein conformations using IMR, RT-QuIC, and sandwich immunoassays are discussed in **Paradox in Detecting**.

4.3.6 Plasma $A\beta_{1\text{–}42}$ Levels among Technologies, Clinical Validation of Assays for Alzheimer's Disease

Research groups have revealed that biomarkers such as T-Tau, pTau181, $A\beta_{1\text{–}40}$, and $A\beta_{1\text{–}42}$ are also able to predict PD, as shown in Fig. 4.5. However, the clinical sensitivity, specificity, and AUC of assessing PD using T-Tau, pTau181, $A\beta_{1\text{–}40}$, and $A\beta_{1\text{–}42}$ levels are obviously lower than those of α-synuclein and p-α-syn129 levels. These findings suggest that synucleinopathies dominate PD, consistent with the pathological hallmark. However, tauopathy and amyloidosis could have nonnegligible contributions to PD.

Lin et al. reported higher plasma α-synuclein levels in DLB samples ($n = 17$) than in NC samples ($n = 35$), in which IMR was used to assay plasma α-synuclein levels, as shown in Fig. 4.28 [124]. The cutoff value for plasma total α-synuclein levels for discriminating DLB patients from NCs is 104 fg/ml, with values of 0.941, 0.971, and 0.956 for clinical sensitivity, specificity, and AUC, respectively.

The increase in plasma total α-synuclein levels in DLB patients compared with NCs was also documented by Chen et al., who used the IMR assay. DLB patients ($n = 58$) had plasma total α-synuclein levels of 240 fg/ml, whereas NCs ($n = 31$) had levels of 80 fg/ml ($p < 0.005$) [125] (Fig. 4.28).

A careful inspection of the data in Figs. 4.24a and 4.25 revealed that the plasma total α-synuclein levels were the highest in PD patients and the lowest in NCs. DLB patients had levels in between the two groups. However, a significant difference in plasma total α-synuclein levels was not observed between PD patients and DLB patients. These results suggest that although synucleinopathy plays a role in both PD and DLB, synucleinopathy is more dominant in PD patients than in DLB patients. This finding could be an important reason why dementia occurs prior to motor disorders or the 1-year rule for DLB.

4.3.7 Plasma Biomarkers Versus Disease Severity

Motor disorders are the key symptoms of PD. The disorders of a PD patient usually start on one side of the body. For example, tremor, rigidity, reduced arm swing, and slowness are present only on one side. Ultimately, a PD patient will not be able to walk at all. Hence, motor disorders associated with PD progress from one side of the body to the whole body. In 1967, Margaret Hoehn and Melvin Yahr proposed a scale to describe the progression/severity of motor disorders in PD patients [126]. The scale is the Hoehn–Yahr stage (H–Y stage).

Traditionally, this scale includes five stages, as briefly described below.

Stage I: Motor disorders on one side of the body
Stage II: Motor disorders affecting both sides of the body
Stage III: Impaired balance, for example, loss of balance when pulled backward
Stage IV: Difficulty walking or need an assistive device to walk
Stage V: Unable to walk

The relationships between plasma biomarker levels and the H-Y stage have attracted the attention of researchers. Lin et al. used IMR to assay plasma total α-synuclein and p-α-syn129 levels in PD patients [117, 118]. These authors reported that plasma total α-synuclein levels do not correlate with the H-Y stage ($p > 0.05$), nor do plasma p-α-syn129 levels ($p > 0.05$). However, once the five stages are regrouped into the early motor stage (H–Y stages I and II, $n = 76$) and the advanced motor stage (H–Y stages III–V, $n = 46$), the measured plasma p-α-syn129 levels are 11.1 ± 8.9 fg/ml and 17.6 ± 10.2 fg/ml, respectively [105]. A significant difference in the plasma p-α-syn129 level was observed between patients in the early motor stage and the advanced motor stage ($p < 0.05$). However, the plasma total α-synuclein levels did not differ significantly between the early motor stage (1322.8 ± 1136.6 fg/ml) and the advanced motor stage (1423.2 ± 1023.8 fg/ml, $p > 0.05$). The absence of a correlation between plasma total α-synuclein levels and H–Y stage ($r = 0.2490$, $p > 0.05$) in PD patients was reported by Fan et al., who used ECL technology to assay total α-synuclein levels in plasma [122]. The same findings were independently reported by Chang et al. ($r = 0.081$, $p > 0.05$) [127], Chen et al., Lin et al., and Lin et al. ($r = -0.196$, $p > 0.05$) [110, 128, 129], who used IMR to assay total α-synuclein levels in plasma, as well as Ng et al., who used SIMOA to assay plasma α-synuclein levels ($p > 0.05$) [130].

However, Chen et al. and Ren et al. used IMR and SIMOA, respectively, to assay plasma α-synuclein levels. Both methods revealed significant positive correlations between the plasma α-synuclein concentration and modified H-Y stage (Chen: $r = 0.62$, $p < 0.01$; Ren: $r = 0.203$, $p < 0.01$) [119, 120]. The role of plasma α-synuclein levels in determining disease severity in PD patients has not been determined. Nevertheless, most published results show the absence of a correlation between the plasma α-synuclein level and H-Y stage. Hence, p-α-syn129 is likely more crucial than total α-synuclein for the progression of motor disorders in PD patients.

H-Y staging is based purely on mobility, without considering mental, behavioral, or other nonmotor problems such as dementia, depression, fatigue, and sleep disorders, which frequently occur in PD or DLB patients. The Unified Parkinson's Disease Rating Scale (UPDRS) is commonly used in clinical practice to evaluate mental health, activities of daily living, and motor functions. Thus, the UPDRS includes three parts:

UPDRS I: Evaluation of mentation, behavior, and mood

UPDRS II: Self-evaluation of the activities of daily living (ADLs), including handwriting, speech, salivating, cutting food, swallowing, falling, hygiene, dressing, walking, turning in bed, etc.

UPDRS III: Clinician-scored monitored motor evaluation

In 2007, the Movement Disorder Society (MDS) proposed a revision of the UPDRS, termed the MDS-UPDRS, which improved two major limitations of the UPDRS, including the lack of consistent anchors among subscales and the low emphasis on the nonmotor features of PD. At present, both the UPDRS and MDS-UPDRS are used clinically to diagnose PD and DLB.

Lin et al. documented a positive correlation between plasma p-α-syn129 levels and the UPDRS III score of PD patients ($r = 0.27$, $p < 0.005$) [118]. However, the absence of a correlation between plasma total α-synuclein levels and the UPDRS III score was independently demonstrated by Lin et al. and Lin et al. [113, 117]—they used IMR to assay plasma total α-synuclein levels. Ng et al. also presented the independence of the UPDRS III score from the plasma α-synuclein levels assayed with SIMOA ($p > 0.05$) [130]. These findings again demonstrate that p-α-syn129 is more important than total α-synuclein for motor disorders in PD.

Lin et al. proposed a possible explanation for this finding [117, 118]. The H–Y stage and UPDRS III score of PD patients primarily reflect deficits arising from nigrostriatal degeneration, whereas the measured plasma total α-synuclein levels are influenced by whole-brain α-synuclein loading. This finding might suggest that the measured plasma p-α-syn129 level mainly originates from nigrostriatal degeneration in PD patients. However, further studies are needed to clarify the pathophysiology underlying the motor symptoms of PD.

Nevertheless, some research groups have reported a positive correlation between the UPDRS III score and plasma α-synuclein in PD patients. For example, Ren et al. used SIMOA to measure plasma α-synuclein levels and observed a positive correlation with the UPDRS III score ($p < 0.05$) [130]. Fan et al. also reported a positive correlation between the UPDRS III score and the plasma α-synuclein level assayed with ECL ($r = 0.417$, $p < 0.01$) [122]. Dumrikarnlert et al. used ELISA to investigate the relationship between plasma α-synuclein levels and UPDRS III scores [131]. A fair correlation was found ($r = -0.261$, $p < 0.05$). Moreover, Niu et al. used ECL to assay neuronal exosomal α-synuclein concentrations and reported a positive correlation with the UPDRS III score ($p < 0.05$) [132]. Chung et al. also reported a negative correlation between the UPDRS III score and plasma extracellular vesicle α-synuclein concentrations assayed with IMR in PD patients ($r = -0.243$, $p < 0.01$) [133]. These inconsistent results for the correlation between plasma α-synuclein levels and the UPDRS III score might be attributed to the disease duration, age, sex, the coexistence of proteinopathies, or other nonmotor factors. Detailed explorations are needed to clarify the exact roles of total α-synuclein and p-α-syn129 in motor disorders in PD patients.

4.3.8 Plasma Biomarkers Versus Cognitive Impairment

In addition to the progression of motor dysfunction, cognitive impairment often occurs not only in DLB patients but also in PD patients. MMSE or MoCA testing is commonly used as part of the clinical evaluation of cognitive impairment. PD patients are categorized according to their scores as follows: PD with normal cognition (PD-NC), PD with mild cognitive impairment (PD-MCI), and PD dementia (PDD). The MMSE or MoCA scores depend on age, social level, economic level, and education. The scores are somewhat subjective. Physicians prefer objective and

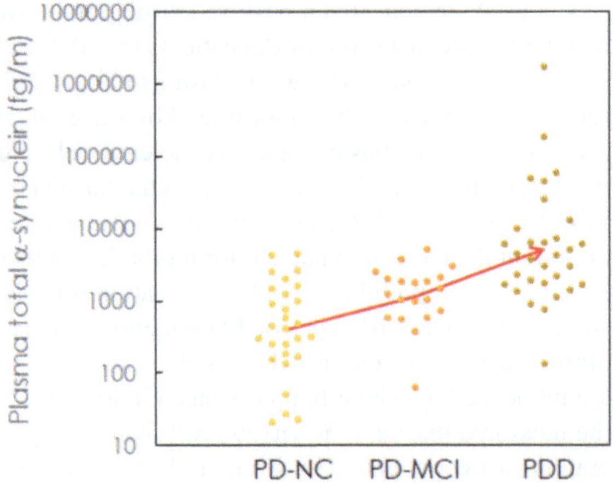

Fig. 4.29 Plasma total α-synuclein measured in PD-NC, PD-MCI, and PDD patients using IMR technology [117]

scientific examinations for assessing cognitive impairment in PD patients. Hence, the correlations between plasma biomarkers and cognitive impairment are attracting the attention of researchers and clinicians.

In 2017, Lin et al. used IMR to assay plasma total α-synuclein levels in PD-NC ($n = 30$), PD-MCI ($n = 21$), and PDD ($n = 29$) patients. Significant differences in the plasma total α-synuclein levels were detected among PD-NC, PD-MCI, and PDD patients, as shown in Fig. 4.29 [117]. Further analysis revealed a negative correlation between the plasma total α-synuclein concentration and the MMSE score ($r = -0.331, p < 0.05$).

The negative correlation between the plasma α-synuclein concentration and the MMSE score was also independently observed by Chen et al. ($r = -0.323, p < 0.005$) and Lin et al. ($r = -0.341, p < 0.005$) [109, 113], who both used IMR to determine the total α-synuclein concentration in the plasma of PD patients. In addition to α-synuclein levels, the MMSE score was negatively correlated with the plasma $A\beta_{1-40}$ levels ($r = 0.276$, $p < 0.05$) and T-Tau levels ($r = -0.322$, $p < 0.005$) [109]. The plasma α-synuclein level was also negatively correlated with the CASI score ($r = -0.328, p < 0.01$) of PD patients [113].

Ren et al. used SIMOA to assay plasma α-synuclein levels but did not find a correlation with the MoCA score [120]. However, Ng et al. used SIMOA to explore the correlation between plasma α-synuclein levels and cognition in PD patients [130]. Higher plasma α-synuclein levels were detected in PD patients with poorer cognition. Although both Ren et al. and Ng et al. used SIMOA to quantify plasma α-synuclein levels, inconsistent correlations with cognitive impairment in PD patients were observed. In addition, Dumrikarnlert et al. used ELISA to investigate the relationship between plasma α-synuclein levels and the UPDRS III score but did not observe a significant correlation [131].

According to the reported results, some researchers observed significant correlations of plasma α-synuclein and p-α-syn129 levels with disease severity, motor disorders, or cognitive impairment in PD patients, whereas others revealed no correlations. These inconsistencies might be attributed to the use of different assay technologies, the measured conformations of α-synuclein, disease duration, education, other potential proteinopathies, etc.

The published results indicate good agreement between plasma α-synuclein or p-α-syn129 levels and the clinical diagnosis of PD or DLB. These findings suggest strong potential for using plasma α-synuclein or p-α-syn129 levels as biomarkers for assessing PD and DLB in clinical practice. For example, the plasma total α-synuclein assay for the assessment of PD in clinical practice is available in Taiwan.

4.3.9 *Plasma Biomarkers Distinguishing PDD from DLB*

One puzzle in clinical practice is the discrimination of PDD from DLB. Clinical symptoms can overlap between the two conditions and include dream enactment behavior (RBD), parkinsonism, cognitive impairment, visual hallucinations, and even fluctuations. The "1-year rule" is currently used for the discrimination of these conditions, but in clinical practice, not all clinicians adhere closely to this rule. However, determining the "1-year rule" precisely is sometimes difficult as historical information lacks precision. Physicians may prefer to have helpful information, such as plasma biomarkers, to assist in discrimination.

In a pilot study, Malaty et al. used IMR to assay plasma biomarkers, including pTau181, total α-synuclein, and $A\beta_{1-42}$, in PDD ($n = 12$) and DLB ($n = 12$) patients [134]. The individual biomarker pTau181 ($p < 0.05$) and the combination of biomarkers $A\beta_{1-42}$xpTau181 ($p < 0.05$) were present at significantly higher levels in DLB patients than in PDD patients. The clinical sensitivity and specificity were within the range of 0.75–0.92. Malaty et al. further combined plasma biomarkers with the H–Y stage, the UPDRS score, or the MoCA score and reported that the pTau181 level + UPDRS score was the best for distinguishing between DLB and PDD patients. Both the clinical sensitivity and specificity were 0.83, and the AUC was 0.88. This result implies that the use of plasma biomarkers in conjunction with established functional clinical measures may increase the diagnostic sensitivity and specificity of those tests over the use of individual tests in distinguishing DLB from PDD. These findings have immediate implications for maximizing the capabilities of diagnostic differentiation, which may lead to better prognoses, earlier patient and family counseling, and better symptomatic management.

4.4 Clinical Validation of Assays for Frontotemporal Dementia

Frontotemporal dementia (FTD) is a rare neurodegenerative disease. The incidence depends on age, i.e., 2.2/100,000 between ages 40 and 49, 3.3/100,000 between ages 50 and 59, and peaks at 8.9/100,000 between ages 60 and 69 [135–137]. Three clinical phenotypes are clearly observed in patients with FTD: behavior variant frontotemporal dementia (bvFTD); primary progressive aphasia (PPA), which is sometimes referred to as language variant FTD (lvFTD); and either FTD with amyotrophic lateral sclerosis (FTD-ALS) or an atypical parkinsonian disorder.

bvFTD is characterized by prominent changes in personality, behavior, and social conduct, which can include apathy, disinhibition (acting inappropriately), loss of empathy, compulsive habits (eating/cleaning), executive function issues (planning/prioritizing), and loss of interest, often with relatively preserved memory in the early stages, leading to socially awkward or disruptive situations. These can significantly disrupt interpersonal relationships. The age of onset is often between 45 and 65 years, but it can develop as early as 20 or as late as 80 years of age.

In lvFTD, the PPA can be further subdivided into semantic dementia (SD), progressive nonfluent aphasia (PNFA), and logopenic variant. SD is characterized by the loss of semantic understanding, resulting in impaired word comprehension, although speech remains fluent and grammatically faultless. PNFA results in a progressive loss of fluency of speech with intact word comprehension. Logopenic progressive aphasia (LPA) is characterized by difficulty with word retrieval, often described as words being "on the tip of the tongue," and trouble repeating sentences and phrases. In summary, the clinical hallmark of lvFTD is speech or language impairment [138, 139].

SD and PNFA have underlying FTD pathology. PNFA is primarily associated with frontotemporal lobar degeneration (FTLD) pathologically, specifically tau pathology and TDP-43 proteinopathies. Pathological markers for semantic dementia (SD) can also include TDP-43 protein aggregates with a regional distribution in the temporal lobes and atrophy in the anterior temporal lobes. Genetic mutations in progranulin (GRN) and microtubule-associated protein tau (MAPT) are associated with some forms of SD, especially those with a family history. Logopenic aphasia is associated with Alzheimer's disease (AD) pathology, including amyloid plaques and tau tangles, in the majority of cases. This pathology is often accompanied by atrophy and/or hypometabolism in specific brain regions, particularly the left posterior temporal and parietal lobes. Thus, the BBMs developed for detection and diagnosis of amnestic variant AD are likely to be useful in LPA as well.

Biomarkers are urgently needed as objective tools to improve clinical diagnosis by neurologists. FTD is associated with three key pathological proteins in neurons: tau protein, TAR DNA-binding protein 43 (TDP-43), and fused in sarcoma (FUS) [140–142]. Filamentous inclusions of tau protein are found in inherited and sporadic FTD patients. Mutations in the tau gene, such as MAPT, cause approximately 5% of FTD cases. Tau protein dysfunction is sufficient to cause neurodegeneration. The mislocalization of TDP-43 to the cytoplasm could lead to the loss of its ability to regulate transcription, splicing, and mRNA stability in the nucleus of patients with FTD, regardless of the presence of a genetic mutation. Mutations in FUS also cause FTD. Disease-causing mutant FUS inhibits intra-axonal protein synthesis and provokes hippocampal synaptic loss and dysfunction without loss of nuclear FUS function or FUS aggregation.

Approximately 50% of FTD patients express an abnormal form of tau protein in the brain. Approximately 50% of FTD patients present TDP-43 accumulation. A small percentage (~5%) of TDP patients present FUS protein accumulation. These abnormalities and the accumulation of tau, TDP-43, and FUS proteins in the brain disrupt normal cellular activities and may cause cell death, ultimately leading to FTD.

Diagnosis is often made clinically with the support of neuropsychological evaluation, MRI, and FDG-PET. FDG-PET in bvFTD typically shows hypometabolism in the frontal and anterior temporal lobes, insula, and anterior cingulate cortex. MRI findings include regional atrophy in the frontal lobes, anterior temporal lobes, and insula, often initially affecting the frontoinsular cortex.

Biomarkers used do not typically include quantification assays of proteins at present. Rather, the majority of blood-based biomarkers are commercially available genetic panels intended to detect mutations in MAPT, GRN, and C9orf72 (which has repeat expansions) as well as mutations in FUS, TARDBP, TBK1, VCP, and CHMP2B.

Since FTD patients are clinically characterized by speech or language impairment and pathological hallmarks, the clinical validation of plasma biomarkers for assessing FTD should be performed by correlating plasma biomarkers with pathological hallmarks and clinical symptoms. The clinical validation of plasma biomarkers is described below.

4.4.1 Plasma Biomarkers Versus Brain Tissue Staining

Figure 4.30a, b show typical immunohistochemical staining for tau protein and TDP neuropathology in the brain [143]. The relationships between plasma biomarkers and NFT pathology are shown in Fig. 4.4. The ability of plasma TDP-43 levels to predict TDP neuropathology in the brain is shown.

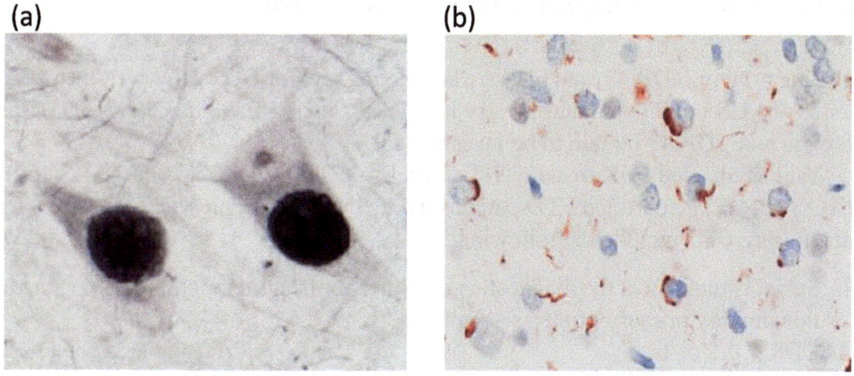

Fig. 4.30 Immunohistochemical staining for tau protein and TDP neuropathology in the brain [143]

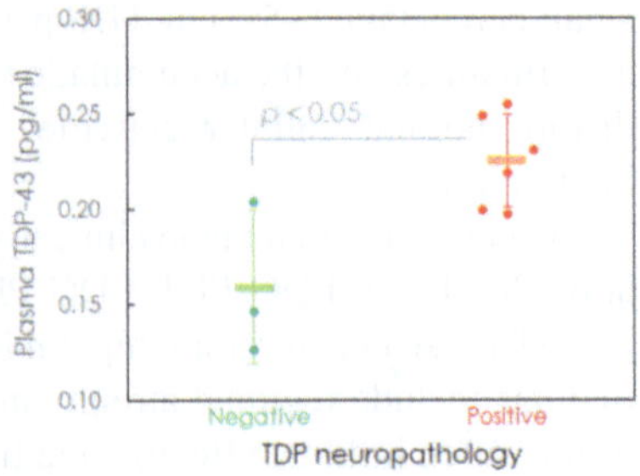

Fig. 4.31 Plasma TDP-43 levels measured with IMR in patients with negative and positive TDP neuropathology [22]

Yang et al. performed a pilot study to investigate the discrimination of TDP neuropathology via plasma TDP-43 levels [22]. They used antemortem plasma samples for the assay of TDP-43 levels with IMR. TDP neuropathology in autopsy samples was analyzed at the Arizona Study of Aging and Neurodegenerative Disorders and Brain and Body Donation Program (AZSAND/BBDP). Antemortem plasma samples were collected 1.5 years before death.

Figure 4.31 shows the TDP-43 levels measured in plasma from patients with negative ($n = 3$) and positive ($n = 5$) TDP neuropathology [22]. In patients with negative TDP neuropathology, the plasma TDP-43 concentration was 0.158 ± 0.040 pg/ml. In patients with positive TDP neuropathology, the plasma TDP-43 concentration was 0.225 ± 0.024 pg/ml, which was significantly higher than that in patients with negative neuropathology ($p < 0.05$). The results in Fig. 4.29 explicitly show the discrimination in TDP neuropathology using plasma TDP-43 levels. Unfortunately, the number of patients included in the pilot study was very limited. Further studies with many more subjects are needed to obtain conclusive evidence. Nevertheless, the pilot study performed by Yang et al. provides promising results to encourage collaboration between biomarker researchers and pathologists.

4.4.2 *Plasma Biomarkers Versus Clinical Diagnosis*

The tau protein, TDP-43, and FUS are key biomarkers for FTD. However, the prevalence of FUS in FTD patients is very low, at less than 5%. Hence, the roles of tau protein and TDP-43 remain to be investigated.

Yang et al. used IMR to assay T-Tau levels in FTD patients and normal controls (NCs) [144]. The enrolled FTD patients were clinically diagnosed. The criteria for diagnosing FTD include the following:

1. Subjects must meet the diagnostic guidelines for frontotemporal lobar degeneration, mainly primary progressive aphasia.
2. CDR = 0.5 or 1.

The exclusion criteria for enrolling patients with FTD include the following:

1. Subjects with cranial metallic implants, cardiac pacemakers, or claustrophobia.

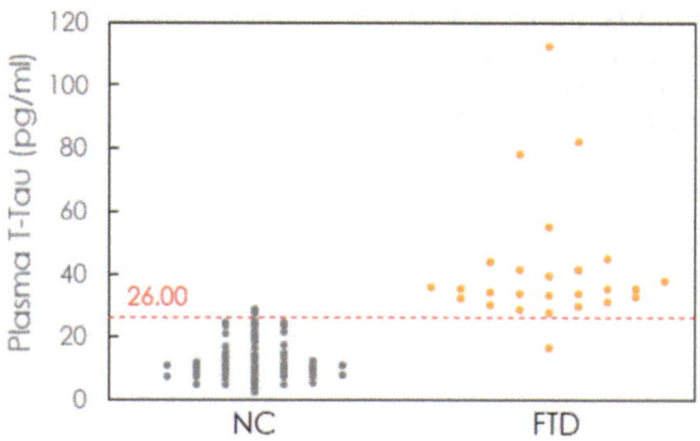

Fig. 4.32 Plasma T-Tau levels measured with IMR in normal control (NC) and FTD patients [144]

2. Subjects with a significant history of depression.
3. A Geriatric Depression Scale score > 8.

In their study, 66 NCs and 26 FTD patients were enrolled. The measured T-Tau levels in the NC and FTD groups were 13.37 ± 7.77 and 41.28 ± 20.13 pg/ml ($p < 0.0001$), respectively, as shown in Fig. 4.32. Through a ROC curve analysis, the cutoff value for the plasma T-Tau level for differentiating FTD patients from NCs was 26.00 pg/ml, with values of 0.962, 0.955, and 0.987 for the clinical sensitivity, specificity, and AUC, respectively. This study clearly revealed higher plasma T-Tau levels in FTD patients than in NCs.

The increase in plasma T-Tau levels in FTD patients assayed with IMR was also reported by Lin et al. [111]. In addition to T-Tau, Lin et al. reported that plasma pTau181 is able to discriminate FTD patients from NCs well ($p < 0.05$).

Foiani et al. used SIMOA to measure plasma T-Tau levels [145]. Higher plasma T-Tau levels were detected in FTD patients (1.96 ± 1.07 pg/ml) than in NCs (1.67 ± 0.50 pg/ml, $p < 0.05$). However, Ill'an-Gala also used SIMOA to measure plasma T-Tau levels but observed poor discrimination between FTD patients and NCs ($p > 0.05$) [146]. Inconsistent results are frequently reported when a particular assay is used. This discrepancy might be due to several factors, such as plasma preparation, assay repeatability, operator-to-operator variation, and batch-to-batch variation. Thus, the authors emphasize the plasma preparation in Chap. 2 and the preclinical performance in Chap. 3. Users of plasma assays are recommended to focus on the details of plasma preparation and the preclinical performance of each assay technology.

Yang et al. also used IMR to assay plasma TDP-43 levels [147]. The measured plasma TDP-43 level in FTD patients ($n = 9$) was 0.419 ± 0.193 pg/ml, whereas it was 0.163 ± 0.097 pg/ml in NCs ($n = 27$). Significant increases in plasma TDP-43 levels have been observed in FTD patients. Further studies in which more subjects (FTD: $n = 10$; NC: $n = 92$) were enrolled again demonstrated higher plasma TDP-43 levels in FTD patients than in NCs, as shown in Fig. 4.33. The cutoff value for the plasma TDP-43 levels for discriminating FTD patients from NCs was 0.237 pg/ml, with values of 0.900, 0.913, and 0.915 for clinical sensitivity, specificity, and AUC, respectively.

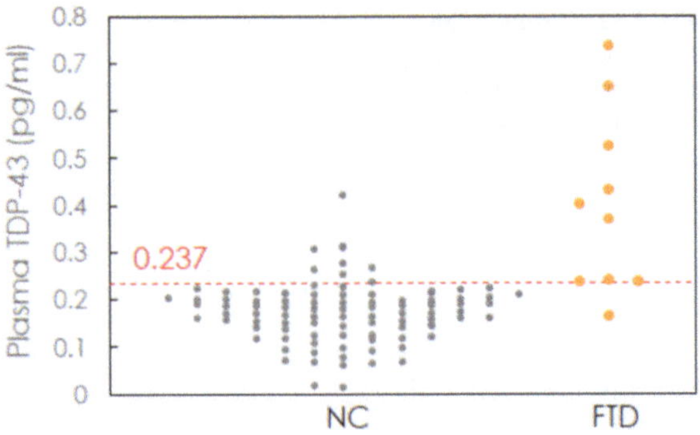

Fig. 4.33 Plasma TDP-43 levels measured with IMR in normal controls (NCs) and FTD patients [147]

In addition, Chatterjee et al. used SIMOA to detect plasma TDP-43 levels in FTD patients and NCs [148]. They also reported a clear increase in plasma extracellular vesicle TDP-43 levels in FTD patients (24.15 pg/ml) compared with those in NCs (10.42 pg/ml, $p < 0.0001$). Foulds et al. developed a homemade ELISA to measure plasma TDP-43 levels in FTD patients. Their results revealed higher plasma TDP-43 levels in FTD patients than in NCs ($p < 0.001$) [149].

Regardless of the use of IMR, SIMOA or ELISA, the measured plasma TDP-43 levels were significantly higher in the FTD patients. As shown in Fig. 4.29, the increased plasma TDP-43 levels revealed increased deposition of TDP-43 in the brain. Undoubtedly, if developed and validated further, plasma TDP-43 levels could be promising as part of assessment of FTD in clinical practice.

References

1. Palmqvist, S., Whitson, H. E., Allen, L. A., Suarez-Calvet, M., Galasko, D., Karikari, T. K., Okrahvi, H. R., Paczynski, M., Schindler, S. E., Teunissen, C. E., Zetterberg, H., Carrillo, M. C., Edelmayer, R. M., Mahinrad, S., McAteer, M. B., Kahale, L. A., Pahlke, S., & Tampi, M. P. (2025). Alzheimer's Association clinical practice guideline on the use of blood-based biomarkers in the diagnostic workup of suspected Alzheimer's disease within specialized care settings. *Alzheimers Dement, 21*, e70535-1-17.
2. Hardy, J., & Selkoe, D. S. (2002). The amyloid hypothesis of Alzheimer's disease: Progress and problems on the road to therapeutics. *Science, 297*, 353–356.
3. Lannfelt, L. (2023). A light at the end of the tunnel – From mutation identification to a potential treatment for Alzheimer's disease. *Upsala Journal of Medical Sciences, 128*, e10316-1-10.
4. Bancher, C., Lassmann, H., Budka, H., Grundke-Iqbal, I., Iqbal, K., Wiche, G., Seitelberger, F., & Wisniewski, H. M. (1987). Neurofibrillary tangles in Alzheimer's disease and progressive supranuclear palsy: Antigenic similarities and differences. Microtubule-associated protein tau antigenicity is prominent in all types of tangles. *Acta Neuropathologica, 74*, 39–46.
5. Vlassenko, A. G., Benzinger, T. L., & Morris, J. C. (2012). PET amyloid-beta imaging in preclinical Alzheimer's disease. *Biochimica et Biophysica Acta, 1822*, 370–379.
6. Okamura, N., Harada, R., Ishiki, A., Kikuchi, A., Nakamura, T., & Kudo, Y. (2018). The development and validation of tau PET tracers: Current status and future directions. *Clinical and Translational Imaging, 6*, 305–316.

7. van Waarde, A., Marcolini, S., de Deyn, P. W., & Dierckx, R. A. J. O. (2021). PET agents in dementia: An overview. *Seminars in Nuclear Medicine, 51*, 196–229.
8. Jie, C. V. M. L., Treyer, V., Schibli, R., & Mu, L. (2021). Tauvid™: The first FDA-approved PET tracer for imaging tau pathology in Alzheimer's disease. *Pharma, 14*, 110-1-12.
9. Jack, C. R., Jr., Knopman, D. S., Jagust, W. J., Shaw, L. M., Aisen, P. S., Weiner, M. W., Petersen, R. C., & Trojanowski, J. Q. (2010). Hypothetical model of dynamic biomarkers of the Alzheimer's pathological cascade. *Lancet Neurology, 9*, 119–128.
10. Jack, C. R., Jr., Bennett, D. A., Blennow, K., Carrillo, M. C., Dunn, B., Haeberlein, S. B., Holtzman, D. M., Jagust, W., Jessen, F., Karlawish, J., Liu, E., Molinuevo, J. L., Montine, T., Phelps, C., Rankin, K. P., Rowe, C. C., Scheltens, P., Siemers, E., Snyder, H. M., & Sperling, R. (2018). NIA-AA research framework: Toward a biological definition of Alzheimer's disease. *Alzheimer's and Dementia, 14*, 535–562.
11. McDade, E., Bednar, M. M., Brashear, H. R., Miller, D. S., Maruff, P., Randolph, C., Ismail, Z., Carrillo, M. C., Weber, C. J., Bain, L. J., Hake, A. M. (2020). The pathway to secondary prevention of Alzheimer's disease. *Alzheimers Dement, 6*(1), e12069.
12. Jack, C. R., Andrews, J. S., Beach, T. G., Buracchio, T., Dunn, B., Graf, A., Hansson, O., Ho, C., Jagust, W., McDade, E., Molinuevo, J. L., Okonkwo, O. C., Pani, L., Rafii, M. S., Scheltens, P., Siemers, E., Snyder, H. M., Sperling, R., Teunissen, C. E., & Carrillo, M. C. (2024). Revised criteria for diagnosis and staging of Alzheimer'sdisease: Alzheimer's association workgroup. *Alzheimer's Dementia, 20*, 5143–5169.
13. Bleecker, M. L., Bolla-Wilson, K., Kawas, C., & Agnew, J. (1988). Age-specific norms for the mini-mental state exam. *The Neurologist, 38*, 1565–1568.
14. Mungas, D., Reed, B. R., Farias, S. T., & Decarli, C. (2009). Age and education effects on relationships of cognitive test scores with brain structure in demographically diverse older persons. *Psychology and Aging, 24*, 116–128.
15. Tripathi, R., Kumar, K., Bharath, S., Marimuthu, P., & Varghese, M. (2014). Age, education and gender effects on neuropsychological functions in healthy Indian older adults. *Dement Neuropsychology, 8*, 148–154.
16. Farran, N., Valmaggia, L., Tadros, N., & Morris, R. (2025). Neuropsychological tests of social cognition in non-Western countries and in individuals from ethnic minoritized groups in Western countries: A systematic review. *Neuropsychology, 39*, 89–136.
17. Jucker, M., & Walker, L. (2013). Self-propagation of pathogenic protein aggregates in neurodegenerative diseases. *Nature, 501*, 45–51.
18. Doraiswamy, P. M., Sperling, R. A., Coleman, R. E., Johnson, K. A., Reiman, E. M., Davis, M. D., Grundman, M., Sabbagh, M. N., Sadowsky, C. H., Fleisher, A. S., Carpenter, A., Clark, C. M., Joshi, A. D., Mintun, M. A., Skovronsky, D. M., & Pontecorvo, M. J. (2012). Amyloid-β assessed by florbetapir F 18 PET and 18-month cognitive decline: A multicenter study. *Neurology, 79*, 1636–1644.
19. Sabri, O., Sabbagh, M. N., Seibyl, J., Barthel, H., Akatsu, H., Ouchi, Y., Senda, K., Murayama, S., Ishii, K., Takao, M., Beach, T. G., Rowe, C. C., Leverenz, J. B., Ghetti, B., Ironside, J. W., Catafau, A. M., Stephens, A. W., Mueller, A., Koglin, N., Hoffmann, A., Roth, K., Reininger, C., & Schulz-Schaeffer, W. J. (2015). Florbetaben PET imaging to detect amyloid beta plaques in Alzheimer's disease: Phase 3 study. *Alzheimer's and Dementia, 11*, 964–974.
20. Tunali, I., Wang, J., Arora, A. K., Kim, M. J., Shcherbinin, S., Pontecorvo, M., & Iaccarino, L. (2025). Development and validation of a ^{18}F-Flortaucipir PET visual stratification nethod. *Journal of Nuclear Medicine, 66*, 612–619.
21. Mueller, A., Bullich, S., Barret, O., Madonia, J., Berndt, M., Papin, C., Perrotin, A., Koglin, N., Kroth, H., Pfeifer, A., Tamagnan, G., Seibyl, J. P., Marek, K., De Santi, S., Dinkelborg, L. M., & Stephens, A. W. (2020). Tau PET imaging with ^{18}F-PI-2620 in patients with Alzheimer disease and healthy controls: A first-in-humans study. *Journal of Nuclear Medicine, 61*, 911–919.
22. Yang, S. Y., & Liu, H. C. (2025). Plasma-to-autopsy study of neurodegenerative diseases. *Journal of Clinical Neuroscience, 9*, 1–8.

23. Iaccarino, L., Burnham, S. C., Tunali, I., Wang, J., Navitsky, M., Arora, A. K., & Pontecorvo, M. J. (2025). A practical overview of the use of amyloid-PET centiloid values in clinical trials and research. *NeuroImage: Clinical, 46*, 103765.
24. Tzen, K. Y., Yang, S. Y., Chen, T. F., Cheng, T. W., Horng, H. E., Wen, H. P., Huang, Y. Y., Shiue, C. Y., & Chiu, M. J. (2014). Plasma Aβ but not tau is related to brain PiB retention in early Alzheimer's disease. *ACS Chemical Neuroscience, 5*, 830–836.
25. Hanseeuw, B. J., Malotaux, V., Dricot, L., Quenon, L., Sznajer, Y., Cerman, J., Woodard, J. L., Buckley, C., Farrar, G., Ivanoiu, A., & Lhommel, R. (2021). Defining a centiloid scale threshold predicting long-term progression to dementia in patients attending the memory clinic: An [^{18}F] flutemetamol amyloid PET study. *European Journal of Nuclear Medicine and Molecular Imaging, 48*, 302–310.
26. Kang, S. K., Heo, M., Chung, J. Y., Kim, D., Shin, S. A., Choi, H., Chung, A., Ha, J. M., Kim, H., & Lee, J. S. (2024). Clinical performance evaluation of an artificial intelligence-powered amyloid brain PET quantification method. *Nuclear Medicine and Molecular Imaging, 58*, 246–254.
27. Fan, L. Y., Tzen, K. Y., Chen, Y. F., Chen, T. F., Lai, Y. M., Yen, R. F., Huang, Y. Y., Shiue, C. Y., Yang, S. Y., & Chiu, M. J. (2018). The relation between brain amyloid deposition, cortical atrophy, and plasma biomarkers in amnesic mild cognitive impairment and Alzheimer's disease. *Frontiers in Aging Neuroscience, 10*, 175-1-10.
28. Schindler, S. E., Bollinger, J. G., Ovod, V., Mawuenyega, K. G., Li, Y., Gordon, B. R., Holtzman, D. M., Morris, J. C., Benzinger, T. L. S., Xiong, C., Fagan, A. M., & Bateman, R. J. (2019). High-precision plasma β-amyloid 42/40 predicts current and future brain amyloidosis. *The Neurologist, 93*, e1647–e1659.
29. Lin, S. Y., Lin, K. J., Lin, P. C., Huang, C. C., Chang, C. C., Lee, Y. C., Hsiao, I. T., Yen, T. C., Huang, W. S., Yang, B. H., & Wang, P. N. (2019). Plasma amyloid assay as a pre-screening tool for amyloid positron emission tomography imaging in early stage Alzheimer's disease. *Alzheimer's Research and Therapy, 11*, 111-1-10.
30. Jiao, F., Yi, F., Wang, Y., Zhang, S., Guo, Y., Du, W., Gao, Y., Ren, J., Zhang, H., Liu, L., Song, H., & Wang, L. (2020). The validation of multifactor model of plasma Aβ42 and total-tau in combination with MoCA for diagnosing probable Alzheimer disease. *Frontiers in Aging Neuroscience, 12*, 212-1-9.
31. Verberk, I. M. W., Thijssen, E., Koelewijn, J., Mauroo, K., Vanbrabant, J., de Wilde, A., Zwan, M. D., Verfaillie, S. C. J., Ossenkoppele, R., Barkhof, F., van Berckel, B. N. M., Scheltens, P., van der Flier, W. M., Stoops, E., Vanderstichele, H. M., & Teunissen, C. E. (2020). Combination of plasma amyloid beta(1-42/1-40) and glial fibrillary acidic protein strongly associates with cerebral amyloid pathology. *Alzheimer's Research and Therapy, 12*, 118-1-14.
32. Wang, P. M., Lin, K. J., Liu, H. C., Andreasson, U., Blennow, K., Zetterberg, H., & Yang, S. Y. (2020). Plasma pyroglutamate-modified amyloid beta differentiates amyloid pathology. *Alzheimer's and Dementia, 12*, e12029-1-9.
33. Janelidze, S., Teunissen, C. E., Zetterberg, H., Allue, J. A., Sarasa, L., Eichenlaub, U., Bittner, T., Ovod, V., Verberk, I. M. W., Toba, K., Nakamura, A., Bateman, R. J., Blennow, K., & Hansson, O. (2021). Head-to-head comparison of 8 plasma amyloid-β 42/40 assays in Alzheimer disease. *JAMA Neurology, 78*, 1375–1382.
34. Keshavan, A., Pannee, J., Karikari, T. K., Rodriguez, J. L., Ashton, N. J., Nicholas, J. M., Cash, D. M., Coath, W., Lane, C. A., Parker, T. D., Lu, K., Buchanan, S. M., Keuss, S. E., James, S. N., Murray-Smith, H., Wong, A., Barnes, A., Dickson, J. C., Heslegrave, A., Portelius, E., Richards, M., Fox, N. C., Zetterberg, H., Blennow, K., & Schott, J. M. (2021). Population-based blood screening for preclinical Alzheimer's disease in a British birth cohort at age 70. *Brain, 144*, 434–449.
35. Chen, T. B., Lin, K. J., Lin, S. Y., Lee, Y. J., Lin, Y. C., Wang, C. Y., Chen, J. P., & Wang, P. N. (2021). Prediction of cerebral amyloid pathology based on plasma amyloid and tau related markers. *Frontiers in Neurology, 12*, 619388-1–619388-9.

36. Benedet, A. L., Brum, W. S., Hansson, O., Karikari, T. K., Zimmer, E. R., Zetterberg, H., Blennow, K., & Ashton, N. J. (2022). The accuracy and robustness of plasma biomarker models for amyloid PET positivity. *Alzheimer's Research and Therapy, 14*, 26-1-11.
37. Donaghy, P. C., Firbank, M., Petrides, G., Lloyd, J., Barnett, N., Olsen, K., Heslegrave, A., Zetterberg, H., Thomas, A. J., & O'Brien, J. T. (2022). The relationship between plasma biomarkers and amyloid PET in dementia with Lewy bodies. *Parkinsonism and Related Disorders, 101*, 111–116.
38. Feizpour, A., Doecke, J. D., Doré, V., Krishnadas, N., Huang, K., Bourgeat, P., Laws, S. M., Fowler, C., Robertson, J., Mackintosh, L., Ayton, S., Martins, R., Rainey-Smith, S. R., Kevin Taddei, K., Ward, L., Stage, E., Bannon, A. W., Masters, C. L., Fripp, J., Villemagne, V. L., & Rowea, C. C. (2024). Detection and staging of Alzheimer's disease by plasma pTau217 on a high throughput immunoassay platform. *EBioMedicine, 109*, 105405.
39. Figdore, D. J., Wiste, H. J., Bornhorst, J. A., Bateman, R. J., Li, Y., Graff-Radford, J., Knopman, D. S., Vemuri, P., Lowe, V. J., Jack, C. R., Jr., Petersen, R. C., & Algeciras-Schimnich, A. (2024). Performance of the Lumipulse plasma Aβ42/40 and pTau181 immunoassays in the detection of amyloid pathology. *Alzheimer's and Dementia, 16*, e12545-1-9.
40. Silva-Spinola, A., Leitao, M. J., Nadal, A., Bastard, N. L., Santana, L., & Baldeiras, I. (2024). Exploring the potential of fully automated LUMIPULSE G plasma assays for detecting Alzheimer's disease pathology. *Alzheimer's Research and Therapy, 16*, 51-1-10.
41. Ashton, N. J., Brum, W. S., Molfetta, G. D., Benedet, A. L., Arslan, B., Jonaitis, E., Langhough, R. E., Cody, K., Wilson, R., Carlsson, C. M., Vanmechelen, E., Montoliu-Gaya, L., Lantero-Rodriguez, J., Rahmouni, N., Tissot, C., Stevenson, J., Servaes, S., Therriault, J., Pascoal, T., Lleo, A., Alcolea, D., Fortea, J., Rosa-Neto, P., Johnson, S., Jeromin, A., Blennow, K., & Zetterberg, H. (2024). Diagnostic accuracy of a plasma phosphorylated tau 217 immunoassay for Alzheimer disease pathology. *JAMA Neurology, 81*, 255–263.
42. Feizpour, A., Doré, V., Krishnadas, N., Bourgeat, P., Doecke, J. D., Saad, Z. S., Triana-Baltzer, G., Laws, S. M., Shishegar, R., Huang, K., Fowler, C., Ward, L., Masters, C. L., Fripp, J., Kolb, H. C., Villemagne, V. L., & Rowe, C. C. (2025). Alzheimer's disease biological PET staging using plasma p217+tau. *Communication and Medicine, 5*, 53-1-8.
43. Cano, A., Capdevila, M., Puerta, R., Arranz, J., Montrreal, L., de Rojas, I., García-González, P., Olivé, C., García-Gutiérrez, F., Sotolongo-Grau, O., Orellana, A., Aguilera, N., Ramis, M., Rosende-Roca, M., Lleó, A., Fortea, J., Tartari, J. P., Lafuente, A., Vargas, L., Pérez-Cordón, A., Muñoz, N., Sanabria, Á., Alegret, M., Morató, X., Tárraga, L., Fernández, V., Marquié, M., Valero, S., Alcolea, D., Boada, M., & Ruiz, A. (2024). Clinical value of plasma pTau181 to predict Alzheimer's disease pathology in a large real-world cohort of a memory clinic. *EBioMed, 108*, 105345-1-15.
44. Janelidze, S., Stomrud, E., Palmqvist, S., Zetterberg, H., van Westen, D., Jeromin, A., Song, L., Hanlon, D., Hehir, C. A. T., Baker, D., Blennow, K., & Hansson, O. (2016). Plasma β-amyloid in Alzheimer's disease and vascular disease. *Scientific Reports, 6*, 26801-1-11.
45. Shin, H. S., Lee, S. K., Kim, S., Kim, H. J., Chae, W. S., & Park, S. A. (2016). The correlation study between plasma Aβ proteins and cerebrospinal fluid Alzheimer's disease biomarkers. *Dementia and Neurocognitive Disorders, 15*, 122–128.
46. Teunissena, C. E., Chiu, M. J., Yang, C. C., Yang, S. Y., Scheltens, P., Zetterberg, H., & Blennow, K. (2018). Plasma amyloid-β (Aβ42) correlates with cerebrospinal fluid Aβ42 in Alzheimer's disease. *Journal of Alzheimer's Disease, 62*, 1857–1863.
47. Álvarez-Sánchez, L., Peña-Bautista, C., Ferré-González, L., Balaguer, A., Baquero, M., Casanova-Estruch, B., & Cháfer-Pericás, C. (2023). Assessment of plasma and cerebrospinal fluid biomarkers in different stages of Alzheimer's disease and frontotemporal dementia. *International Journal of Molecular Sciences, 24*, 1226–1-16.
48. De Rino, F., Rispoli, F., Zuffi, M., Matteucci, E., Gavazzi, A., Salvatici, M., Sansico, D. F., Pollaroli, G., & Drago, L. (2025). Assessment of plasma and cerebrospinal fluid biomarkers in patients with Alzheimer's disease and other dementias: A center-based study. *International Journal of Molecular Sciences, 26*, 4308-1-14.

49. Ledig, C., Schuh, A., Guerrero, R., Heckemann, R. A., Rueckert, D. (2018). Structural brain imaging in Alzheimer's disease and mild cognitive impairment: biomarker analysis and shared morphometry database. *Scientific Reports, 8*(1), 11258.
50. Chiu, M. J., Chen, Y. F., Chen, T. F., Yang, S. Y., Yang, F. P., Tseng, T. W., Chieh, J. J., Chen, J. C., Tzen, K. Y., Hua, M. S., Horng, H. E. (2014). Plasma tau as a window to the brain-negative associations with brain volume and memory function in mild cognitive impairment and early Alzheimer's disease. *Human Brain Mapping, 35*, 3132–3142.
51. Liu, H. C., Chen, H. H., Ho, C. S., Chang, J. F., Lin, C. C., Chiu, M. J., Chen, T. F., Hu, C. J., Yan, S. H., Sun, Y., & Yang, S. Y. (2021). Investigation of the number of tests required for assaying plasma biomarkers associated with Alzheimer's disease using immunomagnetic reduction. *Neurology and Therapy, 10*, 1015–1028.
52. Mattsson, N., Zetterberg, H., Janelidze, S., Insel, P. S., Andreasson, U., Stomrud, E., Palmqvist, S., Baker, D., Hehir, C. A. T., Jeromin, A., Hanlon, D., Song, L., Shaw, L. M., Trojanowski, J. Q., Weiner, M. W., Hansson, O., & Blennow, K. (2016). Plasma tau in Alzheimer disease. *Neurology, 87*, 1827–1835.
53. Deters, K. D., Risacher, A. L., Kim, A., Nho, K., West, J. D., Blennow, K., Zetterberg, H., Shaw, L. M., & Trojanowski, J. Q. (2017). Plasma tau association with brain atrophy in mild cognitive impairment and Alzheimer's disease. *Journal of Alzheimer's Disease, 58*, 1245–1254.
54. Pai, M. C., Wu, C. C., Hou, Y. C., Jeng, J. S., Tang, S. C., Lin, W. C., Lu, C. H., Chiu, M. J., Chen, T. F., Yan, S. H., Hu, C. J., & Yang, S. Y. (2022). Evidence of plasma biomarkers indicating high risk of dementia in cognitively normal subjects. *Scientific Reports, 12*, 1192-1-7.
55. Dittrich, A., Ashton, N. J., Zetterberg, H., Blennow, K., Zettergren, A., Simrén, J., Skillbäck, T., Shams, S., Machado, A., Westman, E., Schöll, M., Skoog, I., & Kern, S. (2023). Association of chronic kidney disease with plasma NfL and other biomarkers of neurodegeneration: The H70 birth cohort study in Gothenburg. *The Neurologist, 101*, e277–e288.
56. Chiu, M. J., Yang, S. Y., Horng, H. E., Yang, C. C., Chen, T. F., Chieh, J. J., Chen, H. H., Chen, T. C., Ho, C. S., Chang, S. F., Liu, H. C., Hong, C. Y., & Yang, H. C. (2013). Combined plasma biomarkers for diagnosing mild cognition impairment and Alzheimer's disease. *ACS Chemical Neuroscience, 4*, 1530–1536.
57. Álvarez-Sánchez, L., Peña-Bautista, C., Baquero, M., & Cháfer-Pericás, C. (2022). Novel ultrasensitive detection technologies for the identification of early and minimally invasive Alzheimer's disease blood biomarkers. *Journal of Alzheimer's Disease, 86*, 1337–1369.
58. Mielke, M. M., Hagen, C. E., Wennberg, A. M. V., Airey, D. C., Savica, R., Knopman, D. S., Machulda, M. M., Roberts, R. O., Jack, C. R., Jr., Petersen, R. C., & Dage, J. L. (2017). Association of plasma total tau level with cognitive decline and risk of mild cognitive impairment or dementia in the Mayo Clinic study on aging. *JAMA Neurology, 74*, 1073–1080.
59. Fossati, S., Cejudo, J. R., Debure, L., Pirraglia, E., Sone, J. Y., Li, Y., Chen, J., Butler, T., Zetterberg, H., Blennow, K., & de Leon, M. J. (2019). Plasma tau complements CSF tau and pTau in the diagnosis of Alzheimer's disease. *Alzheimer's and Dementia: Diagnosis, Assessment and Disease Monitoring, 11*, 483–492.
60. Jiao, B., Liu, H., Guo, L., Liao, X., Zhou, Y., Weng, L., Xiao, X., Zhou, L., Wang, X., Jiang, Y., Yang, Q., Zhu, Y., Zhou, L., Zhang, W., Wang, J., Yan, X., Tang, B., & Shen, L. (2021). Performance of plasma amyloid β, total tau, and neurofilament light chain in the identification of probable Alzheimer's disease in South China. *Frontiers in Aging Neuroscience, 13*, 749649-1-10.
61. Zabala-Findlay, A., Penny, L. K., Lofthouse, R. A., Porter, A. J., Palliyil, S., Harrington, C. R., Wischik, C. M., & Arastoo, M. (2023). Utility of blood-based tau biomarkers for mild cognitive impairment and Alzheimer's disease: Systematic review and meta-analysis. *Cells, 12*, 1184-1-16.
62. Yang, C. C., Chiu, M. J., Chen, T. F., Chang, H. L., Liu, B. H., & Yang, S. Y. (2018). Assay of plasma phosphorylated tau protein (threonine 181) and total tau protein in early-stage Alzheimer's disease. *Journal of Alzheimer's Disease, 61*, 1323–1332.

63. Chiu, M. J., Yang, S. Y., Chen, T. F., Lin, C. H., Yang, F. C., Chen, W. P., Zetterberg, H., & Blennow, K. (2021). Synergistic association between plasma $A\beta_{1-42}$ and p-tau in Alzheimer's disease but not in Parkinson's disease or frontotemporal dementia. *ACS Chemical Neuroscience, 12*, 1376–1383.
64. Chen, L., Niu, X., Wang, Y., Lv, S., Zhou, X., Yang, Z., & Peng, D. (2022). Plasma tau proteins for the diagnosis of mild cognitive impairment and Alzheimer's disease: A systematic review and meta-analysis. *Frontiers in Aging Neuroscience, 14*, 942629-1-10.
65. Liu, H. C., Lin, W. C., Chiu, M. J., Lu, C. H., Lin, C. Y., & Yang, S. Y. (2020). Development of an assay of plasma neurofilament light chain utilizing immunomagnetic reduction technology. *PLoS One, 15*, e0234519-1-17.
66. Chao, S. P., Lin, W. C., Lu, C. H., Wu, H. C., Liu, H. C., & Hu, C. J. (2024). Changes in plasma biomarkers differentiate clinical stages of Alzheimer's disease using immunomagnetic reduction assays. *Journal of Alzheimer's Disease, 102*, 459–469.
67. Arranz, J., Zhu, N., Rubio-Guerra, S., Rodriguez-Baz, I., Ferrer, R., Carmona-Iragui, M., Barroeta, I., Illan-Gala, I., Santos-Santos, M., Fortea, J., Lleo, A., Tondo, M., & Alcolea, D. (2024). Diagnostic performance of plasma pTau217, pTau181, $A\beta_{1-42}$ and $A\beta_{1-40}$ in the LUMIPULSE automated platform for the detection of Alzheimer disease. *Alzheimer's Research and Therapy, 16*, 139-1-15.
68. Quaresima, v., Pilotto, A., Trasciatti, C., Tolassi, C., Parigi, M., Bertoli, D., Mordenti, C., Galli, A., Rizzardi, A., Caratozzolo, S., Benussi, A., Ashton, N. J., Blennow, K., Zetterberg, H., Giliani, S., Brugnoni, D., & Padovani, A. (2024). Plasma pTau181 and amyloid markers in Alzheimer's disease: A comparison between Lumipulse and SIMOA. *Neurobiology of Aging, 143*, 30–40.
69. Martínez-Dubarbie, F., Guerra-Ruiz, a., López-García, S., Lage, C., Fernández-Matarrubia, M., Nevado-Cáceres, A., Rivera-Sánchez, M., Valera-Barrero, A., Pozueta-Cantudo, A., García-Martínez, M., Corrales-Pardo, A., Bravo, M., López-Hoyos, M., Irure-Ventura, J., de Lucas, E. M., Drake-Pérez, M., Cahuana-Santamaría, N. H., García-Unzueta, M. T., Sánchez-Juan, P., & Rodríguez-Rodríguez, E. (2025). Diagnostic performance of plasma pTau217 in a memory clinic cohort using the Lumipulse automated platform. *Alzheimer's Research and Therapy, 17*, 68-1-16.
70. Thijssen, E. H., Verberk, I. M. W., Kindermans, J., Abramian, A., Vanbrabant, J., Ball, A. J., Pijnenburg, Y., Lemstra, A. W., van der Flier, W. M., Stoops, E., Hirtz, C., & Teunissen, C. E. (2022). Differential diagnostic performance of a panel of plasma biomarkers for different types of dementia. *Alzheimer's and Dementia, 14*, e12285-1-11.
71. Palmqvist, S., Janelidze, S., Stomrud, E., Zetterberg, H., Karl, J., Zink, K., Bittner, T., Mattsson, N., Eichenlaub, U., Blennow, K., & Hansson, O. (2019). Performance of fully automated plasma assays as screening tests for Alzheimer disease-related β-amyloid status. *JAMA Neurology, 76*, 1060–1069.
72. Li, C. Y., Fan, L. Y., Lin, C. H., Hu, C. J., & Chiu, M. J. (2025). Ultrasensitive assays detect different conformations of plasma β amyloids. *ACS Omega., 10*, 7256–7263.
73. Lannfelt, L. (2023). A light at the end of the tunnel - from mutation identification to a potential treatment for Alzheimer's disease. *Upsala Journal of Medical Sciences, 128*(1), e10316
74. Chen, Z., Singh, P. K., Calvano, M., Norris, E. H., & Strickland, S. (2023). A possible mechanism for the enhanced toxicity of beta-amyloid protofibrils in Alzheimer's disease. *Proceedings of the National Academy of Sciences, 120*, e2309389120-1-3.
75. Yang, S. Y., Liu, H. C., & Chen, W. P. (2020). Immunomagnetic reduction detects plasma $A\beta_{1-42}$ levels as a potential dominant indicator predicting cognitive decline. *Neurology and Therapy, 9*, 435–442.
76. Hsieh, C. H., Ko, C. A., Liang, C. S., Yeh, P. K., Tsai, C. K., Tsai, C. L., Lin, G. Y., Lin, Y. K., Tsai, M. C., & Yang, F. C. (2024). Longitudinal assessment of plasma biomarkers for early detection of cognitive changes in subjective cognitive decline. *Frontiers in Aging Neuroscience, 16*, 1389595-1-11.

77. Zilka, N., Filipcik, P., Koson, P., Fialova, L., Skrabana, R., Zilkova, M., Rolkova, G., Kontsekova, E., & Novak, M. (2006). Truncated tau from sporadic Alzheimer's disease suffices to drive neurofibrillary degeneration in vivo. *FEBS Letters, 580*, 3582–3588.
78. Gu, J., Xu, W., Jin, N., Li, L., Zhou, Y., Chu, D., Gong, C. X., Iqbal, K., & Liu, F. (2020). Truncation of tau selectively facilitates its pathological activities. *The Journal of Biological Chemistry, 295*, 13812–13828.
79. Liu, P., Smith, B. R., Montonye, M. L., Kemper, L. J., Leinonen-Wright, K., Nelson, K. M., Higgins, L. A., Guerrero, C. R., Markowski, T. W., Zhao, X., Petersen, A. J., Knopman, D. S., Petersen, R. C., & Ashe, K. H. (2020). A soluble truncated tau species related to cognitive dysfunction is elevated in the brain of cognitively impaired human individuals. *Scientific Reports, 10*, 3869-1-18.
80. Zabala-Findlay, A., Penny, L. K., Lofthouse, R. A., Porter, A. J., Palliyil, S., Harrington, C. R., Wischik, C. M., Arastoo, M. (2023). Utility of blood-based tau biomarkers for mild cognitive impairment and Alzheimer's disease: Systematic review and meta-analysis. *Cells, 12*(8), 1184
81. Hanger, D. P., Hughes, K., Woodgett, J. R., Brion, J. P., & Anderton, B. H. (1992). Glycogen synthase kinase-3 induces Alzheimer's disease-like phosphorylation of tau: Generation of paired helical filament epitopes and neuronal localisation of the kinase. *Neuroscience Letters, 147*, 58–62.
82. Lovestone, S., Reynolds, C. H., Latimer, D., Davis, D. R., Anderton, B. H., Gallo, J. M., Hanger, D., Mulot, S., Marquardt, B., Stabel, S., Woodgett, J. R., & Miller, C. C. J. (1994). Alzheimer's disease-like phosphorylation of the microtubule-associated protein tau by glycogen synthase kinase-3 in transfected mammalian cells. *Current Biology, 4*, 1077–1086.
83. Le, X. F., Claret, F. X., Lammayot, A., Tian, L., Deshpande, D., LaPushin, R., Tari, A. M., & Bast, R. C., Jr. (2003). The role of cyclin-dependent kinase inhibitor p27Kip1 in anti-HER2 antibody-induced G1 cell cycle arrest and tumor growth inhibition. *The Journal of Biological Chemistry, 278*, 23441–23450.
84. Watanabe, K., Uemura, K., Asada, M., Maesako, M., Akiyama, H., Shimohama, S., Takahashi, R., & Kinoshita, A. (2015). The participation of insulin-like growth factor-binding protein 3 released by astrocytes in the pathology of Alzheimer's disease. *Molecular Brain, 8*, 82-1-14.
85. Dorsey, E. R., Constantinescu, R., Thompson, J. P., Biglan, K. M., Holloway, R. G., Kieburtz, K., Marshall, F. J., Ravina, B. M., Schifitto, G., Siderowf, A., & Tanner, C. M. (2007). Projected number of people with Parkinson disease in the most populous nations, 2005 through 2030. *Neurology, 68*, 384–386.
86. Del Tredici, K., & Braak, H. (2016). Review: Sporadic Parkinson's disease: Development and distribution of α-synuclein pathology. *Neuropathology and Applied Neurobiology, 42*, 33–50.
87. Munhoz, R. P., Moro, A., Silveira-Moriyama, L., & Teive, H. A. (2015). Non-motor signs in Parkinson's disease: A review. *Arquivos de Neuro-Psiquiatria, 73*, 454–462.
88. Gomperts, S. N. (2016). Lewy body dementias: Dementia with Lewy bodies and Parkinson disease dementia. *Continuum (Minneap Minn)., 22*, 435–463.
89. Menšíková, K., Matěj, R., Colosimo, C., Rosales, R., Tučková, L., Ehrmann, J., Hraboš, D., Kolaříková, K., Vodička, R., Vrtěl, R., Procházka, M., Nevrlý, M., Kaiserová, M., Kurčová, S., Otruba, P., & Kaňovský, P. (2022). Lewy body disease or diseases with Lewy bodies? *NPJ Parkinsons Disease, 8*, 3-1-11.
90. Martin, W. R. W., Younce, J. R., Campbell, M. C., Racette, B. A., Norris, S. A., Ushe, M., Criswell, S., Davis, A. A., Alfradique-Dunham, I., Maiti, B., Cairns, N. J., Perrin, R. J., Kotzbauer, P. T., & Perlmutter, J. S. (2023). Neocortical Lewy body pathology parallels Parkinson's dementia, but not always. *Annals of Neurology, 93*, 184–195.
91. Kung, M. P., Stevenson, D. A., Plössl, K., Meegalla, S. K., Beckwith, A., Essman, W. D., Mu, M., Lucki, I., & Kung, H. F. (1997). [99mTc]TRODAT-1: A novel technetium-99m complex as a dopamine transporter imaging agent. *European Journal of Nuclear Medicine, 24*, 372–380.

92. Huang, W. S., Lin, S. Z., Lin, J. C., Wey, S. P., Ting, G., & Liu, R. S. (2001). Evaluation of early-stage Parkinson's disease with 99mTc-TRODAT-1 imaging. *Journal of Nuclear Medicine, 42*, 1303–1308.
93. de la Fuente-Fernández, R. (2012). Role of DaTSCAN and clinical diagnosis in Parkinson disease. *The Neurologist, 78*, 696–701.
94. Havelund, J. F., Heegaard, N. H. H., Færgeman, N. J. K., & Gramsbergen, J. B. (2017). Biomarker research in Parkinson's disease using metabolite profiling. *Meta, 7*, 42-1-18.
95. Okochi, M., Walter, J., Koyama, A., Nakajo, S., Baba, M., Iwatsubo, T., Meijer, L., Kahle, P. J., & Haass, C. (2000). Constitutive phosphorylation of the Parkinson's disease associated alpha-synuclein. *The Journal of Biological Chemistry, 275*, 390–397.
96. Smith, W. W., Margolis, R. L., Li, X., Troncoso, J. C., Lee, M. K., Dawson, V. L., Dawson, T. M., Iwatsubo, T., & Ross, C. A. (2005). α-synuclein phosphorylation enhances eosinophilic cytoplasmic inclusion formation in SH-SY5Y cells. *The Journal of Neuroscience, 25*, 5544–5552.
97. Oh, Y., Kim, Y. M., Mouradian, M. M., & Chung, K. C. (2011). Human polycomb protein 2 promotes α-synuclein aggregate formation through covalent SUMOylation. *Brain Research, 1381*, 78–89.
98. Oftedal, L., Maple-Grødem, J., Tysnes, O. B., Alves, G., & Lange, J. (2023). Seed amplification assay as a diagnostic tool in newly-diagnosed Parkinson's disease. *Journal of Parkinson's Disease, 13*, 841–844.
99. Fang, C., Lv, L., Mao, S., Dong, H., & Liu, B. (2020). Cognition deficits in Parkinson's disease: Mechanisms and treatment. *Parkinsons Disease, 24*, 2076942-1-11.
100. Gonzalez-Latapi, P., Bayram, E., Litvan, I., & Marras, C. (2021). Cognitive impairment in Parkinson's disease: Epidemiology, clinical profile, protective and risk factors. *Behavioral Science, 11*, 74-1-22.
101. Auning, E., Rongve, A., Fladby, T., Booij, J., Hortobágyi, T., Siepel, F. J., Ballard, C., & Aarsland, D. (2011). Early and presenting symptoms of dementia with Lewy bodies. *Dementia and Geriatric Cognitive Disorders, 32*, 202–208.
102. McKeith, I. (2009). DLB and PDD: The same or different? Is there a debate? *International Psychogeriatrics, 21*, 220–224.
103. Gonzalez, M. C., Tovar-Rios, D. A., Alves, G., Dalen, I., Williams-Gray, C. H., Camacho, M., Forsgren, L., Bäckström, D., Lawson, R. A., Macleod, A. D., Counsell, C. E., Paquet, C., DeLena, C., D'Antonio, F., Pilotto, A., Padovani, A., Blanc, F., Fąlup-Pecurariu, C., Lewis, S. J. G., Rejdak, K., Papuc, E., Hort, J., Nedelska, Z., O'Brien, J., Bonanni, L., Marquié, M., Boada, M., Pytel, V., Abdelnour, C., Alcolea, D., Beyer, K., Tysnes, O. B., Aarsland, D., & Maple-Grødem, J. (2023). Cognitive and motor decline in dementia with Lewy bodies and Parkinson's disease dementia. *Movement Disorders Clinical Practice, 10*, 980–986.
104. Slaets, S., Van Acker, F., Versijpt, J., Hauth, L., Goeman, J., Martin, J. J., De Deyn, P. P., & Engelborghs, S. (2015). Diagnostic value of MIBG cardiac scintigraphy for differential dementia diagnosis. *International Journal of Geriatric Psychiatry, 30*, 864–869.
105. Sakakibara, R., Tateno, F., Aiba, Y., Ogata, T., Kishi, M., Terada, H., Inaoka, T., Nakatsuka, T., & Matsuoka, K. (2018). MIBG myocardial scintigraphy identifies premotor PD/DLB during a negative DAT scan period: Second report. *Movement Disorders Clinical Practice, 6*, 46–50.
106. Braak, H., & Tredici, K. D. (2017). Neuropathological staging of brain pathology in sporadic Parkinson's disease: Separating the wheat from the chaff. *Journal of Parkinson's Disease, 7*, S73–S87.
107. Koeglsperger, T., Rumpf, S. L., Schließer, P., Struebing, F. L., Brendel, M., Levin, J., Trenkwalder, C., Höglinger, G. U., & Herms, J. (2023). Neuropathology of incidental Lewy body & prodromal Parkinson's disease. *Molecular Neurodegeneration, 18*, 32-1-18.
108. Mastrangelo, A., Caldera, S., Mastenbroek, S. E., Vittoriosi, E., Janelidze, S., Serrano, G. E., Atri, A., Shill, H., Driver-Dunckley, E., Mehta, S., Adler, C. H., Mammana, A., Magliocchetti, F., Baiardi, S., Beach, T. G., Hansson, O., & Parchi, P. (2025). Quantification of Lewy body

pathology by cerebrospinal fluid endpoint dilution RT-QuIC in a neuropathological autopsy cohort of clinically heterogeneous participants. *Acta Neuropathologica, 149*, 67-1-13.

109. Chen, N. C., Chen, H. L., Li, S. H., Chang, Y. H., Chen, M. H., Tsai, N. W., Yu, C. C., Yang, S. Y., Lu, C. H., & Lin, W. C. (2020). Plasma levels of α-synuclein, Aβ-40 and T-tau as biomarkers to predict cognitive impairment in Parkinson's disease. *Frontiers in Aging Neuroscience, 12*, 112-1-8.
110. Chen, C. H., Lee, B. C., & Lin, C. H. (2020). Integrated plasma and neuroimaging biomarkers associated with motor and cognition severity in Parkinson's disease. *Journal of Parkinson's Disease, 10*, 77–88.
111. Chen, P. C., Yu, C. C., Chen, Y. S., Lu, C. H., Chan, S. H., Chou, K. H., Chen, M. H., Chen, M. H., & Lin, W. C. (2021). The potential effects of oxidative stress-related plasma abnormal protein aggregate levels on brain volume and its neuropsychiatric consequences in Parkinson's disease. *Oxidative Medicine and Cellular Longevity, 2021*, 3666327-1-10.
112. Huang, C. C., Chen, P. H., Tsai, C. C., Chiang, H. F., Hsieh, C. C., Chen, T. L., Liao, W. H., Chen, Y. L., & Wang, J. J. (2024). Diffusion and structural MRI as potential biomarkers in people with Parkinson's disease and cognitive impairment. *European Radiology, 34*, 126-1-10.
113. Lin, W. C., Lee, P. L., Lu, C. H., Lin, C. P., & Chou, K. H. (2021). Linking stage-specific plasma biomarkers to gray matter atrophy in Parkinson disease. *AJNR. American Journal of Neuroradiology, 42*, 1444–1451.
114. Yu, C. C., Lu, C. Y., Chen, M. H., Chen, Y. S., Lu, C. H., Lin, Y. Y., Chou, K. H., & Lin, W. C. (2021). Brain atrophy mediates the relationship between misfolded proteins deposition and cognitive impairment in Parkinson's disease. *Journal of Personalized Medicine, 11*, 702-1-15.
115. Ishii, R., Tokuda, T., Tatebe, H., Ohmichi, T., Kasai, T., Nakagawa, M., Mizuno, T., & El-Agnaf, O. M. A. (2015). Decrease in plasma levels of α-synuclein is evident in patients with Parkinson's disease after elimination of heterophilic antibody interference. *PLoS One, 10*, e0123162-1-15.
116. Gibb, W. R., & Lees, A. J. (1988). The relevance of the Lewy body to the pathogenesis of idiopathic Parkinson's disease. *Journal of Neurology, Neurosurgery, and Psychiatry, 51*, 745–752.
117. Lin, C. H., Yang, S. Y., Horng, H. E., Yang, C. C., Chieh, J. J., Chen, H. H., Liu, B. H., & Chiu, M. J. (2017). Plasma α-synuclein predicts cognitive decline in Parkinson's disease. *Journal of Neurology, Neurosurgery, and Psychiatry, 88*, 818–824.
118. Lin, C. H., Liu, H. C., Yang, S. Y., Yang, K. C., Wu, C. C., & Chiu, M. J. (2019). Plasma pS129-α-synuclein is a surrogate biofluid marker of motor severity and progression in Parkinson's disease. *Journal of Clinical Medicine, 8*, 1601-1-11.
119. Albillos, S. M., Montero, O., Calvo, S., Solano, B., Trejo, J. M., & Cubo, E. (2021). Can plasma α-synuclein help us to differentiate Parkinson's disease from essential tremor? *Tremor Other Hyperkin Movement, 11*, 1–8.
120. Ren, J., Pan, C., Wang, Y., Xue, C., Lin, H., Xu, J., Wang, H., Zhang, W., Xu, P., Chen, Y., & Liu, W. (2022). Plasma α-synuclein and phosphorylated tau 181 as a diagnostic biomarker panel for de novo Parkinson's disease. *Journal of Neurochemistry, 161*, 506–515.
121. Okuzumi, A., Hatano, T., Matsumoto, G., Nojiri, S., Ueno, S. I., Imamichi-Tatano, Y., Kimura, H., Kakuta, S., Kondo, A., Fukuhara, T., Li, Y., Funayama, M., Saiki, S., Taniguchi, D., Tsunemi, T., McIntyre, D., Gerardy, J. J., Mittelbronn, M., Kruger, R., Uchiyama, Y., Nukina, N., & Hattori, N. (2023). Propagative α-synuclein seeds as serum biomarkers for synucleinopathies. *Nature Medicine, 29*, 1448–1455.
122. Fan, Z., Pan, Y. T., Zhang, Z. Y., Yang, H., Yu, S. Y., Zheng, Y., Ma, J. H., & Wang, X. M. (2020). Systemic activation of NLRP3 inflammasome and plasma α-synuclein levels are correlated with motor severity and progression in Parkinson's disease. *Journal of Neuroinflammation, 17*, 11-1-10.

123. Lin, W. C., Chiu, P. Y., Chiu, M. J., Hu, C. J., Wang, P. N., Chen, T. F., Yang, F. C., Huang, L. K., Lu, C. H., Liu, H. C., & Yang, S. Y. (2025). Assay of neurodegenerative biomarkers in human plasma of patients with Parkinson's disease. *Journal of Biotechnology and Biomedicine, 8*, 159–169.
124. Lin, C. H., Yang, S. Y., Horng, H. E., Yang, C. C., Chieh, J. J., Chen, H. H., Liu, B. H., & Chiu, M. J. (2018). Plasma biomarkers differentiate Parkinson's disease from atypical parkinsonism syndromes. *Frontiers in Aging Neuroscience, 10*, 123-1-9.
125. Chen, Y. T., Orimo, S., Wei, C. Y., Hung, G. U., Yang, S. Y., & Chiu, P. Y. (2022). Synuclein motor dysfunction composite scale for the discrimination of dementia with Lewy bodies from Alzheimer's disease. *Frontiers in Aging Neuroscience, 14*, 920591-1-8.
126. Hoehn, M. M., & Yahr, M. D. (1967). Parkinsonism: Onset, progression and mortality. *The Neurologist, 17*, 427–442.
127. Chang, C. W., Yang, S. Y., Yang, C. C., Chang, C. W., & Wu, Y. R. (2019). Plasma and serum alpha-synuclein as a biomarker of diagnosis in patients with Parkinson's disease. *Frontiers in Neurology, 10*, 1388-1-7.
128. Lin, W. C., Lu, C. H., Chiu, P. Y., & Yang, S. Y. (2020). Plasma total α-synuclein and neurofilament light chain: Clinical validation for discriminating Parkinson's disease from normal control. *Dementia and Geriatric Cognitive Disorders, 49*, 401-1-9.
129. Lin, W. T., Shaw, J. S., Cheng, F. Y., & Chen, P. H. (2021). Plasma total tau predicts executive dysfunction in Parkinson's disease. *Acta Neurologica Scandinavica, 1*, 1-1-8.
130. Ng, A. S. L., Tan, Y. J., Lu, Z., Ng, E. Y. L., Ng, S. Y. E., Chia, N. S. Y., Setiawan, F., Xu, Z., Tay, K. Y., Prakash, K. M., Au, K. L., Tan, E. K., & Tan, L. C. S. (2019). Plasma alpha-synuclein detected by single molecule array is increased in PD. *Annals of Clinical Translational Neurology, 6*, 615–619.
131. Dumrikarnlert, C., Wachirutmangur, L., Udomphanthurak, S., Rattanabannakit, C., Srivanitchapoom, P., & Senanarong, V. (2023). Plasma alpha synuclein as a potent biomarker of diseases with synucleinopathies. *Siriraj Medical Journal, 75*, 864–870.
132. Niu, M., Li, Y., Li, G., Zhou, L., Luo, N., Yao, M., Kang, W., & Liu, J. (2020). A longitudinal study on α-synuclein in plasma neuronal exosomes as a biomarker for Parkinson's disease development and progression. *European Journal of Neurology, 27*, 967–974.
133. Chung, C. C., Chan, L., Chen, J. H., Hung, Y. C., & Hong, C. T. (2021). Plasma extracellular vesicle α-synuclein level in patients with Parkinson's disease. *Biomology, 11*, 744-1-10.
134. Malaty, G. R., Decourt, B., Shill, H. A., & Sabbagh, M. N. (2024). Biomarker assessment in Parkinson's disease dementia and dementia with Lewy bodies by the immunomagnetic reduction assay and clinical measures. *Journal of Alzheimer's Disease Reports, 8*, 1361–1371.
135. Ratnavalli, E., Brayne, C., Dawson, K., & Hodges, J. R. (2002). The prevalence of frontotemporal dementia. *Neurology, 58*, 1615–1621.
136. Onyike, C. U., & Diehl-Schmid, J. (2013). The epidemiology of frontotemporal dementia. *International Review of Psychiatry, 25*, 130–137.
137. Chen, X., Chen, Y., Ni, B., & Huang, C. (2024). Research trends and hotspots for frontotemporal dementia from 2000 to 2022: a bibliometric analysis. *Frontiers in Neurology, 15*, 1399600-1-15.
138. Banović, S., & Sinanović, O. (2021). Speech and language abilities of persons with frontotemporal dementia. *Psychiatria Danubina, 33*(Suppl 13), s183–s188.
139. Samra, K., MacDougall, A. M., Bouzigues, A., Bocchetta, M., Cash, D. M., Greaves, C. V., Convery, R. S., van Swieten, J. C., Seelaar, H., Jiskoot, L., Moreno, F., Sanchez-Valle, R., Laforce, R., Graff, C., Masellis, M., Tartaglia, M. C., Rowe, J. B., Borroni, B., Finger, E., Synofzik, M., Galimberti, D., Vandenberghe, R., de Mendonça, A., Butler, C. R., Gerhard, A., Ducharme, S., Le Ber, I., Tiraboschi, P., Santana, I., Pasquier, F., Levin, J., Otto, M., Sorbi, S., Rohrer, J. D., & Russell, L. L. (2023). Language impairment in the genetic forms of behavioural variant frontotemporal dementia. *Journal of Neurology, 270*, 1976–1988.
140. Rohrer, J. D., & Zetterberg, H. (2014). Biomarkers in Frontotemporal Dementia. *Biomarkers in Medicine, 8*(4), 519–521.

141. Swift, I. J., Synofzik, M., Otto, M., Sogorb-Esteve, A., Heller, C., Graff, C., Galimberti, D., Todd, E., Heslegrave, A. J., van der Ende, E. L., Van Swieten, J. C., Zetterberg, H., & Rohrer, J. D. (2021). Fluid biomarkers in frontotemporal dementia: Past, present and future. *Neurology, Neurosurgery & Psychiatry, 92*, 204–215.
142. Praschan, A. G. N., Newhouse, A., & Chemali, Z. (2023). Biomarkers in frontotemporal dementia: Current landscape and future directions. *Biomarkers in Neuropsychiatry, 8*, 100065-1-14.
143. Tartaglia, M. C., & Mackenzie, I. R. A. (2022). Recent advances in frontotemporal dementia. *Canadian Journal of Neurological Sciences*, 1–10.
144. Yang, S. Y., Chiu, M. J., Chen, T. F., Lin, C. H., Jeng, J. S., Tang, S. C., Lee, Y. F., Yang, C. C., Liu, B. H., Chen, H. H., Wu, C. C. (2017). Analytical performance of reagent for assaying tau protein in human plasma and feasibility study screening neurodegenerative diseases. *Science Reports, 7*, 9304
145. Foiani, M. S., Woollacott, I. O., Heller, C., Bocchetta, M., Heslegrave, A., Dick, K. M., Russell, L. L., Marshall, C. R., Mead, S., Schott, J. M., Fox, N. C., Warren, J. D., Zetterberg, H., & Rohrer, J. D. (2018). Plasma tau is increased in frontotemporal dementia. *Journal of Neurology, Neurosurgery, and Psychiatry, 89*, 804–807.
146. Ill'an-Gala, I., Lleo, A., Karydas, A., Staffaroni, A. M., Zetterberg, H., Sivasankaran, R., Grinberg, L. T., Spina, S., Kramer, J. H., Ramos, E. M., Coppola, G., Joie, R. A., Rabinovici, G. D., Perry, D. C., Gorno-Tempini, M. L., Seeley, W. W., Miller, B. L., Rosen, H. J., Blennow, K., Boxer, A. L., & Rojas, J. C. (2021). Plasma tau and neurofilament light in frontotemporal lobar degeneration and Alzheimer disease. *The Neurologist, 96*, e671–e683.
147. Yang, S. Y., Liu, H. C., Lin, C. Y., Chiu, M. J., Chen, T. F., Lin, C. H., Chen, H. H. (2020). Development of assaying plasma TDP-43 utilizing immunomagnetic reduction. *Journal of Neurological Disorders, 8*, 7
148. Chatterjee, M., Ozdemir, S., Fritz, C., Mobius, W., Kleineidam, L., Mandelkow, E., Biernat, J., Doğdu, C., Peters, O., Cosma, N. C., Wang, X., Schneider, L. S., Priller, J., Spruth, E., Kuhn, A. A., Krause, P., Klockgether, T., Vogt, I. R., Kimmich, O., Spottke, A., Hoffmann, D. C., Fliessbach, K., Miklitz, X., McCormick, C., Weydt, P., Falkenburger, B., Brandt, M., Guenther, R., Dinter, E., Wiltfang, J., Hansen, N., Bahr, M., Zerr, I., Floel, A., Nestor, P. J., Duzel, E., Glanz, W., Incesoy, E., Burger, K., Janowitz, D., Perneczky, R., Rauchmann, B. S., Hopfner, F., Wagemann, O., Levin, J., Teipel, S., Kilimann, I., Goerss, D., Prudlo, J., Gasser, T., Brockmann, K., Mengel, D., Zimmermann, M., Synofzik, M., Wilke, C., Selma-Gonzalez, J., Turon-Sans, J., Santos-Santos, M. A., Alcolea, D., Rubio-Guerra, S., Fortea, J., Carbayo, A., Lleo, A., Rojas-Garcia, R., Illan-Gala, I., Wagner, M., Frommann, I., Roeske, S., Bertram, L., Heneka, M. T., Brosseron, F., Ramirez, A., Schmid, M., Beschorner, R., Halle, A., Herms, J., Neumann, M., Barthelemy, N. R., Bateman, R. J., Rizzu, P., Heutink, P., Dols-Icardo, O., Hoglinger, G., Hermann, A., & Schneider, A. (2024). Plasma extracellular vesicle tau and TDP-43 as diagnostic biomarkers in FTD and ALS. *Nature Medicine, 30*, 1771–1783.
149. Foulds, P., McAuley, E., Gibbons, L., Davidson, Y., Pickering-Brown, S. M., Neary, D., Snowden, J. S., Allsop, D., Mann, D. M. (2008). TDP-43 protein in plasma may index TDP-43 brain pathology in Alzheimer's disease and frontotemporal lobar degeneration. *Acta Neuropathologica, 116*, 141–146

Chapter 5
Clinical Applications of Blood Tests

Abstract In 2024, the Alzheimer's Association recommended blood-based biomarker tests in the revised guidelines of ATN criteria for the clinical diagnosis of Alzheimer's disease. Blood-based biomarkers formally play a role in diagnostic practice. Several potential uses of blood-based biomarkers for Alzheimer's disease have been demonstrated, such as risk evaluation, pre-screening for amyloid PET, prediction of disease progression, monitoring of treatments/interventions, etc. Depending on the intended uses, adequate biomarkers are suggested. Such clinical applications using blood-based biomarkers are feasible for Parkinson's disease, dementia with Lewy bodies, frontotemporal dementia, and poststroke dementia. The authors have been using blood-based biomarkers in routine practice for many years. Valuable experiences in solving clinical puzzles using blood-based biomarkers are presented in this chapter. Examples are given in this chapter to help readers learn when blood-based biomarker tests are needed for clients, what kinds of biomarkers should be measured, and how the test results should be correctly interpreted. The contents also outline the prospects and challenges of blood-based biomarkers in real-world practice.

Keywords Clinical practice · Blood-based biomarkers · Alzheimer's disease · Parkinson's disease · Frontotemporal dementia · Poststroke dementia

5.1 Potential Roles of Blood Tests in Clinical Practice

Among neurodegenerative diseases, Alzheimer's disease (AD) is the most advanced with the use of plasma biomarkers in terms of clinical utility and test validation. Similar to other types of neurodegenerative diseases, AD was diagnosed based on clinical symptoms until the advent of, approval, and reimbursement of positron emission tomography, CSF assessment, and now blood-based biomarkers (BBBMs). Over time there has been growing interest in exploring the potential use of pathological proteins in the diagnosis of AD. However, many years are needed to develop

C. S.-Y. Yang, M. N. Sabbagh, *Assessing Neurodegenerative Diseases: High-Sensitivity Technologies That Assay Pathological Biomarkers in Blood*,
https://doi.org/10.1007/978-3-032-23863-4_5

clear recommendations for the use of plasma biomarkers in clinical practice. The milestones in the use of plasma biomarkers for the diagnosis of AD started in 2010.

In 2010, Jack et al. proposed a hypothetical model of dynamic biomarkers of the Alzheimer's pathological cascade, in which the time-cost $A\beta_{1-42}$ and T-Tau levels in cerebrospinal fluid (CSF) were simulated [1]. CSF Aβ biomarkers become abnormal first, before other neurodegenerative biomarkers and cognitive symptoms, whereas neurodegenerative biomarkers such as tau protein become abnormal later and correlate with clinical symptom severity. The hypothetical model was confirmed by Bateman et al. two years later. Using data from the Dominantly Inherited Alzheimer's Network (DIAN) cohort, a group of patients that have a high risk of developing young-onset AD because of presenilin mutations, their group reported that concentrations of $A\beta_{1-42}$ in CSF appear to be abnormal 25 years before expected symptom onset [2]. Aβ deposition as observed by amyloid positron emission tomography (PET) can be detected 15 years before the expected symptom onset. Increased concentrations of tau protein in CSF and an increase in brain atrophy can be detected 15 years before the expected symptom onset. Cerebral hypometabolism and impaired episodic memory can be observed 10 years before the expected symptom onset. Global cognitive impairment, as measured by the Mini-Mental State Examination (MMSE) and the Clinical Dementia Rating (CDR) score, can be detected 5 years before the expected symptom onset, and patients may meet the diagnostic criteria for dementia at an average of 3 years after the expected symptom onset. Thus, the roles of pathological biomarkers in the CSF in the development of AD have been identified. Importantly the findings were significant in that biomarkers that signal pathological changes of AD precede the onset of symptoms and that the clinical syndrome is the advanced stage of the disease. Notably, CSF $A\beta_{1-42}$ and tau protein are critical biomarkers.

With the accumulation of evidence that correlates CSF $A\beta_{1-42}$ and tau protein with the development of AD in the following years, the Alzheimer's Association formally announced the research framework toward a biological definition of Alzheimer's disease in 2018 [3]. In this research framework, the diagnosis is not based on the clinical symptoms of AD; instead, the definition of AD is shifted to a biological construct. The research framework focuses on the diagnosis of AD with biomarkers in living persons. CSF biomarkers are grouped into those of Aβ deposition (A: $A\beta_{1-42}$ or $A\beta_{1-42}/A\beta_{1-40}$), pathological tau (T: pTau), and neurodegeneration (N: T-Tau), termed the ATN criteria. Remarkably, the Alzheimer's continuum is determined with only positive $A\beta_{1-42}$ or $A\beta_{1-42}/A\beta_{1-40}$. Once $A\beta_{1-42}$ or $A\beta_{1-42}/A\beta_{1-40}$ is not positive, the assessments of CSF biomarkers would be normal or indicate pathological changes not related to AD. This proposal for the use of the ATN criteria was a profound shift away from defining AD by clinical phenotype to defining AD pathologically. The precedent for shifting from a clinical syndrome to a biological definition has happened with other conditions as well [4].

Since 2018, with the use of ultrasensitive assay technologies in clinical studies and trials, it has become increasingly common and accepted, and many results that use plasma biomarkers related to the development of AD have been reported. Not only CSF but also plasma biomarkers have been proposed as part of the diagnosis of AD [5]. In 2020s, several studies have reported on plasma biomarkers that augment the diagnosis of AD. Plasma $A\beta_{1-42}$, pTau, and nonphosphorylated mid-region fragments of tau protein are key plasma biomarkers for revealing the status of amyloid and tau pathologies in the revised criteria for the diagnosis and staging of AD.

In 2024, the Alzheimer's Association (AA) revised the criteria to define Alzheimer's disease (AD) as a biological continuum identified by specific biomarkers, rather than by the clinical symptoms alone. The ATN criteria were expanded to include an "X" for additional pathologies, forming the ATX(N) framework where "X" could include pathologies such as synuclein, TDP-43, inflammation, and vascular brain injury [5].

In the revised criteria, biomarkers were stratified into tiers (Core 1 and Core 2). Core 1 biomarkers represent abnormal changes early in the disease process and are likely to be sufficient for diagnosis. Examples include amyloid PET scans, approved cerebrospinal fluid (CSF) assays, and highly accurate $A\beta_{1-42}$, plasma (blood-based) pTau217 tests. Core 2 biomarkers become abnormal later and are intended to be used to stage biological severity when combined with Core 1 biomarkers. Examples include tau PET imaging and specific phosphorylated tau (pTau) blood/CSF markers (e.g., pTau217, pTau181, MTBR-Tau243, non-phosphorylated mid-region tau fragments).

The Alzheimer's Association has put forth position statements as guidance documents to suggest clinical practice guidelines for the use of blood-based biomarkers in the diagnostic workup of suspected AD within specialized care settings [6]. Two key recommendations are given in these guidelines. First, blood-based biomarker tests with ≥90% clinical sensitivity and ≥75% clinical specificity can be used as triage tests. Second, blood-based biomarker tests with ≥90% clinical sensitivity and specificity can serve as substitutes for amyloid PET imaging or CSF AD biomarker testing in patients with cognitive impairment presenting for specialized care for memory disorders. Ultimately, plasma biomarkers, particularly $A\beta_{1-42}$, pTau217, and nonphosphorylated mid-region fragments of tau protein, are useful in the triage or confirmatory diagnosis of AD in clinical practice.

The use of blood biomarkers in the triage/confirmatory diagnosis of AD has opened a new era for early care for patients with neurodegenerative diseases, incrementally decreasing the barriers associated with identifying patients with early-stage AD. In addition, blood tests are powerful tools for managing brain health with the speculation that they could be predictive and might open a window of opportunity to detect pathological changes before symptom onset. However, the issues associated with caring for patients with neurodegenerative diseases are not limited to the demand for convenient tests for early-stage diagnosis. Other issues such as the prediction of cognitive decline in cognitively unimpaired (CU) individuals, monitoring of disease progression in early-stage patients, and monitoring responses to interventions/treatments are also urgent and could potentially be addressed with the aid of plasma biomarkers. Researchers and clinicians are searching for solutions involving blood tests for these issues.

In fact, many clinical studies and trials have demonstrated that the clinical uses of plasma biomarkers are not limited to the triage/confirmatory diagnosis of

AD. Potential applications such as the prediction of cognitive decline in CU individuals, monitoring of disease progression in those with amnestic mild cognitive impairment (aMCI), and monitoring responses to interventions/treatments, among others, using plasma biomarkers are feasible. Such potential applications of plasma biomarkers in clinical practice have also been proposed for Parkinson's disease (PD), dementia with Lewy bodies (DLB), amyotrophic lateral sclerosis (ALS), and frontotemporal dementia (FTD). In this chapter, we provide a comprehensive review of potential applications for plasma biomarkers in clinical practice with the aim of allowing blood tests to contribute more to the prevention, diagnosis, and management of neurodegenerative diseases.

5.2 Clinical Utility of Plasma Biomarkers in Alzheimer's Disease

Owing to their high convenience, high safety, low cost, and high availability, assays of pathological biomarkers in blood are expected to become more broadly used in clinical practice. The intended uses of plasma biomarkers in patients with Alzheimer's disease (AD) could include initial evaluation for AD, preliminary evaluation prior to ordering amyloid positron emission tomography (PET) findings, potential prediction of disease progression, and ideally monitoring of response to interventions/treatments. We aim to show the feasibility of these clinical uses for plasma biomarkers. Importantly, expectations of biomarkers have been amplified to achieve all aspects of diagnosis, progression, and theragnosis when in fact individual tests are unlikely to be able to perform in all manners.

5.2.1 Risk Evaluation for AD

According to the clinical diagnosis based on the NIA-AA or IWG guidelines [7–10], the development of AD progresses through various stages of severity. The NIA-AA Six-Stage Clinical Framework (for A+ individuals) contextualizes the continuity from biomarker-only change to clinical manifestations. This framework focuses on individuals with biomarker evidence (A+) of AD pathology.

Stage 1: Preclinical/asymptomatic (biomarker-positive). AD pathology (amyloid/tau) is present in the brain, but the person is cognitively normal.
Stage 2: Mild cognitive impairment (MCI).
Stage 3: Mild dementia.
Stage 4: Moderate dementia.
Stage 5: Moderately severe dementia.
Stage 6: Severe dementia.

Clinically, the physician or practitioner focuses on discriminations between cognitively unimpaired (CU) individuals, patients with amnestic mild cognitive impairment (aMCI), and patients with Alzheimer's disease dementia (loss of independence stemming from cognitive impairment) (AD). Recently, the demand to discriminate between positive and negative amyloid pathology in CU individuals, i.e., preclinical AD, has increased and promoted the development of preclinical treatments. Hence, plasma biomarkers able to stage CU individuals with negative amyloid pathology, CU individuals with positive amyloid pathology (preclinical AD), and individuals with aMCI and AD are preferred for the evaluation of risk for AD in clinical practice. Various cutoff values for biomarker levels for identifying disease stages are needed.

In terms of assay technologies, various plasma biomarkers, such as $A\beta_{1-42}$, $A\beta_{1-40}$, T-Tau, pTau181, and pTau217, alone or in combination, have been validated to be useful for such discrimination. In addition to the clinical accuracy of this discrimination, the consistency among studies and sites matters in clinical practice. Without limited variation among studies, e.g., <20% or lower, the measured levels of plasma biomarkers in CU individuals and individuals with preclinical AD, aMCI, or AD would be very heterogeneous from site to site. Thus, it would be difficult to have universal cutoff values to discriminate these stages using plasma biomarkers without limited variations. As discussed in "**Clinical Validation of Assays for Alzheimer's Disease - Plasma Biomarkers vs. Neuropsychological Tests**" in Chap. 4, the assays using magnetic particle-based chemiluminescent enzyme immunoassay (MP-CLEIA) show study-to-study variation of 5% in plasma $A\beta_{1-40}$ and $A\beta_{1-42}$ in CU individuals enrolled in three independent studies. Assays using immunomagnetic reduction (IMR) showed study-to-study variation of 8.1%, 4.1%, and 19.4% in plasma $A\beta_{1-40}$, $A\beta_{1-42}$, and T-Tau, respectively, in CU individuals enrolled in 30 independent studies. IMR also showed study-to-study variation of 18.2% in the plasma level of pTau181 in CU individuals enrolled in nine independent studies. Assays using single molecule array (SIMOA) revealed study-to-study variation of 18.4%, 61.5%, and 45.7% in plasma $A\beta_{1-40}$, $A\beta_{1-42}$, and T-Tau, respectively, in CU individuals enrolled in 13 independent studies. SIMOA also showed 130% variation in plasma pTau181 levels in CU individuals enrolled in three independent studies. Reducing variation for plasma pTau181 is important since emerging evidence suggests that a normal value has clinical relevance. pTau181 as a BBBM is highly effective at ruling out Alzheimer's amyloid pathology, showing excellent negative predictive values (NPV), often over 90% (e.g., 97.9% for pTau181 test using ME-ECL technology) in primary care settings. This means a negative result strongly suggests no significant amyloid plaque buildup. This high NPV helps clinicians rule out early Alzheimer's disease, reducing the need for invasive PET scans or spinal taps and guiding patients to appropriate care pathways for other causes of cognitive decline.

Thus, MP-CLEIA appears to show greater consistency among studies in terms of assaying plasma biomarkers. However, the number of reported studies is limited to only three. Moreover, the results involving the use of MP-CLEIA are not complete enough to discriminate CU individuals and individuals with preclinical AD, aMCI,

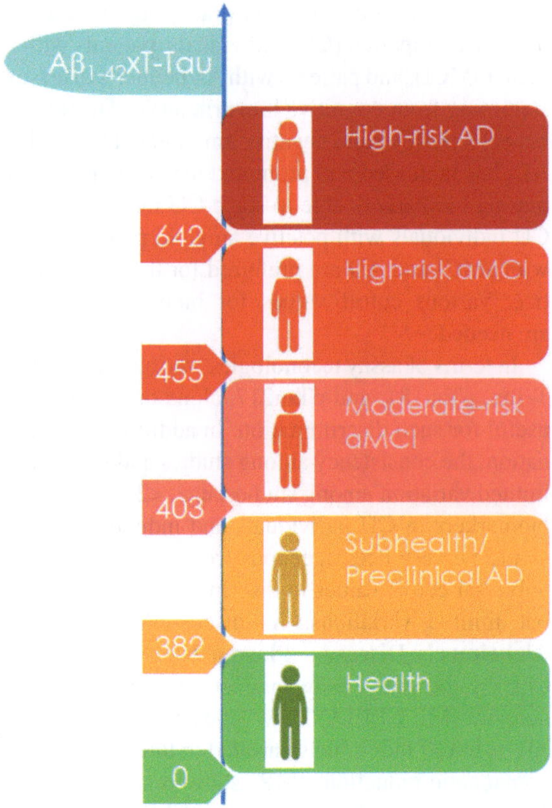

Fig. 5.1 Plasma $A\beta_{1-42}$xT-Tau levels assayed by IMR for discriminating the staging risk of health, subhealth/preclinical AD, aMCI, and AD [11–17]

or AD. More studies using MP-CLEIA are needed to support such discrimination in the future. Fortunately, IMR shows reasonable variation in the measured plasma biomarker levels among studies (<20%). The correlation between plasma $A\beta_{1-42}$xT-Tau and disease severity according to published results from studies in 2013–2025, in which more than 2000 participants were enrolled, involving the use of IMR is shown in Fig. 5.1 [11–17].

When the plasma $A\beta_{1-42}$xT-Tau concentration ranges from 0 to 382 $(pg/ml)^2$, the subject does not have a risk of amnestic mild cognitive impairment (aMCI) or AD. This population is referred to as the healthy group. When the plasma $A\beta_{1-42}$xT-Tau concentration ranges from 382 to 403 $(pg/ml)^2$, clinical studies have revealed that this population is cognitively normal but very prone to positive amyloid PET findings, i.e., preclinical AD. Studies have also shown that 26% of individuals in this population progress to aMCI within 1–2 years, although they are cognitively normal at baseline. The details are discussed in "**Prediction of aMCI Development in Cognitively Unimpaired Subjects**." Hence, they are referred to as the sub-healthy groups.

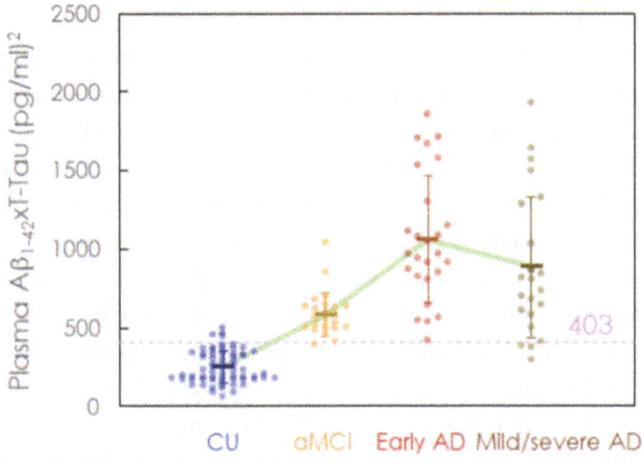

Fig. 5.2 Plasma $A\beta_{1-42}$xT-Tau levels assayed by IMR in the CU, aMCI, early AD and mild or severe AD groups

A plasma $A\beta_{1-42}$xT-Tau concentration greater than 403 (pg/ml)^2 indicates aMCI. Notably, the frequencies of clinical symptoms and the disease severity could vary among aMCI patients. Hence, plasma $A\beta_{1-42}$xT-Tau levels differ between mild-risk and high-risk aMCI patients. In the case of mild-risk aMCI, the plasma $A\beta_{1-42}$xT-Tau concentration ranges from 403 to 455 (pg/ml)^2. In the case of high-risk aMCI, the plasma $A\beta_{1-42}$xT-Tau concentration ranges from 455 to 642 (pg/ml)^2. Once the plasma $A\beta_{1-42}$xT-Tau concentration is greater than 642 (pg/ml)^2, the population is at risk of early AD dementia.

Interestingly, plasma $A\beta_{1-42}$xT-Tau levels are reduced in some patients with mild or severe AD, as shown in Fig. 5.2. The average values of plasma $A\beta_{1-42}$xT-Tau in the CU, aMCI, early AD, and mild/severe AD groups are indicated by the green line. These values increase from the CU group to the aMCI and early AD groups and then decreases in the mild/severe AD group. Some mild/severe AD patients have a plasma $A\beta_{1-42}$xT-Tau concentration lower than 403 (pg/ml)^2, as indicated by the pink dashed line in Fig. 5.2. False-negative results could be obtained for mild or severe AD by assaying plasma $A\beta_{1-42}$xT-Tau using IMR. One possible reason for the low levels of plasma $A\beta_{1-42}$xT-Tau in mild or severe AD is the so-called burn-out effect, in which abundant neurons vanish because of death, resulting in the reduced release of $A\beta_{1-42}$ and T-Tau.

If an individual has a plasma $A\beta_{1-42}$xT-Tau concentration lower than 382 (pg/ml)^2, no action is necessary. If an individual has a plasma $A\beta_{1-42}$xT-Tau concentration between 382 and 403 (pg/ml)^2, the subject is subhealthy and is at risk of preclinical AD, and an annual blood examination is recommended. In addition, the subject should adopt suitable interventions to prevent progression to aMCI. If an individual has a plasma $A\beta_{1-42}$xT-Tau level greater than 403 (pg/ml)^2, the subject should visit a neurologist/neuropsychologist for detailed examinations such as amyloid PET or magnetic resonance imaging for confirmatory diagnosis of aMCI. Lifestyle interventions together with drug treatments would be recommended for the subject by neurologists/neuropsychologists. After receiving advice from a neurologist/neuropsychologist, it would be recommended for the subject to undergo blood tests every 3–6 months to monitor and manage the risk of developing AD.

The predictive utility of blood-based biomarkers (BBBMs) continues to emerge. In 2025, Salvado et al. included 2916 cognitively unimpaired participants from 12 international independent observational cohorts in the USA, Europe, Australia, and Canada with available plasma pTau217 levels and CSF or PET Aβ biomarkers. Performance comparisons between mass spectrometry and immunoassay-based pTau217 measurements were also performed ($n = 964$) to determine the Aβ status, as measured by CSF or Aβ PET biomarkers. In the study, participants had a age of 66.9±9.9 years; 971 (33.3%) were Aβ-positive by either CSF or PET, 1667 (57.2%) were women, and 1108 (38.1%) carried at least 1 *APOE* ε4 allele. As a stand-alone test, plasma pTau217 achieved a positive predictive value (PPV) of 79% (95% CI, 74–84) and an overall accuracy of 81% (95% CI, 80–82). In a two-step workflow, the PPV and accuracy significantly increased to 91% (95% CI, 86–95). These data suggest that plasma ptau217 predicts Aβ pathology in preclinical participants [18].

5.2.2 Risk Evaluation of Subjective Cognitive Decline

Among CU individuals, clinicians frequently receive complaints about cognitive decline from the individuals themselves, particularly memory decline. In the clinic, this population is referred to as patients with subjective cognitive decline (SCD). The suspicious symptoms are asserted by the subjects themselves, not by clinicians. Hence, it is possible for the suspicious symptoms to not be true. Strictly speaking, SCD is not officially diagnosed as cognitive impairment without medically professional assessments. Individuals with SCD are often declared to be cognitively normal by clinicians. However, most individuals with SCD tend to be dismissed by clinicians. Thus, individuals with SCD could be troubled by subjective cognitive loss, which causes great mental pressure and is usually a very large burden to carry. In serious cases, subjects could have trouble at work because of decreased self-confidence. Hence, individuals with SCD suffer all the time, not only in daily life but also at work. Thus, it would be useful to have scientific evidence to support the diagnosis of SCD and help individuals with SCD receive timely medical assistance.

The data suggest that the self-report of a memory complaint in the context of SCD can be predictive. To investigate the association between subjective memory complaints (SMCs) and long-term risk of cognitive impairment, 1107 cognitively normal subjects were followed for 2 decades. SMCs were assessed shortly after baseline and repeatedly over time with the yes/no question, "Do you feel you have more problems with memory than most?" Cognitive status 18 years later (normal or impaired with mild cognitive impairment or dementia) was determined by an expert panel. Using logistic regression, the association between SMCs over time and the risk of cognitive impairment, adjusting for demographics, baseline cognition, and characteristics that differed between those with and without SMCs, was investigated. At baseline, 8.0% of participants ($n = 89$) endorsed SMCs. Baseline SMCs were associated with increased risk of cognitive impairment 18 years later (adjusted odds ratio = 1.7, 95%; confidence interval = 1.1–2.8). The association between

SMCs and cognitive impairment was greatest at the last SMC assessment time point (18 years before diagnosis: adjusted odds ratio = 1.7 [1.1–2.9]; 14 years before diagnosis: adjusted odds ratio = 1.6 [0.9–2.7]; 10 years before diagnosis: adjusted odds ratio = 1.9 [1.1–3.1]; 4 years before diagnosis: adjusted odds ratio = 3.0 [1.8–5.0]). SMCs are associated with cognitive impairment nearly two decades later among older women [19].

The source of cognitive complaints predicts diagnostic conversion differentially among nondemented older adults. A study was conducted to compare whether different sources of SMC (i.e., subjective and informant) predict diagnostic conversion in nondemented older adults. Participants from NACC had a baseline diagnosis of NC ($n = 4414$; mean age, 73 ± 8 years) or MCI ($n = 1843$; mean age, 74 ± 8 years). Multinomial logistic regression related baseline cognitive complaint (no complaint, self only, informant only, or both self and informant) to diagnostic outcome (reversion, stable, or conversion). Among NC participants, self-complaint only (odds ratio = 2.1; 99% confidence interval = 1.5–2.9; $p < 0.001$), informant complaint only (odds ratio = 2.2; 99%; confidence interval = 1.2–3.9; $p < 0.001$), and both self and informant complaint (odds ratio = 4.2; 99% confidence interval = 2.9–6.0; $p < 0.001$) were associated with diagnostic conversion compared with no complaint. Cognitive complaints are related to conversion among nondemented older adults. Complaint from both (i.e., mutual complaint) sources was most predictive of diagnostic outcome, followed by informant complaint [20].

SMC/SCD might predict Aβ pathology in the brain. The AIBL group recruited 120 subjects which were healthy controls (HC) or patients with mild cognitive impairment (MCI). 67 participants including 47 HC and 20 MCI had Aβ scans available for analysis. HC with positive Aβ (HCAβ+) acknowledged a progressive memory decline compared to HC with negative Aβ (HCAβ-), while HCAβ+ individuals acknowledged progressive memory change, suggesting they are aware of memory changes not yet detectable via neuropsychological measures. Qualitative analysis of SMCs can inform the earliest clinical manifestations of Alzheimer's disease [21].

SMC/SCD might be particularly predictive in ApoE ε4 carriers. The relationship of self-reported SMC among ApoE ε4 carriers and noncarriers from the Nurses' Health Study was assessed. In both groups, increasing subjective cognitive concern score predicted worse baseline memory and faster rates of subsequent memory decline, after adjustment for age, education, and depression. The relation with baseline memory appeared statistically stronger in ApoE ε4 carriers (P-interaction = 0.03). ApoE ε4 carriers with self-assessed cognitive concerns appear to have worse memory and possibly accelerated memory decline [22].

SMC/SCD might predict progression. To investigate whether people with SMC but no objective deficits are at increased risk of developing MCI or dementia, a meta-analysis of 28 studies found that there were 29,723 unique individuals (14,714 with SMC and 15,009 without SMC) with a mean age of 71.6 years, followed on average for 4.8 years for the development of dementia. The annual conversion rate (ACR) of SMC to dementia was 2.33% (95% confidence interval = 1.93%–2.78%), a relative risk of 2.07 (95% confidence interval = 1.76–2.44) compared with those without SMC ($n = 15{,}009$). From 11 studies the ACR of developing MCI was 6.67% (95% confidence interval = 4.70–8.95%). In long-term studies over 4 years, 14.1% (9.67–19.1%) of people with SMC developed dementia, and 26.6% (95% confidence interval = 5.3–39.7) went on to develop MCI. Older people with SMC but no objective impairments are twice as likely to develop dementia as individuals without SMC. Approximately 2.3% and 6.6% of older people with SMC will progress to dementia and MCI per year [23].

Thus, plasma biomarkers could provide such evidence in SCD to determine at risk status. Owing to the difficulties in determining SCD in the clinic, studies on plasma biomarkers in individuals with SCD are very limited. In 2024, Bernard et al. used SIMOA to measure plasma biomarkers in CU individuals with different degrees of SCD. The results showed that a greater degree of SCD in CU individuals was predictive of a greater plasma T-Tau/Aβ_{1-42} ratio [24].

Additionally, in 2024, Hsieh et al. used IMR to measure plasma biomarkers in CU individuals with and without SCD [25]. Non-SCD and SCD subjects are matched for age, education year, body mass index, and results of cognitive tests such as the CDR and MMSE. The authors reported that plasma Aβ_{1-42}, but neither T-Tau nor pTau181, was the dominant index for discriminating SCD and non-SCD in CU individuals. Greater levels of plasma Aβ_{1-42} were detected in SCD subjects (17.2 ± 1.1 pg/ml) than in non-SCD subjects (16.4 ± 0.9 pg/ml, $p < 0.05$). Thus, the cutoff value of plasma Aβ_{1-42} (16.9 pg/ml) for discriminating between SCD and non-SCD in CU individuals (subhealthy) was added to the risk evaluation in Fig. 5.1,

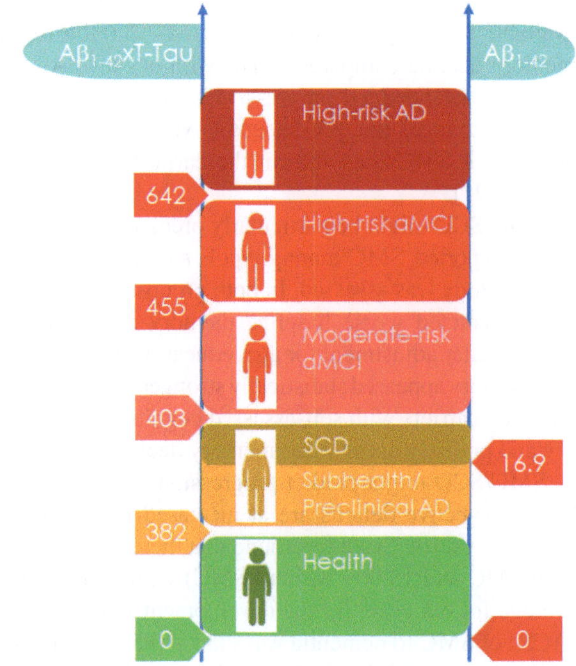

Fig. 5.3 Plasma Aβ_{1-42}xT-Tau levels assayed by IMR for discriminating the staging risks for health, subhealth/preclinical AD, SCD, aMCI, and AD

as shown in Fig. 5.3. Moreover, the baseline levels of plasma $A\beta_{1-42}$ in SCD subjects predict cognitive decline in the following 1.7 years.

Remarkably, the dominant $A\beta_{1-42}$ detected by IMR is soluble oligomers or protofibrils, which are key proteins in the development of AD. The greater levels of plasma $A\beta_{1-42}$ detected with IMR by Hsieh et al. reveal the possibility that SCD could be a prior stage to amnestic MCI (aMCI), which is characterized by early and mild cognitive impairment anchored in memory by clinicians. These findings are consistent with the framework proposed by the Alzheimer's disease continuum in which SCD is an indicator of a transitional stage in the development of AD. More clinical evidence validating the use of plasma biomarkers in the diagnosis of SCD in CU individuals, particularly to determine if they are predictive of progression, are needed. With this potential success, we could easily identify individuals in the transitional state from health to aMCI or early AD. Treatments or interventions could be applied in such a population at the extremely early stage of AD to prevent disease progression.

5.2.3 Risk Evaluation of Cognitive Impairment in Cognitively Normal Subjects with Risk Factors

The impact of plasma biomarkers on the diagnosis of Alzheimer's disease has gradually increased from the late stages (mild or severe AD) to include the early stages (early AD or aMCI) and the transitional stage (SCD). Generally, AD should be identified as early as possible so that suitable interventions or treatments could be more effective. Recently, a shift in focus has emerged, i.e., the demand for assessments of subtle cognitive impairment of at-risk populations with normal cognition, mostly non-SCD CU individuals. Many computer-based platforms that are AI-assisted are being developed to detect the cognitive changes. The AI landscape offers newer digital platforms such as MoCA Cognition's XpressO, Linus Health's DCR, Cogniguard, and Cogmi that digitize, streamline, and enhance these assessments, using AI for deeper insights, personalized feedback, and user-friendly tracking on devices, moving beyond traditional paper tests to improve early detection and management of memory/brain health issues.

To maximize the detectability of subtle cognitive changes, risk assessment has been proposed using aggregated risk by odds ratio from published data. Kivipelto et al. and Norton et al. listed the risk factors for the development of cognitive impairment, as shown in Table 1 [26, 27]. These risk factors include family history, age, education, sex, comorbidities, ApoE genotype, disabilities, the presence of sleep apnea, smoking, cardiac disease, and brain injury. The odds ratio for each risk factor is also given. Among these risk factors, age and the ApoE genotype are more predominant.

If the sum of the odds ratios is greater than 5, the individual is at moderate risk of developing cognitive impairment. For example, the odds ratio for an individual with hypertension (odds ratio = 2.2) and obesity (odds ratio = 2.3) who frequently smokes (odds ratio = 2.3) is 6.8, which indicates a moderate risk of cognitive impairment. If an individual suffers from risk factors with a sum of individual odds ratios greater than 15, he or she is at great risk of developing cognitive impairment in the near future.

Future aggregate risk tables might include anosmia (loss of smell), anosognosia (lack of self-awareness of condition), and the head turning sign. The "head turning sign" (HTS) in Alzheimer's is a behavioral cue where a patient, when asked a question they can't answer, turns their head to their caregiver for help or cues, acting as a subtle request for the answer. It's a simple, highly specific potential screening marker for dementia, particularly Alzheimer's, indicating cognitive impairment, dependency, and correlation with lower cognitive scores and AD biomarkers like tau, suggesting it's an excuse behavior for memory gaps [28].

In addition to consulting the risk factors in Table 5.1, we would like to have biological information to assist in evaluating the risk of developing cognitive impairment in non-SCD CU individuals which would increase or diminish the aggregated risk. Many research groups have used IMR to correlate these risk factors with plasma biomarker levels in a cognitively unimpaired population (CU individuals) without SCD (non-SCD). The results are summarized as follows [29–31].

First, the age-dependent plasma $A\beta_{1-42}$xT-Tau concentration determined by IMR was investigated. There were 897 CU individuals enrolled. The ages of the enrolled CU individuals ranged from 20 to 91 years. The percentage of female participants was 60.8%. A plot of the plasma $A\beta_{1-42}$xT-Tau levels versus age is shown in Fig. 5.4.

Table 5.1 Risk factors for developing cognitive impairment in CU oindividuals [26, 27]

Risk factor	Odds ratio
A first degree relative	3.0
History of head injury with loss of consciousness	2.0
Age >65 years	1.0
Age >75 years	4.0
Age >85 years	16.0
Education <7 years	3.6
Female sex	1.5
Hypertension (systolic blood pressure >140 mm-hg)	2.2
Obesity (body mass index >30 kg/m^2)	2.3
Cholesterol >6.5 mmol/l	1.9
ApoE 4 carrier	4.0
History of stroke	2.4
History of myocardial infarction	2.5
Untreated type 2 diabetes mellitus	2.0
Low physical activity (sedentary)	1.7
Continuation of smoking	2.3
Obstructive sleep apnea	2.2

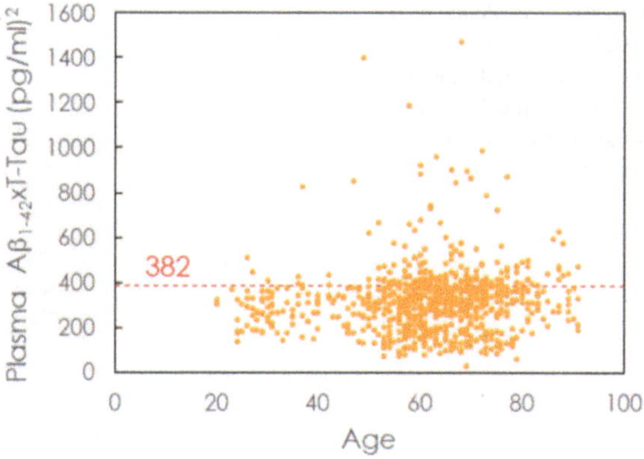

Fig. 5.4 Plasma $A\beta_{1-42}$xT--Tau levels assayed by IMR in non-SCD CU individuals aged from 20 to 91 years [29, 30]

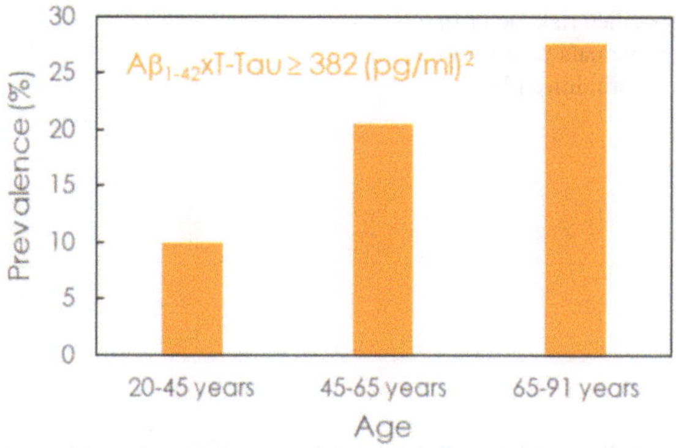

Fig. 5.5 Prevalence of the specified risk factor in CU individuals aged 20–45 (young adults), 45–65 (middle-aged adults), and 65–91 (aged adults) years [30]

The Pearson correlation coefficient for the data in Fig. 5.4 is 0.094 ($p < 0.05$), which reveals the independent effect of plasma $A\beta_{1-42}$xT-Tau on age in CU individuals.

Furthermore, the prevalence of a plasma $A\beta_{1-42}$xT-Tau concentration greater than 382 (pg/ml)^2 among individuals aged 20–45 (young adults), 45–65 (middle-aged adults), and 65–91 (aged adults) years was analyzed, as shown in Fig. 5.5. The prevalence increased from 9.89% in young adults to 20.4% in middle-aged adults and 27.6% in aged adults. Obviously, the risk of developing cognitive impairment in CU individuals increases with age. When young adults were used as a reference, the odds ratio of developing cognitive impairment was 2.1 in middle-aged adults and 2.8 in aged adults.

Second, the effect of the ApoE ε4 genotype on the prevalence of the risk factor, i.e., a plasma $A\beta_{1-42}$xT-Tau concentration greater than 382 (pg/ml)^2, in CU individuals aged more than 45 years was examined. The number of enrolled CU individuals was 188; 158 were ε4 noncarriers, and the others were ε4 carriers. The prevalence of the risk factor in the ApoE ε4 noncarrier and carrier groups is shown in Fig. 5.6. A total of 27.8% of ε4 noncarriers were at risk, and 37.5% of ε4 carriers were at risk. The odds ratio was 1.5. From the plasma biomarker point of view, the ApoE ε4

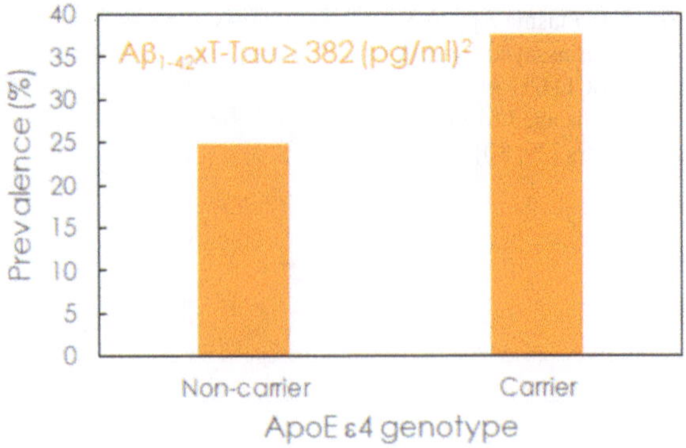

Fig. 5.6 Prevalence of the specified risk factor in CU individuals aged more than 45 years in the ApoE ε4 noncarrier and carrier groups

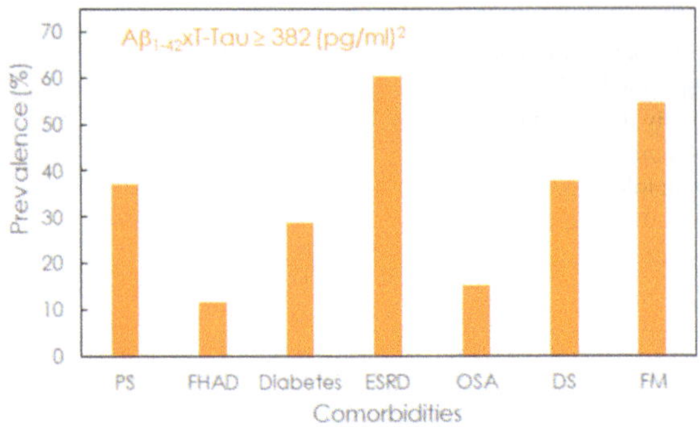

Fig. 5.7 Prevalence of the specified risk factor in CU individuals with various comorbidities [31]

genotype clearly increases the risk of developing cognitive impairment in non-SCD CU individuals. Also demonstrated by Jack et al., the ApoE 4 carrier status is a known risk factor for AD and is a proxy marker of amyloid and is age-dependent [32].

Third, the prevalence of this risk factor in non-SCD CU individuals with comorbidities was investigated, as shown in Fig. 5.7. The comorbidities of interest included poststroke (PS), family history of AD (FHAD), type 2 diabetes mellitus (diabetes), end-stage renal disease (ESRD), obstructive sleep apnea (OSA), Down's syndrome (DS), and fibromyalgia (FM). The prevalence ranged from 11.4% to 60.2%, depending on the comorbidity. Notably, the prevalence of the risk factor for developing cognitive impairment was greater than 25% among CU individuals with PS, diabetes, ESRD, DS, or FM. In particular, 60.2% of the population at risk of developing cognitive impairment had ESRD; this proportion is incredibly high.

Given the strong evidence shown in Figs. 5.4, 5.5, 5.6, and 5.7, the effects of the risk factors listed in Table 5.1 on brain health in CU individuals cannot be ignored and should be applied in clinical practice with the BBBMs for subjects with risk factors to assess their brain health.

Here, we provide a typical example to show how blood-based biomarker (BBBM) tests assist in the diagnosis of aMCI and AD in clinical practice.

The patient's name is Alex, aged 66 years. "Alex" is not a real name but a character's name in the example. Hereafter, each example has a character's name, not the real name of the patient; he had been educated for 14 years, and his mother was an AD patient. Therefore, he had a family history of AD (ADFH). Alex used to smoke one pack every 2 days, i.e., half pack per day. The ApoE genotype was determined to be ε3/ε3. A neuropsychologist performed brief tests; the FAST score was 2, and the MMSE score was 26.

According to the prior examinations, he had risk factors in terms of age, ADFH, continuation of smoking, and MMSE score. However, the ApoE genotype was not an increased risk factor for developing aMCI/AD. His FAST score did not indicate objective impairment. Remarkably, the MMSE score of 26 is between that of individuals with normal cognition and mild cognitive impairment. Hence, the findings of the prior examinations resulted in difficulty in making a timely diagnosis for Alex.

In this case, the regular process included performing detailed neuropsychological tests, such as ADL, IADL, and WIS-III determination, as well as MRI. In most settings, the patient would be observed for protracted periods of time without intervention or further assessment. In addition, a longitudinal follow-up, e.g., 3–6 months, was performed to monitor the changes in his cognition and behaviors. It takes much time and effort to obtain a confirmatory diagnosis. During the period of the detailed examinations, usually 3–6 months, Alex and his family members continued to be concerned about cognitive changes.

To obtain a timely diagnosis for Alex, the neurologist strongly suggested the use of IMR blood tests. Alex agreed with the neurologist and underwent the blood tests immediately. Seven days later, Alex obtained the blood test results. The measured plasma $A\beta_{1-42}$xT-Tau was 468.5 $(pg/ml)^2$, which revealed the risk of aMCI for Alex, as shown in Fig. 5.3. Thus, the neurologist immediately performed the detailed neuropsychological tests and MRI.

The detailed examinations by the neurologist indicated that Alex had clinical symptoms such as memory loss, cognitive impairment, episodic confusion, and short-term memory impairment suggestive of amnestic mild cognitive impairment. In addition, MRI was performed to exclude the possibility of brain ischemia. The results of detailed examinations clearly revealed that Alex suffered from aMCI. The diagnosis through detailed examinations was consistent with that through IMR blood tests. Alex needed medical treatment immediately.

In this typical case, blood tests provided significant indicators to the neurologist that detailed examinations were needed immediately. Specifically, with the aid of blood tests, the neurologist could make suitable decisions in a timely manner for Alex in clinical practice. Thus, blood tests truly assist neurologists in the timely and correct diagnosis of aMCI and AD.

The case illustrates that clinical diagnosis without the use of biomarkers is imprecise. The literature confirms this.

5.2.4 *Preliminary Evaluations of Amyloid PET in aMCI and AD*

Currently, the gold standard for the biologically confirmatory diagnosis of aMCI and AD is amyloid PET. Amyloid PET, to be approved, had to demonstrate validity and correlation to autopsy. This was completed for florbetapir, florbetaben, flutemetamol, and all three radiotracers are FDA approved and have been used in clinical practice since the national coverage determination was lifted. However, the cost and radiological exposure associated with amyloid PET are the most concerning issues in clinical practice, and often the tests are not reimbursed, which limits utility due to expense. Hence, the uptake of amyloid PET for the diagnosis of aMCI and AD is limited, creating a demand for screening evaluations before amyloid PET. Such preliminary evaluations should be low cost, low risk, and highly available and should be significantly correlated with amyloid PET. This is the rationale for the development of BBBMs.

Blood tests are low cost, low risk, and highly available. In general, the cost of blood tests is 1/6–1/4 that of amyloid PET. The last issue to address for blood tests to serve as a preliminary evaluation before amyloid PET is how the results correlate with those of amyloid PET.

According to the published results shown in Fig. 4.6 in Chap. 4, the significant discriminations in status indicated by amyloid PET using plasma biomarkers have been validated. The first plasma biomarker used before amyloid PET is $A\beta_{1-42}/A\beta_{1-40}$. More recently, other biomarkers, such as pTau217, pTau181, $A\beta_{1-42}$xT-Tau, $A\beta_{pE3-40}$, $A\beta_{pE3-40}$/T-Tau, $A\beta_{1-42}$ combined with the ApoE genotype, and pTau217/$A\beta_{1-42}$, have been developed and marketed, some with laboratory-development-test (LDT) coverage only. The accuracy of discriminating positive/negative amyloid PET using plasma biomarkers could be greater than 90%. Therefore, assays of plasma biomarkers are useful as preliminary evaluations before amyloid PET in the diagnosis of aMCI and AD.

Here we provide two typical examples of preliminary evaluations of amyloid pathology using plasma biomarkers in aMCI patients using the model proposed by Lin et al. in 2019 [33]. This model is specific for the prediction of amyloid pathology in aMCI. The model considers not only protein biomarkers but also genomic biomarkers. Both the plasma $A\beta_{1-42}$ concentration and the ApoE genotype are parameters that predict amyloid deposition in the brain in aMCI patients, as shown in Fig. 5.8. In fact, genotyping of ApoE is a requisite step risk stratification when considering amyloid-targeted therapies such as lecanemab. The remainder of Fig. 5.8 is briefly illustrated as follows.

First, aMCI patients were divided into two groups by genotype, i.e., ApoE ε4: ApoE ε4 carriers (ε2/ε4, ε3/ε4, ε4/ε4) and ApoE ε4 noncarriers (ε2/ε2, ε2/ε3, ε3/ε3). Second, among the ApoE ε4 noncarriers, aMCI patients were further divided into two groups: ApoE ε2 carriers (ε2/ε2, ε2/ε3) and ApoE ε2 noncarriers (ε3/ε3). The cutoff value for plasma $A\beta_{1-42}$ for discriminating between positive and negative amyloid pathology in ApoE ε4 carriers (ε2/ε4, ε3/ε4, ε4/ε4) was 18.53 pg/ml. In ApoE ε4 noncarriers and ApoE ε2 noncarriers (ε3/ε3), the cutoff value for plasma $A\beta_{1-42}$ for differentiating between positive and negative amyloid pathology was 16.01 pg/ml. The aMCI patients who were ApoE ε4 noncarriers and ApoE ε2 carriers (ε2/ε2, ε2/ε3) were found to be negative for amyloid pathology.

As a case example, the first patient, Batty, was a female, aged 68 years; she had received 16 years of education. She was clinically diagnosed as an aMCI patient, with a CDR = 0.5 and MMSE = 27. Her ApoE genotype was determined to be ε3/

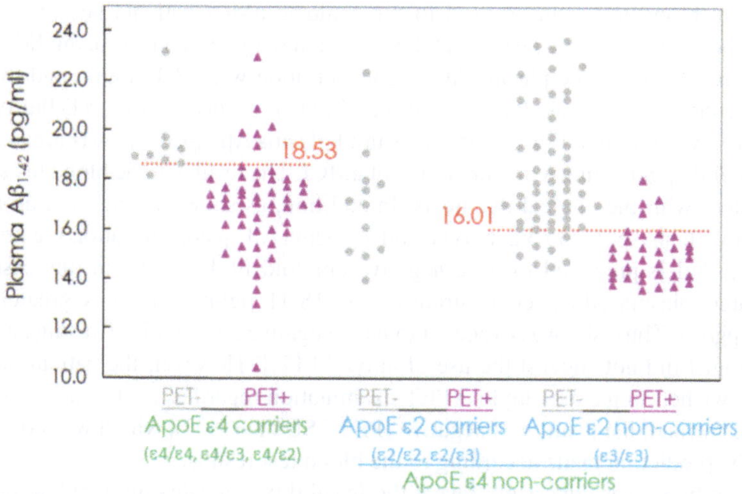

Fig. 5.8 Discrimination of positive (PET+) and negative (PET-) amyloid pathology using ApoE genotype and plasma $A\beta_{1-42}$ in aMCI patients. The amyloid pathology was confirmed with amyloid PET. The plasma $A\beta_{1-42}$ was assayed by IMR

ε4. Both her clinical symptoms and genotype present clear risks of the patient having amyloid deposition in the brain. However, amyloid PET is expensive and carries radiological risks. Furthermore, the hospital she visited in her healthcare system lacked a PET machine. The patient needed to be triaged to the neighboring medical center for amyloid PET. Cautiously, the neurologist suggested that the patient should undergo blood tests first. By IMR, the plasma $A\beta_{1-42}$ concentration and T-Tau concentration were determined to be 16.60 pg/ml and 26.25 pg/ml, respectively. $A\beta_{1-42}$xT-Tau was 435.7 (pg/ml)^2. According to Fig. 5.3, individuals with plasma $A\beta_{1-42}$xT-Tau greater than 403 (pg/ml)^2 are at risk of aMCI. Thus, the plasma biomarker testing results were consistent with the clinical diagnosis for aMCI. In addition, as shown in Fig. 5.8, an aMCI patient with an ApoE genotype of ε3/ε4 and plasma $A\beta_{1-42}$ concentration lower than 18.12 pg/ml is predicted to have positive amyloid deposition. The patient in this case example had a plasma $A\beta_{1-42}$ concentration of 16.60 pg/ml. With the information from the plasma Aβ, the neurologist referred the patient to the neighboring medical center for amyloid PET examination. The examination was positive for amyloid pathology (PET+). High consistency between the results of blood tests and amyloid PET was demonstrated.

The second patient was also a female patient, Christine, aged 65 years with 12 years of education. Christine had been clinically diagnosed with aMCI, with a CDR = 0.5 and MMSE = 29. Her ApoE genotype was ε3/ε3. The neurologist was reluctant to order amyloid PET examination for the patient because ApoE ε3/ε3 is not a risk factor for amyloid deposition. Moreover, Christine's MMSE score was normal at 29. However, her CDR and clinical symptoms were suggestive of pathology. Such a mix of information regarding the possibility of amyloid deposition in the brain can be confusing for neurologists as the information is conflicting. In this situation, blood tests can be used to pre-evaluate aMCI patients before amyloid PET. The neurologist suggested that the patient undergo blood tests using IMR. The results revealed that the plasma $A\beta_{1-42}$ concentration was 18.41 pg/ml and that the T-Tau concentration was 25.31 pg/ml, resulting in a plasma $A\beta_{1-42}$xT-Tau of 466 $(pg/ml)^2$. According to Fig. 5.3, the measured plasma $A\beta_{1-42}$xT-Tau was greater than 455 $(pg/ml)^2$, indicating a moderate risk of aMCI. The results of the blood tests were consistent with the clinical diagnosis. In addition, as shown in Fig. 5.8, an aMCI patient with an ApoE ε3/ε3 genotype and a plasma $A\beta_{1-42}$ concentration greater than 15.86 pg/ml is predicted to have negative amyloid PET results. In this case, the measured plasma $A\beta_{1-42}$ concentration was 18.41 pg/ml, which is greater than 15.86 pg/ml. Thus, she was expected to have negative amyloid PET results, and the neurologist did not suggest the use of amyloid PET. However, the patient and her family wished to pursue amyloid PET examination regardless. After amyloid PET was performed, the result was negative (PET-, SUVr = 1.04), which was consistent with the prediction made according to the blood test results.

These two cases concretely reveal the feasibility of evaluating amyloid deposition in the brains of aMCI patients by using blood tests. Blood tests can help neurologists make decisions regarding the recommendation of amyloid PET for aMCI patients.

Notably, the model used in these two cases to evaluate patients before amyloid PET involves plasma $A\beta_{1-42}$ and the ApoE genotype. Recently, another model involving plasma pTau217 that inspired the use of blood tests in evaluating amyloid pathology has attracted interest in clinical practice. Several assay technologies, such as SIMOA, ECL, and MP-CLEIA, have been demonstrated to discriminate amyloid PET positivity and negativity by assaying plasma pTau217, as shown in Fig. 4.6 in Chap. 4.

5.2.5 Prediction of aMCI Development in Cognitively Unimpaired Subjects

In addition to risk evaluations for aMCI, the prediction of developing aMCI in CU individuals is a critical issue. Predicting the development of aMCI within a few years using plasma biomarker levels at baseline would be beneficial for managing the health of the brain in CU individuals. Timely interventions could be provided to prevent the development of aMCI. Clinical evidence from longitudinal studies is needed to determine how to use blood tests to predict the possibility of developing aMCI among CU individuals.

Lee et al. performed a longitudinal study to monitor the progression of CU individuals to aMCI [16]. Sixty CU individuals were enrolled. The age at baseline was 71.6 ± 9.8 years. The MMSE scores of all the CU individuals ranged from 29 to 30. The plasma $A\beta_{1-42}$xT-Tau determined by IMR was 357.7 ± 85.5 $(pg/ml)^2$. One and half years later, seven subjects progressed to aMCI, termed the progressive group, and the others remained CU, termed the stable group.

At baseline, the stable group had MMSE scores ranging from 29 to 30, and the progressive group had MMSE scores ranging from 28.0 to 30. The difference in the at-baseline MMSE scores between the two groups was not significant ($p > 0.05$). Age, frequency of ApoE alleles, and years of education did not significantly differ between the two groups either. However, the plasma $A\beta_{1-42}$xT-Tau in the stable group at baseline was 351.1 ± 88.8 $(pg/ml)^2$, which was significantly lower than that in the progressive group (407.3 ± 39.9 $(pg/ml)^2$, $p < 0.05$). Through ROC curve analysis, the cutoff value of plasma $A\beta_{1-42}$xT-Tau at baseline for discriminating between the progressive and stable groups was found to be 382 $(pg/ml)^2$. Further analysis revealed that 26% of CU individuals having a plasma $A\beta_{1-42}$xT-Tau greater than 382 $(pg/ml)^2$ progressed to aMCI one and half a years later. This study revealed the cutoff value of 382 $(pg/ml)^2$, as shown in Fig. 5.3.

Remarkably, once one has a plasma $A\beta_{1-42}$xT-Tau greater than 382 $(pg/ml)^2$, he or she might develop aMCI, although he or she does not have clear clinical symptoms. Coincidently, an autopsy study by Jiang et al. demonstrated that CU individuals with positive amyloid neuropathology in postmortem brain tissue had antemortem plasma $A\beta_{1-42}$xT-Tau levels greater than 380 $(pg/ml)^2$, whereas CU individuals with negative amyloid neuropathology in postmortem brain tissue had antemortem plasma $A\beta_{1-42}$xT-Tau levels lower than 380 $(pg/ml)^2$ [34]. These results suggest that CU individuals whose plasma $A\beta_{1-42}$xT-Tau is greater than 380 $(pg/ml)^2$ have a high risk of preclinical AD and a high probability of developing aMCI.

In clinical practice, it is becoming feasible to identify CU individuals at risk of developing aMCI with the aid of blood tests. This is an area of tremendous investigation because of the possibility of identifying CU patients at risk for progression. Suitable interventions, such as nutrition, physician exercise, sleep improvement, and cognitive training, as well as care for comorbidities, should be given to CU individuals with a plasma $A\beta_{1-42}$xT-Tau greater than 380 $(pg/ml)^2$ to prevent them from developing aMCI in the near future.

5.2.6 *Prediction and Monitoring of Cognitive Decline in aMCI*

The prediction of aMCI development in CU individuals is a key issue in AD. Blood tests make this process feasible in clinical practice. However, the prediction of cognitive decline in aMCI patients is also important and has long been impossible. Fortunately, blood tests now enable this prediction to be achieved.

In 2020, Tsai et al. reported that aMCI patients with greater levels of plasma $A\beta_{1-42}$ assayed by IMR at baseline experienced rapid decreases in MMSE scores 1–1.5 years later [15]. A significant negative correlation ($r = -0.51$, $p < 0.05$) was found between at-baseline plasma $A\beta_{1-42}$ levels and the annual changes in MMSE scores in aMCI patients. These results indicate that the plasma $A\beta_{1-42}$ level is a promising index for predicting cognitive decline in aMCI patients. However, Tsai et al. did not show a cutoff value for the plasma $A\beta_{1-42}$ level at baseline to predict cognitive decline in aMCI patients. Chen et al. completed a longitudinal study to determine the cutoff value [35].

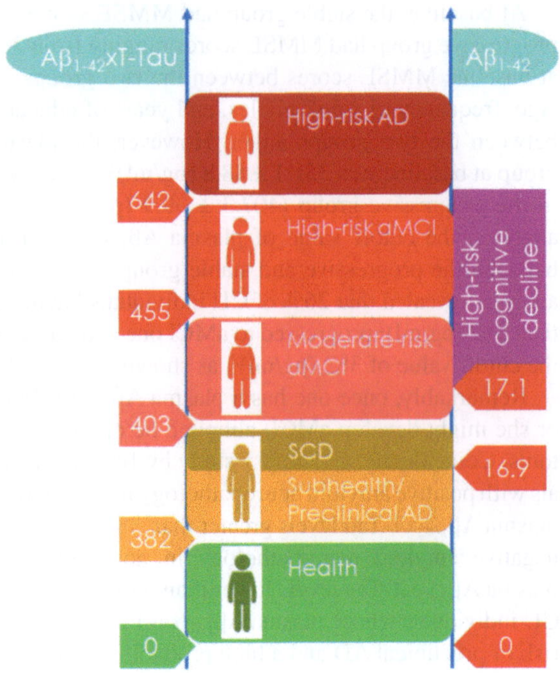

Fig. 5.9 Plasma $A\beta_{1-42}$xT-Tau levels assayed with IMR for discriminating the staging risks of health, subhealth/preclinical AD, SCD, aMCI, and AD also for predicting the cognitive decline in aMCI

In Chen's study, more than 75% of aMCI patients whose plasma $A\beta_{1-42}$ concentration was greater than 17.1 pg/ml at baseline, as determined by IMR, developed cognitive decline 1–1.5 years later [35]. More than 75% of aMCI patients whose plasma $A\beta_{1-42}$ concentration was lower than 17.1 pg/ml at baseline remained stable in terms of cognition 1–1.5 years later. The value of 17.1 pg/ml for plasma $A\beta_{1-42}$ is clearly a suitable cutoff value for predicting cognitive decline in aMCI patients. This cutoff value has been added to Fig. 5.3 as a reference for determining the correlation between plasma biomarkers and clinical significance in AD, as shown in Fig. 5.9.

If an aMCI patient has a plasma $A\beta_{1-42}$ concentration greater than 17.1 pg/ml, the neurologist should monitor for cognitive changes. The most common monitoring method involves routinely performing the MMSE or MoCA for aMCI patients, e.g., every 3–6 months. However, practice effects do occur with MMSE or MoCA when he or she is tested frequently. Hence, the testing results would be less significant for monitoring changes in cognition. As such, the clinician needs to have objective results to help monitor for cognitive decline in MCI patients. Although other examinations, such as amyloid PET or CSF biomarker measurements, are useful for the diagnosis of aMCI, they are not suggested to be performed for a patient so frequently. The most promising examination, which is able to be frequently performed and provide objective results, might be blood tests.

Chu et al. coordinated a longitudinal study to monitor the annual changes in plasma pTau181 levels by IMR in aMCI patients [36]. Thirty-seven aMCI patients were enrolled at baseline (MMSE = 26.2 ± 2.2). One year later, 10 of 37 aMCI patients had progressed to AD, and the others had not. Moreover, 20 age-matched CU individuals were enrolled at baseline (MMSE = 29.4 ± 0.5). All the CU individuals were still CU within 1 year.

The results revealed an increase in the annual changes in plasma pTau181 levels in all the converted aMCI patients, as shown in Fig. 5.10. Ten of the twenty-seven nonconverted aMCI patients had greater plasma pTau181 levels 1 year later, and the others had lower plasma pTau181 levels at follow-up. The annual changes in plasma pTau 181 were significantly lower in the aMCI nonconverter group than in the aMCI converter (−1.07 pg/ml; 95% confidence interval: −1.78 to −0.35 pg/ml; $p < 0.005$) and CU groups (−0.59 pg/ml; 95% confidence interval: −1.16 to −0.03 pg/ml; $p < 0.05$). Thus, an increase in plasma pTau181 levels over time is an objective and significant indication to help neurologists monitor cognitive decline in aMCI patients. However, pTau181 is emerging as a reliable marker for the negative predictive value. Specifically, a normal pTau181 values predicts the absence of AD pathology.

In addition to pTau181, T-Tau in plasma is a helpful biomarker for monitoring the cognitive changes of individuals with aMCI. As shown in Fig. 5.11, the x axis denotes the annual change in the MMSE score ratio, defined as follows:

$$\text{Annual change in MMSE ratio} = \left(\frac{\text{MMSE score on follow} - \text{up} - \text{MMSE score at baseline}}{\text{MMSE score at baseline}} \right) / \text{Period} \tag{5.1}$$

where period denotes the time interval between the follow-up and baseline in years. The y axis in Fig. 5.11 denotes the annual change in the T-Tau ratio, defined as follows:

$$\text{Annual change in T} - \text{Tau ratio} = \left(\frac{\text{T} - \text{Tau level on follow} - \text{up} - \text{T} - \text{Tau level at baseline}}{\text{T} - \text{Tau level at baseline}} \right) / \text{Period} \tag{5.2}$$

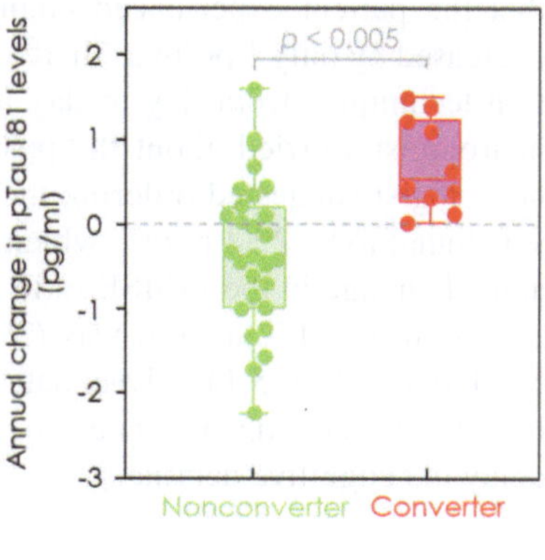

Fig. 5.10 Annual changes in plasma pTau181 levels in aMCI nonconverter and converter groups [36]

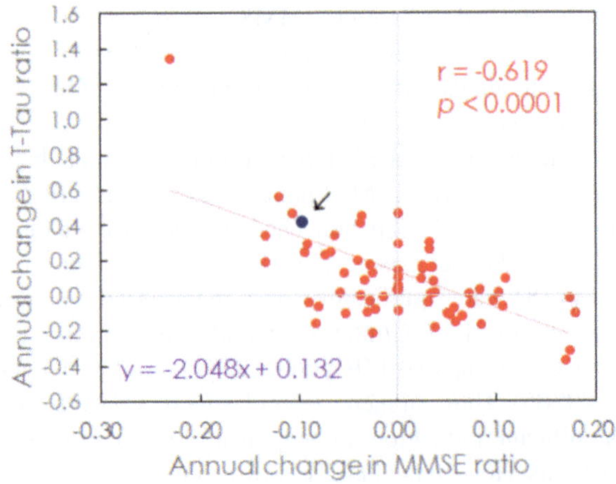

Fig. 5.11 Correlation between annual change in the T-Tau ratio and annual change in the MMSE score ratio in aMCI. The ratios are calculated using Eqs. (5.1) and (5.2). The plasma T-Tau levels were assayed by IMR

where period denotes the time interval between the follow-up and baseline in years.

A significantly negative correlation ($r = -0.619$, $p < 0.0001$) was found, as shown in Fig. 5.11. The negative correlation is expressed as follows:

$$y = 2.048x + 0.132, \tag{5.3}$$

where y is the annual change in the T-Tau ratio and x is the annual change in the MMSE ratio, as shown in Fig. 5.11. This indicates that the decrease/increase in plasma T-Tau levels reflects the improvement/decline in the cognition of an individual. Changes in plasma T-Tau or pTau181 levels can be objectively monitored as indicators of cognitive changes in aMCI patients. Such follow-up examinations would be very helpful for clinicians to understand whether their aMCI patients are experiencing cognitive improvement or decline.

Here is another case example. The patient, Dyron, a 72-year-old male, was an MCI patient whose plasma T-Tau concentration was 25.2 pg/ml and whose MMSE score was 26 at baseline. Three months later, the neurologist administered the MMSE which revealed a score of 25. Although the 3-month follow-up MMSE score was lower than that at baseline, it would have been difficult to say that the patient experienced cognitive decline because the MMSE score had decreased by only 1 point as there is frequent variance on scoring and administration techniques from day to day and practitioner to practitioner. However, the neurologist worried about the possibility of true cognitive decline. Hence, the neurologist suggested ordering the BBBMs. The 3-month follow-up T-Tau level was found to be 29.4 pg/ml, which was higher than the baseline T-Tau level. The annual change in the MMSE ratio was (25–26)/26/0.4 = −0.096, and the annual change in the T-Tau ratio was (29.4–25.2)/25.2/0.4 = 0.417, as plotted with a black dot in Fig. 5.11. These data indicated that the patient was more likely to have prodromal AD. The neurologist could immediately adjust the treatments to delay the cognitive decline.

5.2.7 Assessment of Non-amnestic Forms of AD

While most clinicians are aware of the typical features of amnestic AD, including repetition of sentences and questions, forgetting details of conversations, and misplacing objects, many are not aware of the non-amnestic forms. There are two that are recognized. The first is logopenic progressive aphasia, and the second is posterior cortical atrophy. Logopenic progressive aphasia is commonly lumped with the primary progressive aphasia (PPA).

Logopenic PPA is characterized by the following symptoms:

Progressive word-finding difficulty: pauses and hesitations while speaking to find words (anomia), sometimes using long descriptions (circumlocution)

Impaired sentence repetition: trouble repeating phrases or sentences, a core feature linked to phonological processing issues; hesitant speech and slowed speech rate where speech becomes slow and halting, especially when trying to be specific despite knowing word meanings; managing to communicate with complex sentences

Phonological errors: may substitute words with similar-sounding but incorrect words (e.g., "telephone" for "lelephone")

Symptoms can progress from word-finding difficulties to significant communication impairment, affecting daily life, and may also involve cognitive issues like memory or spatial difficulties, requiring supportive management.

Pathologically logopenic PPA has amyloid and tau as is typical for amnestic forms by have regional changes of atrophy particularly in the left posterior temporal cortex and inferior parietal lobule.

Assessment has included MRI to detect regional changes of atrophy and neuropsychological assessment which will probe for impaired word retrieval (anomia), phonological errors, and difficulty repeating sentences, while sparing grammar initially, but reveals broader deficits in phonological working memory, attention, calculation, and visuospatial skills, often alongside apathy, differentiating it from other PPA types. Key components involve confrontation naming (e.g., the Boston Naming Test), sentence repetition tests (e.g., Sentence Repetition Task), and probes for phonological errors (e.g., phonemic paraphasias), assessments of category fluency, and assessments of the Controlled Oral Word Association Test (COWAT).

Even for expert clinicians and neuropsychologists, the findings of logopenic PPA may be difficult to discern from other forms of PPA and to diagnose. BBBMs could be particularly useful. pTau217, pTau181, and $A\beta_{1-42}/A\beta_{1-40}$ ratio help confirm underlying AD pathology (plaques/tangles), while neurofilament light chain (NfL) helps differentiate logopenic PPA from other PPA types (semantic/nonfluent impairment).

The other non-amnestic form of AD is called posterior cortical atrophy (PCA). PCA is a rare form of dementia, often considered a visual variant of Alzheimer's, caused by degeneration of the brain's posterior cortex leading to progressive problems with vision and visual processing, such as difficulty reading, judging distance, recognizing objects, and navigating in physical space. Memory and language are spared early on. Symptoms include trouble with spatial awareness, math, and orientation, progressing over time to affect other cognitive functions.

Diagnosis involves detection of posterior atrophy on MRI, neuropsychological assessment, and neuro-ophthalmological assessment. The neuropsychological profile of posterior cortical atrophy (PCA) is marked by early, progressive decline in visuospatial skills, visual perception, reading (alexia), and writing (agraphia), with relative preservation of memory and executive function early in the disease course. Key deficits often include elements of Balint's (simultanagnosia, optic ataxia) and Gerstmann's syndromes (acalculia, finger agnosia).

Even for expert clinicians, PCA can be difficult to diagnose. Here BBBMs could be very useful. Blood-based biomarkers for Alzheimer's disease (AD), like pTau217, pTau181, Aβ ratios, NfL, and GFAP, are showing promise for diagnosing PCA.

5.2.8 Monitoring Interventions/Treatments

AD is a disease of various stages, e.g., preclinical, aMCI, and early AD, depending on the severity of clinical symptoms. Currently, neurologists would like to provide different recommendations for patients at different stages.

For preclinical or at-risk subjects, neurologists prefer to recommend nonpharmaceutical interventions. Most nonpharmaceutical interventions involve lifestyle changes, such as nutritional programs, exercise, sleep improvement, cognitive training, and social activities. One famous study to demonstrate the effects of nonpharmaceutical interventions on delaying the onset of cognitive impairment in preclinical or at-risk subjects is the Finnish Geriatric Intervention Study to Prevent Cognitive Impairment and Disability (FINGER) study [37–39].

FINGER is the first big trial in the world to demonstrate that multidomain lifestyle interventions can improve brain health and prevent cognitive decline. The multimodal lifestyle interventions in the FINGER study include Nordic Nutrition Recommendations, regular physical training, online cognitive stimulation and training, and monitoring of cardiovascular health and social activities. The FINGER study has been conducted for 11 years.

The key finding in the FINGER study is that the total average improvement in the intervention group was 25% greater than that in the control group. In addition, there was an 83% greater improvement in executive function, 150% greater improvement in psychomotor speed, and 40% greater improvement in complex memory tasks in the intervention group than in the control group. In the control group, there was a 30% greater risk of developing cognitive impairment after 2 years than in the intervention group. Lifestyle changes were more significant in the intervention group, and active participation was associated with greater benefits from the intervention.

The FINGER study was the first large, long-term, randomized controlled trial to suggest that a multidomain intervention could improve or maintain cognitive functioning in at-risk elderly people from the general population [38]. The findings of the FINGER study were corroborated by the US POINTER Trial. In this randomized clinical trial of 2111 older adults at risk of cognitive decline and dementia, a structured lifestyle intervention of regular moderate- to high-intensity physical exercise, adherence to the MIND diet, cognitive challenge and social engagement, and cardiovascular health monitoring led to a statistically significant greater improvement in global cognition over 2 years relative to a lower-intensity self-guided intervention (mean composite z-score increase of 0.243 standard deviation per year vs 0.213 standard deviation per year, respectively) [40].

The FINGER study revealed improvements in cognition and behaviors in an at-risk population. To avoid practice effects, consecutive examinations should not be conducted within 6 months. However, it would be better to determine the effects of interventions frequently, e.g., every month. The collection of plasma biomarker information could include instant monitoring of the objective effects of interventions on improvements in brain health during the study. In other words, the BBBMs could be theragnostic. Unfortunately, plasma biomarker information is absent from this study.

In addition to lifestyle changes, many pharmaceutical companies and research groups are developing drugs to treat preclinical subjects preventively before onset. However, it is difficult to efficiently enroll preclinical subjects in clinical studies or studies because of the low availability of participants and high complexity of examinations. As mentioned, preclinical subjects are defined as cognitively unimpaired (CU) individuals who are amyloid-positive. CU individuals who do not have clear clinical symptoms rarely undergo cognitive examinations. It is also difficult to convince CU individuals to see a neurologist for detailed examinations. There are prevention clinics termed under the moniker of Brain Health that are being established, but they are sparse, and most are not well established.

Although we are lucky enough to have data from CU individuals, the determination of amyloid pathology would be performed by using amyloid PET or CSF biomarker tests. In research, because PET or CSF confirmation is expensive, BBBMs have become the new screening measure of eligibility and detection of AD pathology. In particular, fewer than 10% of CU individuals have positive amyloid pathology. The failure rate of enrollment would be very high. This implies high costs in terms of money and time in enrolling CU individuals with positive amyloid pathology. It would be better to have low-cost, convenient, and objective evaluations for the enrollment of preclinical subjects to speed up trials and studies. As shown in Fig. 5.10, CU individuals with a plasma $A\beta_{1-42}$xT-Tau greater than 382 $(pg/ml)^2$ are at risk of preclinical AD. Hence, the use of plasma $A\beta_{1-42}$xT-Tau to evaluate amyloid pathology in CU individuals is advantageous because of its low cost, low risk, high availability, high reliability, and convenience.

The A4 study was the first ATT trial used in a secondary prevention paradigm. The screening methodology was amyloid PET, and the initial screen failure rate was 90%. It did improve over time to 66%. Successive secondary prevention trials include AHEAD-A345 investigating lecanemab, TRAILBLAZER-ALZ3 investigating donanemab, and TRAILRUNNER-ALZ3 investigating remternetug.

For aMCI or early AD patients, neurologists provide drug treatments. Unfortunately, there is no drug to cure AD; however, there are drugs to delay disease progression. Two important antibody drugs, lecanemab from Eisai and Biogen and donanemab from Eli Lilly, eliminate aggregated or deposited amyloid-β in the brain [41–44]. Amyloid plaques in the brain significantly decrease in patients after they receive these drug treatments. The progression of cognitive decline is delayed by 3–5 years after 1.5 years of treatment. In addition, nonpharmaceutical interventions, i.e., combinations of lifestyle changes and drug treatments, are strongly suggested for aMCI patients and early AD patients.

In clinical practice, neurologists would like to know the instant responses of patients to drug treatments and would need routine examinations during drug treatment regimens. However, it is impossible to perform amyloid PET, which requires lumbar punctures, frequently, e.g., every 3 months. Alternative examinations of low cost, low risk, convenience, and high availability are preferred to achieve frequent monitoring during drug treatments. Blood tests could be promising examinations as theragnostic markers [45].

According to the results of a clinical trial of lecanemab initiated by Biogen and Eisai, plasma pTau181 levels were significantly lower in the treated group than in the placebo group ($p < 0.001$) at the end of the study [45]. In a clinical trial of donanemab initiated by Eli Lilly, the plasma level of pTau217 was significantly lower in the treated group than in the placebo group ($p < 0.001$) at the end of the study [46]. An increasing number of results from clinical trials or studies have demonstrated the feasibility of using plasma biomarkers to monitor treatment effects in aMCI or early AD patients.

Here, we provide examples of the application of blood tests to monitor intervent ions/treatments.

The first example is a case report from 6 years ago. Eric, male and aged 56 years, was cognitively normal. He had hypertension, high cholesterol levels, and high blood sugar levels, which are risk factors for cognitive impairment, as listed in Table 5.1. However, Eric was not an individual with SCD. Hence, he did not worry about his cognitive impairment. However, his daughter, Fion, expressed concerns about the possibility of her father having cognitive impairment because of the risk factors. After discussions, Eric was willing to undergo detailed examinations for aMCI/AD.

Eric had normal scores on the bedside cognitive assessments, e.g., CDR = 0 and MMSE = 29. The neurologist suggested amyloid PET which was not financially feasible. Alternatively, the neurologist suggested IMR blood tests. The measured value of plasma $A\beta_{1-42}$xT-Tau was 472 $(pg/ml)^2$, which was much greater than 382 $(pg/ml)^2$. Thus, there was a high suspicion of preclinical AD.

With the significantly abnormal blood test result, the patient became anxious. He could not imagine how terrible his life would be or how difficult his daughter's life would be once he suffered from AD 10 years later. He became deeply anxious and automatically asked the neurologist for treatments.

Since he was cognitively normal, he was not qualified to take medicine for treatments. The neurologist recommended nutritional and exercise programs as described in the brain health literature and published studies. The patient fastidiously adhered to the regimen for 18 months. After the programs, repeat IMR blood tests were performed. The measured value of plasma $A\beta_{1-42}$xT-Tau was 296 $(pg/ml)^2$. A significant decrease in the plasma $A\beta_{1-42}$xT-Tau level was detected after the interventions, revealing a reduction in the risk of cognitive impairment. Furthermore, the plasma $A\beta_{1-42}$xT-Tau level after intervention was lower than 382 $(pg/ml)^2$, indicating that the patient was at less risk for preclinical AD. He subsequently became less anxious but maintained engagement in the nutrition and exercise programs.

The second example originates from the clinical study conducted by Liu et al. [47] Thirty patients with obstructive sleep apnea (OSA) and normal cognition were enrolled in the study. As listed in Table 5.1, OSA is a risk factor for cognitive impairment. OSA patients need suitable interventions/treatments to prevent them from developing cognitive impairment. However, monitoring the effects of interventions/treatments on cognitive changes is difficult because OSA patients are cognitively normal at baseline, e.g., MMSE = 28.7 ± 1.2 at baseline in this study. It is impossible to examine the significant improvements in cognition due to interventions/treatments in subjects who are cognitively normal at baseline. Instead, biological information such as plasma biomarkers might be useful for monitoring the effects of interventions/treatments on OSA patients.

In the clinic, OSA patients are frequently treated with continuous positive airway pressure (CPAP). In this study, 20 OSA patients received 3 months of CPAP treatment (CPAP group), whereas the others did not (placebo group). IMR blood tests were performed on each of the enrolled OSA patients. Three months later, IMR blood tests were performed again for each OSA patient. The patients in the CPAP group had decreased plasma $A\beta_{1-42}$xT-Tau levels (−44.35 ± 88.92 pg/ml), as shown in Fig. 5.12. However, the patients in placebo group had significantly greater plasma $A\beta_{1-42}$xT-Tau levels (48.73 ± 34.02 pg/ml) than patients in the CPAP group ($p < 0.05$). Notably, OSA patients who received CPAP treatments had decreased plasma $A\beta_{1-42}$xT-Tau levels. This indicates that the risk of developing cognitive impairment can be mitigated in the future. This example demonstrates the application of blood tests in monitoring the effects of interventions/treatments on the prevention of cognitive impairment in cognitively normal subjects.

The third example is the application of blood tests to monitor the effects of olfactory stimulation with cognitive training on patients with mild to moderate dementia. Lin et al. enrolled 18 participants with mild to moderate dementia who were equally randomly assigned to the control group (placebo) and olfactory stimulation with cognitive training group (OS) [48]. The intervention was a 12-week program with one 30-min session twice a week. Lin et al. employed IMR for measuring plasma biomarkers and several cognitive and psychological tests to measure basal and postintervention values.

The results revealed no significant differences between the groups after the intervention, as measured using the Mini-Mental State Examination, Loewenstein Occupational Therapy Cognitive Assessment (LOTCA), Top International Biotech Smell Identification Test, or Geriatric Depression Scale (GDS). However, the

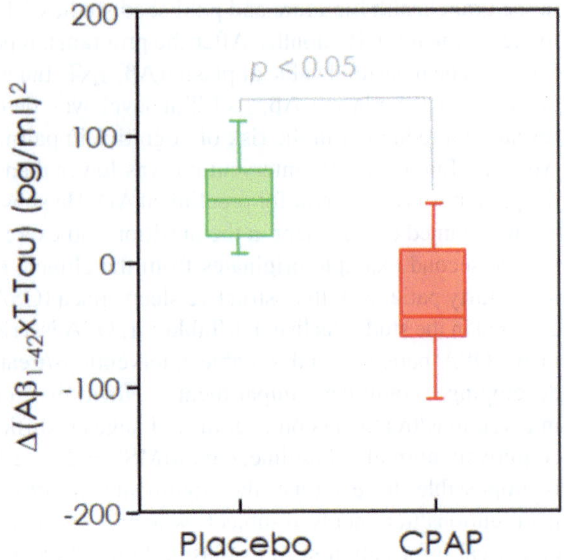

Fig. 5.12 Changes in plasma $A\beta_{1-42}$xT-Tau levels in OSA patients with CPAP (CAP) and without (placebo) 3-month CPAP treatment. The plasma $A\beta_{1-42}$ and T-Tau levels were assayed by IMR [47]

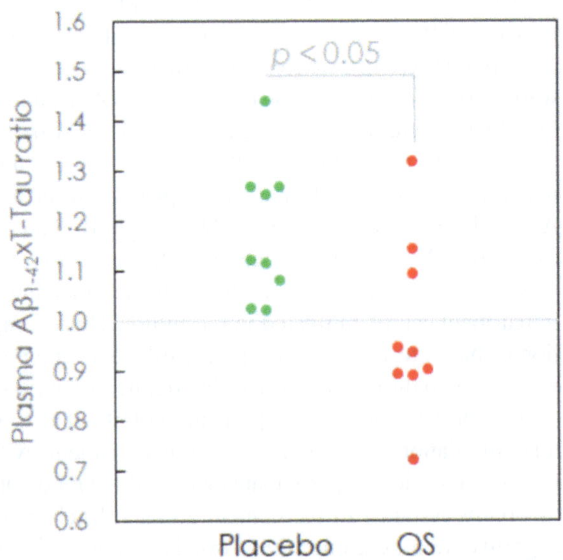

Fig. 5.13 Changes in plasma $A\beta_{1-42}$xT-Tau levels in patients mild to moderate dementia with (OS) and without (placebo) the 12-week olfactory stimulation with cognitive training. The plasma $A\beta_{1-42}$ and T-Tau levels were assayed by IMR [48]

postintervention-to-preintervention biomarker ratios significantly differed between the placebo and OS groups, as shown in Fig. 5.13.

A biomarker ratio of 1 indicates that the postintervention biomarker level is equal to the preintervention biomarker level. In the case of a biomarker ratio <1, the measured biomarker level decreased after OS intervention, and when the biomarker ratio is greater than 1, the measured biomarker level increased after OS intervention.

As shown in Fig. 5.3, the plasma $A\beta_{1-42}$xT-Tau ratio was 1.18 ± 0.14 in the placebo group and 0.98 ± 0.18 in the OS group. A significant difference in the plasma $A\beta_{1-42}$xT-Tau ratio was obtained ($p < 0.05$) between the placebo and OS groups. These results reveal the possibility of suppressing increases in plasma $A\beta_{1-42}$xT-Tau levels with the intervention. Thus, the risk of disease progression is promisingly reduced with the aid of olfactory stimulation with cognitive training.

Compared with cognitive and psychological tests, blood tests for monitoring changes in plasma biomarkers are more sensitive for identifying disease progression. Neurologists could use blood tests to effectively manage/treat patients. At present, the BBBMs are used for screening and diagnosis, but the future will include detection and monitoring of treatment responses as well as monitoring the progression of the disease.

5.2.9 Correlations of Brain Atrophy

In aMCI or AD patients, brain atrophy continues to occur as a marker of neurodegeneration. Cognitive and behavioral decline follow brain atrophy, resulting in the clinical progression of aMCI to AD dementia. However, effective interventions/treatments are being investigated to reduce the rate of brain atrophy or preserve brain structure, and many drugs are using measurements of atrophy as a marker of efficacy [49–52]. Magnetic resonance imaging is the standard method for observing brain atrophy using volumetric software. However, magnetic resonance imaging might not be available in regional hospitals or clinics. Also, there is great variability in the strength of the magnet ranging from 0.5 T to 7 T, with the standard practice being either 1.5 or 3 T. This results in variability in image resolution. Patients may need to visit larger medical centers for examination by magnetic resonance imaging, which may not be convenient for patients to do frequently. Therefore, blood tests are the most convenient and more widely available examinations for evaluating the occurrence of brain atrophy.

As shown in Fig. 4.9 in Chap. 4, plasma T-Tau levels are negatively correlated with brain volume. This correlation reveals the feasibility of monitoring the brain volume in aMCI and AD patients during interventions/treatments. For example, as we found, if the increases in the plasma T-Tau level of an aMCI and AD patient continue to be unchanged or to become faster during interventions/treatments, this indicates that the patient's brain is becoming smaller and that the neurologist needs to modify the recommendations to prevent further atrophy of the brain. However, if the increasing rate of the plasma T-Tau level of an aMCI or AD patient is slowed down, this indicates that interventions/treatments are protecting the brain from damage. Hence, blood tests for inspecting brain atrophy every 1–3 months are strongly recommended for aMCI and AD patients in clinical practice.

In summary, the clinical applications of plasma biomarkers in the diagnosis of AD at various stages, such as preclinical AD, SCD, aMCI, early AD, and amyloid pathology, are demonstrated. In addition, the ability of plasma biomarkers to predict the development of cognitive impairment in CU individuals and cognitive decline in aMCI patients is illustrated. Ultimately, monitoring the effects of interventions/treatments using plasma biomarkers in at-risk individuals, aMCI patients, early AD patients, and mild to moderate AD patients is feasible. Therefore, plasma biomarkers can be practically utilized in clinical practice on a routine basis in the management of Alzheimer's disease.

5.3 Clinical Utility in Parkinson's Disease and Dementia with Lewy Bodies

The gold standard for diagnosing Parkinson's disease (PD) is the observation of motor signs and symptoms characterized by asymmetric tremor, bradykinesia, cogwheel rigidity, and gait disturbance. However, movement disorders can be attributed to other causes, such as tauopathies, including progressive supranuclear palsy, heavy metal toxicity, and autonomic dysfunction [53–55]. Dementia with Lewy bodies (DLB) can also be categorized as a movement disorder. Although the "1-year rule" helps to differentiate DLB from PD dementia, it relies on informant history, which can inherently be imprecise. It is not easy to exactly identify the onset of cognitive impairment or motor disorders in DLB. Therefore, "1 year" is extremely difficult to precisely determine in the clinic. Other neurological diseases, such as cerebral autosomal dominant arteriopathy with subcortical infarcts and leukoencephalopathy (CADASIL), are accompanied by movement disorders [56–59].

In some cases of very early-stage PD, nonmotor symptoms such as rapid eye movement, insomnia, anosmia, constipation, facial stiffness, etc., which are not easily recognized, can occur. Observations of motor disorders make it difficult to achieve precise and early PD diagnosis. Biological diagnosis could be a powerful way to help neurologists make the right decisions when caring for their patients. With the clinical validations of plasma biomarkers described in Chap. 4, assays of plasma biomarkers would be promising for biological diagnosis.

Regarding the application of plasma biomarkers in the management of PD, several issues are of interest. Is it possible to evaluate the severity of PD using plasma biomarkers? Could we use plasma biomarkers to determine the degree of cognitive impairment in PD? Do PD patients with different degrees of motor disorders have different levels of plasma biomarkers? How can plasma biomarkers be used to predict the progression of motor disorders in PD? As PD patients are treated, could plasma biomarkers be used as monitoring indices? In this section, we illustrate how to use plasma biomarkers to address these issues of interest in clinical practice.

5.3.1 Assessment of PD

The primary pathological biomarker of PD is α-synuclein. The results listed in Table 4.4 in Chap. 4 demonstrate the clinical sensitivity and specificity of using plasma α-synuclein to assist in the diagnosis of PD in clinical practice. Assays utilizing IMR and IP/RT-QuIC show that the agreement between plasma α-synuclein diagnosis and clinical diagnosis is greater than 90%. This indicates that testing plasma α-synuclein could be used to confirm the diagnosis of PD in clinical practice. The high agreement between plasma α-synuclein diagnosis and clinical diagnosis using IMR and IP/RT-QuIC is mainly due to the presence of α-synuclein fibrils in plasma. Notably, fibrils, not monomers, of α-synuclein are critical to the development of PD. Measurements of fibrils of α-synuclein are essentially strongly associated with PD development.

The movement symptoms of PD often start with rest tremors. However, tremors, so-called essential tremors, are common among elderly people. Although the symptoms of essential tremors differ from those of PD tremors, elderly people are usually confused due to a lack of awareness of the distinction between these two tremors, and it can be difficult to differentiate them. Additionally, medications can cause tremors. Thus, the neurologist relies on the characterization of the tremor, which can be imprecise. This strengthens the argument for a biological identification of PD using BBBMs. Here, an example is given.

Fion, aged 78 years, has been suffering from tremors for 2 months. She worried about developing PD. Just in case, her daughter brought her to visit a neurologist, who carefully documented the clinical symptoms and assessed the clinical signs, including assessing for bradykinesia, checking the gait, and assessing for cogwheel rigidity. The neurologist informed the patient that she was not suffering from PD although she had tremors. For example, the tremors disappeared when she was at rest. However, the patient was not convinced by the neurologist because of her clear tremors. She asked the neurologist for detailed examinations such as dopamine scan. Due to the limited availability of a dopamine scan, Fion remained on a waiting list for 1–1.5 months. Fion complained to the neurologist about the long waiting time and asked the neurologist for an alternative examination that could be performed immediately. Hence, the neurologist recommended blood tests for α-synuclein, which were performed.

One week later, at the follow-up visit, Fion was informed that the level of plasma α-synuclein was below the cutoff value, i.e., 116 fg/ml, as determined by IMR, indicating that it was not consistent with PD. Her neurologist informed her that she was not suffering from PD but was suffering from essential tremors. She received treatments for essential tremors from the neurologist.

George, aged 46 years, had symptoms of gait impairment, bradykinesia, and rigidity. His UPDRS III score was 12, and his H-Y stage was 2. He clearly had typical symptoms of PD and was diagnosed with clinical PD. However, he was relatively young to have a neurodegenerative disease. Conservatively, the neurologist asked him to undergo blood testing to ensure that he had PD and considered genetic testing as well.

The patient's plasma α-synuclein concentration was below the cutoff value, i.e., 116 fg/ml by IMR. The neurologist suspected that the patient might have something other than PD. Hence, the neurologist strongly recommended neuroimaging examinations for George. The results of the dopamine scan were negative, but the cortex was abnormally thin. The neurologist ordered an assessment of the expression of the Notch3 gene, which revealed that the patient had a mutation. Following detailed neurological and cardiovascular examinations, the patient was diagnosed with CADASIL, not PD.

Another male, Henry, aged 63 years, had a family history of PD. His father died from PD. Although Henry did not have any clinical symptoms of PD, he underwent genetic analysis because of the family history. The results revealed that Henry had a mutation in the SNCA gene, which is a common genetic cause of PD. Therefore, he was at risk for PD and always worried about developing PD.

To monitor for emergent PD, Henry visited the family clinician regularly and underwent blood tests every year. A few years later, his plasma α-synuclein level was found to be slightly greater than the cutoff value of 116 fg/ml by IMR. The family clinician immediately triaged the patient to a neurologist. Through careful examinations by a neurologist, the patient was found to have very early-stage PD requiring treatment. Fortunately, Henry was at a very early stage of PD, and it was not difficult to receive the treatments. With the early diagnosis and early treatments, he continues to enjoy a high-quality life.

Therefore, assays of plasma α-synuclein are very helpful for neurologists to easily diagnose or exclude PD for outpatients.

5.3.2 *Monitoring of Disease Progression in PD*

When a patient is diagnosed with PD, a neurologist also identifies the severity of the disease. The severity of PD involves several aspects, mainly including motor disorders; mental, behavioral, or other nonmotor problems; and cognitive impairment. The H-Y stage and UPDRS are used to quantify motor severity, while MMSE and MoCA scores are typically used to assess cognitive symptoms of PD. All the diagnoses are symptomatic observations, whose results strongly depend on the clinical experience of a neurologist. It is not easy to objectively clarify the changes in these scores as the disease progresses, especially when the time interval between two clinical visits is short. It would be helpful to have supportive biological information for the neurologist to determine the severity PD progression.

As demonstrated in Chap. 4, the levels of various plasma biomarkers are related to the H-Y stage and the UPDRS III, MMSE, and MoCA scores. These findings reveal the possibility of using plasma biomarkers as supportive biological information for monitoring PD progression in clinical practice. For example, the plasma p-α-syn129 level positively correlates with the H-Y stage. Thus, PD patients with

Table 5.2 Plasma biomarkers for monitoring the declines in the H-Y stage and UPDRS III, MMSE, and MoCA scores in PD patients [60–65]

		Plasma biomarkers	
Domain	Decline	Increasing level	Decreasing level
Motor	H-Y stage	p-α-syn129	–
	UPDRS III	p-α-syn129, NfL	
Cognition	MMSE	$A\beta_{1-40}$	α-synuclein, T-Tau
	MoCA		
Brain structure	Volume, thickness, surface	NfL, T-Tau, pTau181	–

greater plasma p-α-syn129 levels would have later H-Y stages, indicating the progressive loss of motor ability.

Plasma biomarkers for monitoring the progression of PD are listed in Table 5.2. The main clinical domains of disease progression are motor and cognitive improvement or decline. The clinical diagnosis in the motor domain includes the H-Y stage and UPDRS III score. The clinical diagnosis in the cognitive domain includes the MMSE and MoCA scores in PD. The associations among plasma biomarkers, clinical domains, and clinical diagnoses, such as the H-Y stage and UPDRS III, MMSE, and MoCA scores, in PD patients are linked in Table 5.2 [60–65].

Notably, the key plasma biomarker for monitoring motor progression in PD patients is p-α-syn129, whereas total α-synuclein is an important plasma biomarker for monitoring cognitive progression in PD patients. Obviously, p-α-syn129 and total α-synuclein play different roles in PD. More studies are expected to clarify the functions of p-α-synuclein and total α-synuclein in the pathogenesis of PD.

As shown in Table 5.2, the plasma NfL increases as the UPDRS III score increases in PD patients. Since NfL is usually released as nerve cells are damaged, this observation might result from damage to nerve cells in the motor domains of the brain. Hence, monitoring the plasma NfL level is a promising method for monitoring nerve cell damage in the motor domain of the brain in PD patients.

Amyloid-β and T-Tau can be used to monitor cognitive progression in PD patients. Researchers have shown that amyloid-β and T-Tau are able to significantly discriminate PD patients with dementia from PD patients with normal cognition ($p < 0.05$). These findings suggest that Aβ and T-Tau might contribute to the cognitive impairment in PD patients, although PD is essentially a synucleinopathy, not an amyloidosis or a tauopathy. The potential participation of Aβ and T-Tau in the cognitive progression of PD reveals the complexity of the pathogenesis of PD as cognitive impairment develops. Combinations of treatments involving synucleinopathy, amyloidosis, and tauopathy should be considered for PD patients with cognitive impairment. During the combined treatments, plasma biomarkers of total α-synuclein, p-α-syn129, Aβ, and T-Tau are convenient and efficient parameters for routinely monitoring treatment effects.

In addition to the progression of motor disorders and cognitive impairment, brain atrophy is a key clinical issue in PD. The gold standard for assessing brain atrophy is volumetric magnetic resonance imaging (MRI). However, it is difficult to perform

MRI examinations frequently, e.g., once per month, in PD patients. Instead, blood tests could be performed every month to care for these patients. As described in Chap. 4, many clinical results support the feasibility of monitoring brain atrophy in PD patients by measuring plasma NfL, T-Tau, and pTau181 levels. The levels of these plasma biomarkers significantly and continuously increase as brain atrophy continues to occur. Although the elevations in the levels of these plasma biomarkers cannot reveal the exact regions of atrophy in the brain, blood tests are convenient for the instant monitoring of indices reflecting structural changes in the brain in patients with PD and help neurologists obtain information about brain damage in PD patients.

5.3.3 Prediction of Decline in Movement Disorders

In PD patients, a decline in motor disorders is common in the clinic. However, the rates of decline differ from one patient to another, depending on many factors. Neurologists prefer to know how rapid the decline will be for a PD patient and to provide complete treatments to the patient to delay the decline.

Line et al. demonstrated the feasibility of predicting motor decline in PD patients [60]. Plasma p-α-syn129 assayed by IMR at baseline is a promising biomarker for predicting future motor decline.

A total of 112 PD patients were enrolled at baseline. The plasma p-α-syn129 concentration of each PD patient was assayed using IMR technology. A longitudinal follow-up study was conducted to observe motor decline. After a mean follow-up of 1275 ± 766 days, 36 of 122 patients (29.5%) with PD presented with a sustained increase of at least 3 points in the MDS-UPDRS part III score. Baseline p-α-syn129 levels correlated with the decline in motor function. Lin et al. analyzed the receiver operating characteristic (ROC) curve and identified the cutoff value based on the Youden's J index. They reported that among PD patients, those with baseline p-α-syn129 concentrations greater than 8.46 fg/ml were at a greater risk of motor decline by at least 3 points in the MDS-UPDRS III than those with p-α-syn129 concentrations lower than 8.46 fg/ml. Kaplan–Meier analysis revealed a clear difference between patient subgroups whose p-α-syn129 levels were above and below the cutoff value, as shown in Fig. 5.14. Obviously, the plasma p-α-syn129 level at baseline is a powerful predictor of motor symptom progression in PD patients.

5.3.4 Monitoring of Treatment Responses in PD

Most patients with Parkinson's disease (PD) are treated with medication. Medications are taken to indirectly or directly increase dopamine levels in the brain. Hence, the medications are for modulating symptoms, not curing nerve cells in the brain. Currently, there are two main drugs used: levodopa and dopamine agonists. The

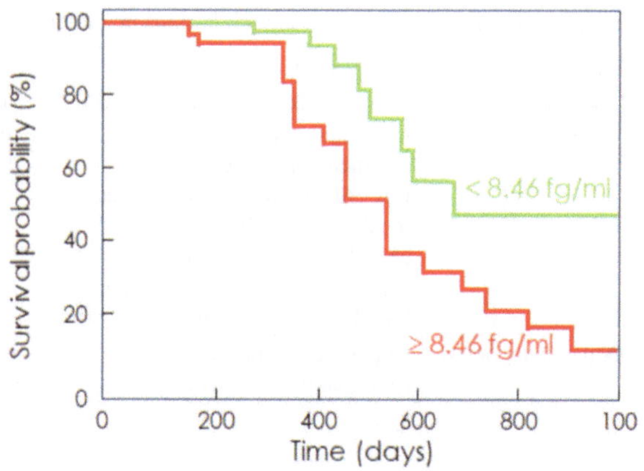

Fig. 5.14 Survival probability without motor symptom progression in PD patients with high or low plasma p-α-syn129 levels at baseline during follow-up

mechanism, dosage, and side effects of these two drugs are listed in Table 5.3. Depending on the severity of PD, neurologists provide individual drugs or combinations of the two drugs to a patient.

To enhance the treatment effects, anticholinergics, amantadine, monoamine oxidase-B inhibitors, catechol-O-methyltransferase inhibitors, or adenosine A2A antagonists can sometimes be given to a PD patient. The mechanism, dosage, and side effects of these drugs are also listed in Table 5.3.

In addition to medications, surgical therap therapies so-called deep brain stimulation (DBS), and High Intensity Focused Ultrasound (HIFUS) are used to treat patients with PD. DBS involves implanting electrodes within areas of the brain. An impulse generator battery, which is similar to a heart pacemaker and approximately the size of a stopwatch, is also implanted under the collarbone or in the abdomen to activate electrical stimulation to targeted areas in the brain that control movement. Hence, DBS can help relieve symptoms but not cure or stop disease progression. DBS is typically performed for PD patients who experience disabling tremors, wearing-off spells, and medication-induced dyskinesia.

DBS is a surgical therapy that can have side effects, such as bleeding in the brain, brain ischemia, infection, trouble breathing, nausea, heart problems, and seizures. Electrical stimulation can have additional side effects, such as headache, confusion and trouble concentrating, hardware complications, or temporary pain and swelling at the implantation site. Patients undergoing treatment by DBS need to frequently visit neurologists for safety issues, especially during the initial period.

High-intensity focused ultrasound (HIFUS) has emerged as a surgical alternative to DBS. Focused ultrasound (FUS) is a noninvasive, permanent, one-time procedure creating a brain lesion, ideal for severe unilateral tremor, but less adjustable and currently limited to one side. The choice depends on symptom profile (tremor vs. other symptoms), preference for hardware/invasiveness, and need for adjustability, with DBS offering flexibility and FUS offering immediate, incisionless relief.

Sometimes, to relieve muscle stiffness or joint pain through movement and to solve difficulties in patients' daily lives, such as dressing, swallowing, or speech,

Table 5.3 Medications for patients with Parkinson's disease

Medication	Mechanism	Dosage	Side effects
Levodopa	Levodopa is absorbed by nerve cells in the brain and turned into the chemicaldopamine It is often combined with other medication, such as benserazide or carbidopa, to prevent the levodopa from being broken down in the bloodstream before it enters the brain	Tablet or liquid	Feeling sick Tiredness Dizziness
Dopamineagonists	Dopamineagonists act as a substitute for dopamine in the brain	Tablet or skin patch	Feeling sick Tiredness Dizziness Compulsive behaviors
Anticholinergics	These are the oldest type of Parkinson's medication and help restore the balance between dopamine and another brain chemical, acetylcholine. They are primarily used for managing severe tremors and rigidity, particularly in younger patients, as they can cause significant cognitive side effects in older adults	Spray	Being unable to pee Blurred vision Confusion Dry eyes Dry mouth Fast heart rate High body temperature Pupils that are very dilated Skin that's flushed and dry Slowed digestion Vomiting Weakened muscle strength
Amantadine	This medication works on dopamine and glutamate pathways and can be used alone in early, mild Parkinson's or to treat involuntary movements (dyskinesias) that may develop as a complication of long-term levodopa use	Tablet	Bladder pain Bloody or cloudy urine Blurred vision Confusion Difficult, burning, or painful urination Dizziness or lightheadedness Fainting Falls Frequent urge to urinate Lower back or side pain Seeing, hearing, or feeling things that are not there Swelling of the hands, feet, or lower legs
Monoamine oxidase-B inhibitors	These drugs block the effects of monoamine oxidase-B, an enzyme that breaks down dopamine	Tablet	Feeling sick Headaches Abdominal pain High or low blood pressure

(continued)

Table 5.3 (continued)

Medication	Mechanism	Dosage	Side effects
Catechol-O-methyltransferase inhibitors	These drugs prevent levodopa from being broken down by the enzyme catechol-O-methyltransferase	Tablet	Feeling sick Diarrhea Abdominal pain
Adenosine A2A antagonists	A newer class of medication (e.g., istradefylline) that blocks the brain chemical adenosine to boost dopamine signaling and help manage "off" periods when added to levodopa/carbidopa treatment	Tablet	Dyskinesia Constipation Nausea Dizziness Hallucinations Diarrhea Insomnia Decreased appetite Upper respiratory tract infections Skin rash Elevated liver enzymes Hyperglycemia Increased urea levels

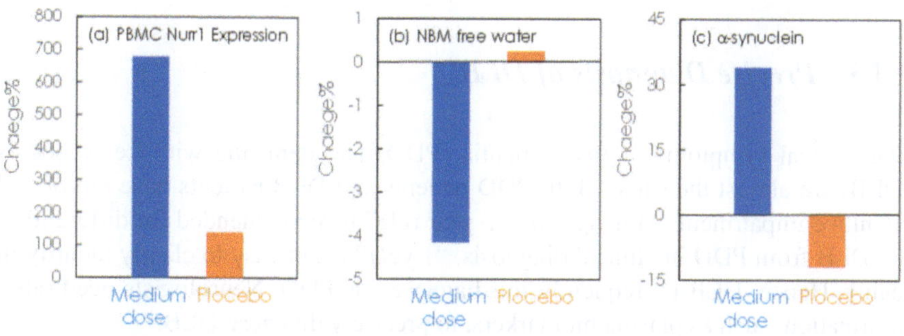

Fig. 5.15 Changes in (**a**) peripheral blood mononuclear cell mRNA expression of nuclear receptor-related-1 (PBMC Nurr1), (**b**) free water in the nucleus basalis of Meynert (NBM), and (**c**) serum α-synuclein level through lithium carbonate treatment at a medium dose (15 mg/day for 24 weeks) [66]

supportive therapies such as physiotherapy, occupational therapy, and speech and language therapy can be performed for PD patients. In some cases, dietitians provide dietary advice to PD patients to improve or avoid constipation, dizziness, and unintentional weight loss.

All the therapies mentioned above are passive and do not affect the recovery of nerve cells. Researchers would like to develop therapies that are able to protect or rescue nerve cells to cure PD. For example, lithium has multiple neuroprotective effects, including suppressing microglial activation, reducing inflammation and oxidative stress, and enhancing autophagy and mitochondrial biogenesis and function. Lithium has been shown to be effective in animal models of Parkinson's disease (PD). Guttuso et al. conducted a pilot study on the effects on PD patients [66]. According to the results shown in Fig. 5.15, compared with the placebo, the medium-dose therapy (MDT), i.e., lithium carbonate at 15 mg/day for 24 weeks, activated cells significantly in PD patients (change: 679% in MDT; 139% in placebo) and decreased the free-water levels in the thalamus, the nucleus basalis, and the posterior substantia nigra, which is the opposite of the known longitudinal free-water change in PD patients (change: −0.98 to −4.16% in MDT; 0.25–1.8% in placebo). Furthermore, the expression of total α-synuclein was recovered by monitoring its levels in serum using an IMR assay (change: 34% in the MDT group; −12% in the placebo group) through medium-dose therapy.

Pharmaceutical companies are developing not only drugs to cure PD but also vaccines to prevent PD. For example, by injecting UB-312, antibodies selectively targeting pathological oligomeric and fibrillar α-synuclein forms are induced [67]. The results of a phase I trial revealed that the CSF p-α-syn129 concentration in PD patients injected with UB-312 but without a titer continuously increased when the IMR was assayed for 42 weeks. However, PD patients injected with UB-312 and with a titer showed continuous decreases in the CSF p-α-syn129 concentration. These examples demonstrate how to apply assays of total α-synuclein and p-α-syn129 to monitor responses to treatment in PD patients.

5.3.5 Precise Diagnosis of DLB

The clinical symptoms of PD dementia (PDD) and dementia with Lewy bodies (DLB) are almost the same. Both PDD patients and DLB patients have motor and cognitive impairment. Although the "1-year rule" is recommended for differentiating DLB from PDD in clinical diagnosis, "1 year" is not easy to clearly identify in reality. Hence, DLB is frequently misdiagnosed as PDD. Neurologists need other information, such as plasma biomarkers, to precisely diagnose DLB.

As shown in Figs. 4.26 and 4.28 in Chap. 4, compared with those in the NCs, both PDD patients and DLB patients have significantly greater levels of total α-synuclein in the plasma. The levels of plasma total α-synuclein overlap between DLB and PDD. The consistency in the plasma total α-synuclein levels between DLB and PDD patients is reasonable because DLB and PDD patients suffer from neurodegeneration-derived motor disorders. The motor disorders in DLB and PDD likely have the same pathogenic mechanism. However, the origins of cognitive impairment differ between DLB and PDD. Hence, by taking pTau181 into account, as discussed in **Plasma Biomarkers Distinguishing PDD from DLB** in Chap. 4, the combination of the plasma pTau181 level and UPDRS score is able to discriminate DLB from PDD well, with an AUC of 0.88. This is a promising method for precisely diagnosing DLB.

Therefore, we suggest a flow chart for evaluating DLB with the aid of plasma biomarkers. First, the patient must show clear motor disorders and cognitive impairment. Second, the plasma total α-synuclein concentration must be elevated, e.g., ≥ 104 fg/ml, as shown in Figs. 4.26 and 4.28 in Chap. 4. This result confirms the presence of neurodegeneration-derived motor disorders. Third, the combination of the plasma pTau181 level and UPDRS score can be used to determine whether the patient is suffering from DLB or PDD. Notably, the regular examinations for PDD and DLB by neurologists cannot be ignored in clinical practice.

5.4 Biological Assessments of Frontotemporal Dementia

Frontotemporal dementia (FTD) frequently presents with a presenile presentation (age less than 65 years), occurring in males and females equally. Health professionals estimate the prevalence of FTD to be between 15/100,000 and 22/100,000 which makes it relatively uncommon. At present, 1.2–1.8 million people suffer from FTD worldwide. Therefore, FTD is among the more common neurodegenerative diseases.

The clinical symptoms of FTD differ among patients, and the disease is subcategorized into behavioral variant (bvFTD) and language variant (lvFTD). The most common symptoms of bvFTD include loss of inhibition, apathy, loss of empathy, compulsive behaviors, changes in diet or mouth-centered behaviors (Kluver–Bucy syndrome), and loss of executive function. The most common symptoms of lvFTD include troubles in grammar and forming words, confusion with complicated sentences, difficulties in choosing and understanding words, and speaking nonsensically. Patients could experience individual or multiple symptoms in combination. In terms of clinical symptoms, FTD can be divided into three main subgroups:

- Behavioral form, termed behavioral variant frontotemporal dementia (bvFTD).
- Language variant (lvFTD), termed primary progressive aphasia (PPA). This can be further divided into progressive nonfluent aphasia (PNFA) and semantic dementia. Logopenic PPA is described above.
- Motor presentation, termed either FTD with amyotrophic lateral sclerosis (FTD-ALS) or atypical parkinsonian disorder.

Neurologists need to carefully examine the clinical symptoms of patients to identify the subtype of FTD. Especially at the early stage, patients with FTD have clinical symptoms similar to those of patients with other neurodegenerative diseases.

Approximately 40% of FTD patients have a family history. This suggests that FTD is highly genetic. More than 30% of FTD patients are familial, with autosomal dominant mutations in three genes accounting for most of the inheritance:

- Progranulin (GRN)
- Chromosome 9 open reading frame 72 (C9orf72)
- Microtubule-associated protein tau (MAPT)

Recently, biological evidence such as abnormal forms and levels of T-Tau, TDP-43, and FET proteins (FUS, EWSR1, and TAF15) has been reported in the majority of FTD patients. The results shown in Figs. 4.32 and 4.33 in Chap. 4 clearly reveal the elevations in plasma T-Tau and TDP-43 in FTD patients. Hence, plasma T-Tau, TDP-43, and FET proteins are promising biomarkers for assessing FTD. Until now, the diagnosis involved detection of genetic markers, FDG-PET, and clinical characterization.

Neuroanatomically, FTD involves the deterioration of the frontal and temporal lobes in the brain. Recent studies have revealed more widespread cortical, subcortical, cerebellar, and brainstem involvement. Neuroanatomic examinations can be performed with magnetic resonance imaging or PET/CT.

Depending on where the deterioration starts in the brain, FTD affects one's behavior, speech, and understanding of others. FTD is a progressive disease that is not curable or treatable, but some symptoms can be relieved.

Here is an example.

Ivan, aged 67 years, male, presents with difficulty talking fluently to his family. One day, his daughter brought Ivan to a neurological clinic. His daughter informed the neurologist that her father sometimes forgot the date. Whenever she asked her father, "What day is today?" he did not say anything. Did he have problems with memory? Did he have Alzheimer's disease?

The neurologist asked the patient some quick questions. During the examination, the neurologist wanted the patient to remember three kinds of fruits: apple, watermelon, and peach. A few seconds later, the neurologist asked more questions. Ten minutes later, the neurologist asked, "What are the three kinds of fruits I wanted you to remember?" Not surprisingly, Ivan could not answer the neurologist. This indicates the possibility of having a short-term memory problem, which is a typical symptom of Alzheimer's disease.

Then, the neurologist asked the daughter, "Does your father usually speak slowly, sometimes not fluently?" She responded, "My father was a vice president at an insurance company. He had to frequently communicate with his colleagues and customers. Therefore, he was talkative." The daughter kept saying, "However, your question reminded me that my father talks more slowly now." Thus, the neurologist suspected that the patient might be suffering from FTD, not Alzheimer's disease. The neurologist recommended Ivan with blood tests for TDP-43. The testing results revealed irregularly high levels of plasma TDP-43, i.e., > 0.237 pg/ml, as determined by IMR technology. The neurologist then ordered magnetic resonance imaging and fluorodeoxyglucose positron emission tracer (FDG-PET) examinations. With all the diagnostic details, the patient was identified as having FTD, not Alzheimer's disease.

The family was misinterpreting the anomia as being a cardinal feature of AD. The inability to answer questions could be attributed to several possible causes. Loss of memory could be one of them. An alternative cause could be speech disorders. In this case, the patient could remember the three kinds of fruits in his brain, but he had a language problem preventing him from answering the question. Therefore, the patient did not have problems in memory, but in language, which was evidenced by the changes in daily communications. Such behaviors frequently occur among FTD patients. Short-term memory disorders cannot be identified in patients based solely on the inability to answer such questions. More careful observations should be made with respect to patients' behaviors as outpatients.

Unfortunately, there is no clinical cure or treatment for FTD. Certain medicines and speech therapy can help manage the clinical symptoms of FTD, particularly the management of the behavioral symptoms. In some cases, speech therapy teaches FTD patients to use communication aids. We hope that researchers can successfully develop curative or therapeutic drugs to help FTD patients in the near future.

5.5 Differential Evaluations Among Neurodegenerative Diseases

In the clinic, the first step in the diagnosis of neurodegenerative diseases involves behavioral observations and cognitive tests. This is the so-called phenotype examination. The results strongly depend on the professional experiences of neurologists or neuropsychologists and on the education, age, and social economics of the examined individual. In addition, early-stage dementia of various types could have common phenotypes, as illustrated in Table 5.4. For example, all the neurodegenerative diseases show changes in mood and personality, movement difficulties, or cognitive impairment. Both aMCI/AD patients and PDD patients have memory deficits and deterioration. PDD and DLB patients have movement difficulties. It is difficult to achieve an objective and precise diagnosis via phenotype examinations. In the clinic, several challenging situations prevent a precise diagnosis. For example, there

Table 5.4 Phenotypes of neurodegenerative diseases, including aMCI/AD, PD/PDD, DLB and FTD

Disease	aMCI/AD	PD/PDD	DLB	FTD
Age of onset (years)	> 65	50–85	50–85	45–65
MMSE score	17.8 ± 4.5	17.5 ± 5.8		21.41 ± 8.11
Cognitive impairment	✓	✓	✓	✓
Inability to concentrate	✓	✓		✓
Get lost	✓	–		–
Memory deficits	✓ (Early onset)	✓ (51.2%)		Less prominent in early onset
Changes in mood and personality	✓	✓	✓	✓
Loss of semantic understanding	✓	✓		✓
Loss of speech fluency	✓	✓		✓
Movement	✓	✓	✓	✓
Sleep deterioration	✓ (Fragmented sleep)	✓ (Sleepbehavior disorder)	✓ (Sleepbehavior disorder)	Rare

can be confusion between DLB and PDD, early DLB and aMCI, FTD and AD, etc., for neurologists and neuropsychologists in making a clearly decisive diagnosis. Hence, misdiagnosis can occur occasionally via phenotype examinations. Occasionally, misdiagnosis can cause irreparable mistakes or major disasters for subjects and their families.

To avoid the misdiagnosis of neurodegenerative diseases, clinicians would like to perform CSF biomarker assays or neuroimaging for patients at early stages. However, owing to the invasiveness of lumbar puncture and the high cost and low availability of neuroimaging, most subjects do not want or cannot undergo CSF biomarker assays or neuroimaging. Hence, the precise diagnosis of neurodegenerative diseases is difficult to achieve in clinical practice.

With the successful development of blood tests for the diagnosis of neurodegenerative diseases, the challenges associated with achieving precise diagnosis are overcome because blood tests are low cost, low risk, highly convenient, and highly available. In particular, blood tests are available for identifying versatile biomarkers associated with different types of neurodegenerative diseases. A panel consisting of versatile plasma biomarkers associated with various types of neurodegenerative diseases could be helpful for achieving precise diagnosis in routine practice.

Hereafter, we demonstrate the feasibility of using a panel of versatile plasma biomarkers to differentiate among types of neurodegenerative diseases, such as AD, PD, DLB, and FTD.

5.5.1 Differentiation Between DLB and PDD

DLB and PDD are associated with common clinical symptoms of motor disorders and cognitive impairment. Although the "1-year rule" is used in clinical practice, determining the exact time interval between the onset of motor disorders and cognitive impairment is difficult. Moreover, the plasma total α-synuclein level does not significantly differ between PDD patients and DLB patients, although both PDD patients and DLB patients had higher plasma total α-synuclein levels than CU individuals, as shown in Fig. 5.16. Hence, the symptomatic diagnosis and plasma total α-synuclein assay fail to differentiate DLB from PDD. The misdiagnosis of DLB as PDD frequently occurs in clinical practice.

Giovanni et al. reported that the plasma pTau181 level can significantly discriminate DLB (3.83 ± 0.79 pg/ml) from PDD (3.03 ± 0.65 pg/ml; $p < 0.01$) [68]. This suggests that tauopathy could be attributed to DLB but not to PDD. Furthermore, the combinations of clinical examination scores and plasma biomarkers demonstrate the ability to differentiate DLB from PDD when the plasma pTau181 level is combined with the UPDRS score (clinical sensitivity = 0.83, specificity = 0.83, AUC = 0.88) or when the plasma total α-synuclein level is combined with the UPDRS score (clinical sensitivity = 0.83, specificity = 0.83, AUC = 0.85). The combinations of plasma biomarkers (pTau181 or total α-synuclein) with the UPDRS

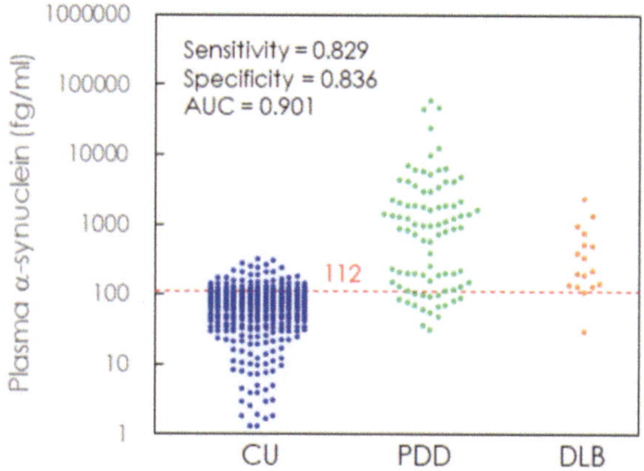

Fig. 5.16 Plasma total α-synuclein levels discriminate PDD/DLB from CU, but not PDD from DLB

score proposed by Giovanni et al. can facilitate the diagnosis by precisely differentiating DLB from PDD.

In addition to the proposal by Giovanni et al., Kannarkat et al. suggested that α-synuclein conformations in plasma can distinguish DLB from PD (clinical sensitivity = 0.92, specificity = 0.61, AUC = 0.83) [69]. With this convincing clinical evidence, more effective and precise examinations involving plasma biomarkers can reduce the misdiagnosis of DLB or PDD.

5.5.2 *Differentiation Between Early DLB and aMCI*

DLB is easily confused with not only PDD but also aMCI or early AD. In DLB, most patients experience cognitive impairment before motor disorders, especially at early stages. Early DLB patients can show cognitive symptoms similar to those of aMCI or early AD patients. In addition, DLB has overlapping pathology associated with tauopathy and amyloid deposition ranging from 48% to 88%, but Braak staging tends to be lower (Braak IV), resulting in approximately 30% of DLB patients having abnormal plasma $A\beta_{1-42}$xT-Tau levels, which reveal a high risk of aMCI or early AD. Hence, one-third of DLB patients at early stages could be easily diagnosed with aMCI or early AD, and effective examinations for discriminating aMCI from DLB are needed.

aMCI is characterized by tauopathy and amyloidosis. However, DLB might not be associated with amyloidosis. It is worth comparing the plasma $A\beta_{1-42}$ levels in aMCI and DLB patients. By using IMR technology, the levels of plasma $A\beta_{1-42}$ in aMCI and DLB patients were measured, as shown in Fig. 5.17. Compared with DLB patients, aMCI patients showed significantly greater levels of plasma $A\beta_{1-42}$. Therefore, the use of plasma $A\beta_{1-42}$ to discriminate early aMCI patients from early DLB patients is strongly suggested. An assay of plasma $A\beta_{1-42}$ would be very helpful for increasing the diagnostic accuracy of DLB, especially at early stages.

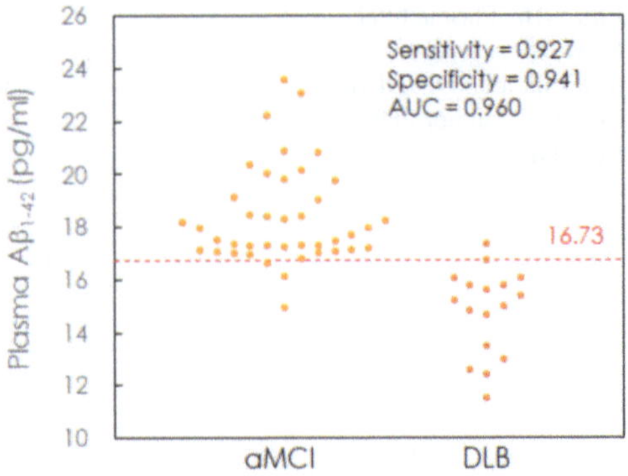

Fig. 5.17 Plasma $A\beta_{1-42}$ levels discriminate aMCI from DLB

In addition to $A\beta_{1-42}$, other plasma biomarkers have been demonstrated to discriminate DLB from aMCI or early AD. Chouliaras et al. reported significantly greater levels of plasma pTau181 in aMCI/AD patients than DLB patients ($p < 0.01$) [70]. Chen et al. reported the discrimination between DLB and AD using the plasma total α-synuclein level ($p = 0.001$) [71]. All the results demonstrate the utility of plasma biomarkers in precisely diagnosing DLB.

5.5.3 Differentiation Between FTD and AD

FTD patients usually have clinical disorders very similar to those of AD patients. For example, AD patients generally have difficulties in managing their finances and show poor judgment and irritability, and these symptoms equally trouble FTD patients. Without the aid of neuroimaging or CSF biomarkers, FTD patients are usually diagnosed with AD. Misdiagnosed patients take medications to treat AD; however, the treatment effects will not meet the expectations. Such situations are deeply frustrating for patients and their families and may even confuse healthcare professionals.

Pathologically, FTD and AD are characterized by the presence of different kinds of proteins. TDP-43 and T-Tau are important pathological biomarkers for FTD, whereas amyloidosis and tauopathy are predominant in AD. TDP-43 (TAR DNA-binding protein-43) is a crucial biomarker being developed for neurodegenerative diseases like ALS and FTD. TDP-43 has also been found as a pathological hallmark of Limbic Associated TDP43 Encephalopathy (LATE); a mimic of AD. Thus, the use of plasma TDP-43 for differentiating FTD from AD is promising. As shown in Fig. 5.18, the plasma TDP-43 level is significantly greater in FTD than in AD and CU. Foulds et al. reported that elevated levels of TDP-43 protein in plasma were detected in 46% of FTD patients, 22% of AD patients, and 8% of control individuals [72]. Clearly, the plasma TDP-43 assay is able to promote differentiation between FTD and AD. The plasma TDP-43 assay is in development by several groups utilizing NULISA or other technologies.

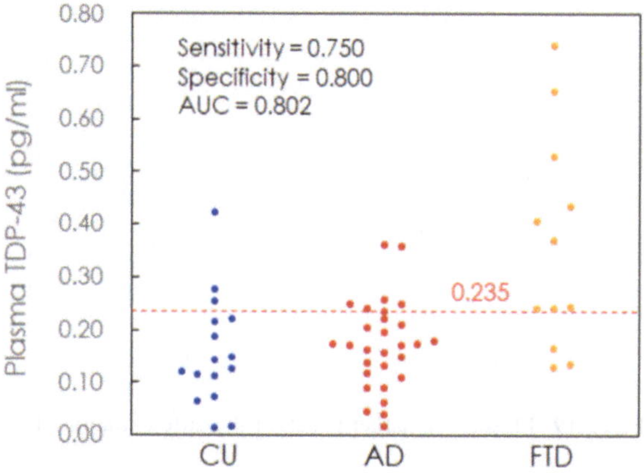

Fig. 5.18 Plasma TDP-43 levels discriminate FTD from AD

TDP-43 has also been shown to be a biomarker for limbic-associated TDP-43 encephalopathy (LATE), which is a major cause of late-onset amnestic dementia, yet no diagnostic biomarker is available [73].

Instead of TDP-43, Gomez-Tortosa et al. revealed the dominant role of non-pathological plasma biomarkers such as NfL in discriminating FTD from AD (clinical sensitivity = 0.88, specificity = 0.64, AUC = 0.78) [74]. When plasma NfL was combined with GFAP, i.e., GFAP/NfL, the clinical sensitivity increased to 0.89, the specificity became 0.78, and the AUC was 0.86 for differentiating FTD from AD.

The ability to discriminate FTD from AD using plasma NfL and GFAP levels was also demonstrated by Thijssen et al., who further proposed that plasma pTau181 might be useful for discriminating between FTD and AD [75]. Clearly, with the aid of plasma biomarkers, the difficulty in precisely diagnosing FTD can be easily overcome.

5.5.4 Differentiation Among FTD, PD, AD, and CU

From the perspective of brain health, many people would like to know of any possibility of developing various types of neurodegenerative diseases. Particularly for new patients or outpatients, comprehensive examinations for various types of neurodegenerative diseases, such as FTD, PD, and AD, would be helpful. With this comprehensive information, healthcare professionals can understand the risks of different types of neurodegenerative diseases well for their patients, and personalized plans for managing or treating brain health could be determined.

However, it is impossible to assay so many biomarkers in plasma because of the high cost. Hence, a biomarker panel that is compact, precise, and economic in routine practice needs to be developed. An example is given below.

First, plasma TDP-43 may be assayed to identify FTD patients, as shown in Fig. 5.19a [76]. Second, three biomarkers—total α-synuclein, $A\beta_{1-42}$, and

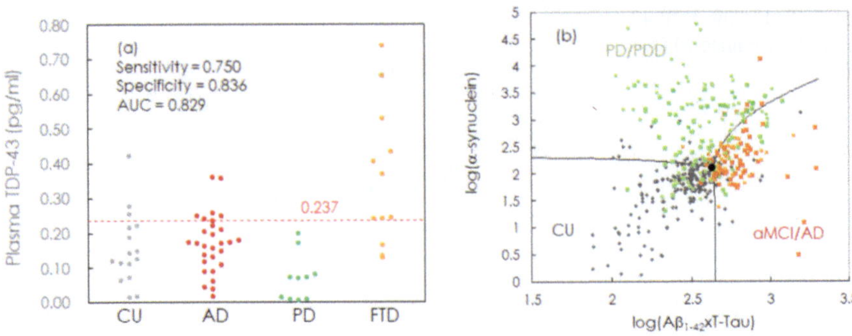

Fig. 5.19 Plasma biomarker panel for the differentiation of FTD, AD, PD, and CU

T-Tau—may be assayed, as shown in Fig. 5.19b. As guided by the solid lines in Fig. 5.19b, AD, PD, and CU can be well discriminated by these results.

Compared with neuroimaging, blood tests are definitely low cost and safe. In the case of the differentiation among AD, PD, FTD, and CU, an individual must undergo an amyloid PET scan, a dopamine scan, and an FDG-PET scan to identify AD, PD, and FTD, respectively. FDG-PET shows high diagnostic performance for differentiating AD from FTD, with meta-analyses reporting sensitivities around 96% and specificities around 84% for AD vs. FTD, though results vary, with some studies showing sensitivities from 80% to 99% and specificities from 63% to 98%. Could you imagine the cost and the risk of undergoing so many kinds of neuroimaging examinations? Moreover, neuroimaging is typically not available in community-based hospitals or private clinics. People usually have to wait for neuroimaging examinations at medical centers for 1–3 months and then take long trips to the medical centers, which is mentally and physically exhausting. Currently, we do not necessarily use high-cost and high-risk neuroimaging. Instead, blood tests can be performed. Blood tests represent revolutionary progress not only in precisely diagnosing neurodegenerative diseases but also in easily caring for brain health.

5.5.5 Differentiation Between Dementia-Derived Depression and Primary Depression

In the 1–3 years before the onset of a neurodegenerative disease, most patients have symptoms of depression, such as apathy, loss of interest in activities and hobbies, social withdrawal, isolation, irritability, and anxiety. This is now termed mild behavioral impairment (MBI), which is a newly described construct akin to MCI. All the symptoms are mental disorders, including neither cognitive impairment nor motor disorders. Often early cognitive symptoms are attributed to depression and mood disorders.

Psychologists administer medications to patients to treat depression. However, medications do not work because patients are not suffering from primary depression but dementia-derived depression. One to three years later, patients start to show cognitive impairment or motor disorders. At that time, psychologists change the diagnosis to that of a neurodegenerative disease. In such cases, 1–3 years are wasted treating patients inappropriately. If efficient examinations to identify dementia-related depression among patients become available, healthcare professionals can provide the right medical policies to patients. Do we have efficient examinations to identify whether patients are suffering from primary or dementia-derived depression? Blood tests have promise for application as such examinations.

As shown in Fig. 5.19, plasma TDP-43, $A\beta_{1-42}$, T-Tau, and total α-synuclein are the key biomarkers for assessing FTD, AD, and PD. Hence, by assaying the four plasma biomarkers, the risk of having FTD, AD, or PD can be determined. When the plasma TDP-43, $A\beta_{1-42}$, T-Tau, and total α-synuclein levels are higher than the cutoff values, healthcare professionals should perform detailed examinations, such as neuroimaging, to confirm the probability of a neurodegenerative disease. In such cases, healthcare professionals should treat patients as having neurodegenerative diseases, not depression. If the levels of the four biomarkers are normal, this reveals a high possibility of primary depression for patients. Further examinations associated with depression, such as assays of brain-derived neurotrophic factor (BDNF) in plasma, should be performed in patients. With solid clinical evidence, healthcare professionals can easily confirm primary depression among patients.

5.6 Prediction of Cognitive Impairment After First-Ever Ischemic Stroke

Neurodegeneration is the main cause of cognitive impairment. However, cognitive impairment also frequently occurs immediately after ischemic stroke, the so-called poststroke cognitive impairment (PSCI). Severe PSCI occurs in 7% of patients 1 year after stroke, whereas mild PSCI occurs in 52% of patients 6 months after stroke. Unfortunately, there is not a good way to predict PSCI in stroke patients. Fifty percent of stroke patients have to live with cognitive impairment within 6–12 months after stroke.

PSCI patients usually suffer from impaired daily activities, poor quality of life, and an increased burden of care. Both PSCI patients and their families are living hard days. If we could predict the development of PSCI in stroke patients, healthcare professionals could develop good plans to prevent PSCI. Some research groups have focused on the prediction of PSCI in stroke patients. One goal is to evaluate the feasibility of using plasma biomarkers for predicting PSCI in stroke patients.

5.6.1 Prediction of PSCI at 12 Months After Stroke

Chi et al. enrolled 55 stroke patients who were aged 20 years or older and had a first-ever ischemic stroke, no previous cognitive impairment, and no speech impairment after stroke [77]. The follow-up period was 12 months. The occurrence of cognitive impairment was defined as a MoCA score less than 23. Patients were examined with the National Institute of Health Stroke Scale (NIHSS), neuroimaging, MoCA test, and plasma biomarkers in the acute phase (<7 days) at 3 months and 12 months after stroke. The plasma biomarkers were assayed using IMR technology.

At 12 months after stroke, 13 patients had PSCI (PSCI+) (23.6%), and 42 patients did not have PSCI (PSCI-). At 3 months, the NIHSS score, stroke etiology, diffusion-weighted image lesion volume, Fazekas score, and periventricular combined with deep white matter findings did not differ between PSCI+ and PSCI- patients. However, the levels of plasma $A\beta_{1-42}$ at 3 months significantly differed between PSCI+ (15.1 pg/ml) and PSCI- (17.2 pg/ml, $p < 0.05$) patients. Through ROC curve analysis, the AUC for discriminating between PSCI+ and PSCI- patients at 12 months using plasma $A\beta_{1-42}$ levels at 3 months was 0.73. The level of plasma T-Tau at 3 months also significantly differed between PSCI+ (16.7 pg/ml) and PSCI- (19.9 pg/ml, $p < 0.05$) patients at 12 months, with an AUC of 0.72. The levels of plasma $A\beta_{1-42}$ and T-Tau at 3 months after stroke can clearly predict cognitive impairment at 12 months after stroke.

Chi et al. combined plasma biomarkers with MoCA scores at 3 months after stroke to differentiate between PSCI+ and PSCI- patients at 12 months after stroke. The AUC was 0.85 in the case of the plasma $A\beta_{1-42}$ level combined with the MoCA score and 0.87 in the case of the plasma T-Tau level combined with the MoCA score. By combining the results of the cognitive tests and plasma biomarkers at 3 months after stroke, the occurrence of cognitive impairment at 12 months after stroke can be predicted.

5.6.2 Prediction of PSCI at 3 Months After Stroke

When a person experiences an ischemic stroke, he or she is immediately sent to the intensive care unit (ICU). Blood is drawn from the patient in the ICU. Hence, it is convenient to perform blood tests during the acute phase. With biomarker information from the acute phase, i.e., within 7 days after stroke, it would be possible to explore suitable indices for predicting PSCI at 3 months after stroke. In such cases, healthcare professionals could plan treatment or prevention strategies for ischemic stroke patients during the acute phase.

Huang et al. enrolled 136 patients within 7 days after ischemic stroke [78]. Huang et al. performed NIHSS, neuroimaging, and MoCA tests and measured plasma biomarkers during the acute phase and 3 months after stroke. Patient cognitive impairment was defined as a MoCA score less than 23. Forty of 136 patients (29.4%) developed cognitive impairment at 3 months after stroke, referred to as PCSI+, whereas 96 of 136 patients (70.5%) were cognitively normal at 3 months after stroke, referred to as PSCI-.

The acute-phase NIHSS score, stroke etiology, diffusion-weighted image lesion volume, Fazekas score, and periventricular combined with deep white matter findings did not differ between PSCI+ and PSCI- patients. The acute-phase $A\beta_{1-42}$ and T-Tau levels did not differ between PCSI+ patients and PSCI- patients either. Surprisingly, compared with PCSI- (4.16 ± 2.18 pg/ml, $p < 0.01$), PSCI+ was associated with significantly lower levels of acute-phase pTau181 (3.19 ± 1.77 pg/ml) as determined by IMR technology. These results demonstrate the feasibility of predicting the onset of cognitive impairment in patients with ischemic stroke at 3 months after stroke using the acute-phase plasma pTau181 level.

5.6.3 Potential Roles of the Glymphatic System in PSCI

In the studies by Chi et al. and Huang et al., IMR technology was applied to assay plasma $A\beta_{1-42}$, T-Tau, and pTau181. Greater levels of plasma $A\beta_{1-42}$, T-Tau, and pTau181 are detected in aMCI and AD patients by IMR. This suggests that greater levels of plasma $A\beta_{1-42}$, T-Tau, and pTau181 are associated with greater risk of neurodegeneration in the brain. However, in the study by Chi et al., patients with lower levels of plasma $A\beta_{1-42}$ and T-Tau at 3 months after stroke were at risk of developing cognitive impairment at 12 months. Moreover, in the study conducted by Huang et al., patients with lower levels of plasma $A\beta_{1-42}$ and T-Tau during the acute phase after stroke were at risk of developing cognitive impairment at 3 months. The results of aMCI/AD studies and ischemic stroke studies seem to contradict each other. A possible reason for this contradiction may be the origin attributed to cognitive impairment.

The cognitive impairment in aMCI and AD patients is due to abnormal neurodegeneration involving the accumulation of $A\beta$ or the truncation of tau protein. Hence, the detected levels of plasma $A\beta_{1-42}$, T-Tau, and pTau181 increased, as determined by IMR. However, the cognitive impairment associated with ischemic stroke may be due to the dysfunction of waste cleaning in the brain.

Recent studies have shown that the glymphatic system is responsible for clearing waste from the nervous system [79, 80]. Through perivascular pathways, the glymphatic system facilitates the exchange of cerebrospinal fluid (CSF) and interstitial fluid (ISF). Glymphatic system activity is predominantly driven by the pressure generated by the continuous production of CSF, arterial pulsation, and slow vasomotion. Most importantly, cerebral arterial pulsation accelerates the rapid movement of solutes in the perivascular space (PVS), resulting in the exchange of solutes between the CSF and ISF. The elasticity of the vascular wall can affect cerebral arterial pulsation. Hence, healthy cerebral arteries are accompanied by good glymphatic activity.

Ischemic stroke causes the loss of blood–brain barrier integrity and damage to cerebral arteries, leading to dysfunction of the glymphatic system. Ischemic stroke clearly reduces CSF circulation because of decreased arterial pulsation, impairing glymphatic waste clearance and exacerbating waste accumulation. Thus, glymphatic system dysfunction hinders the removal of $A\beta_{1-42}$ and tau protein from the brain, resulting in lower levels of plasma $A\beta_{1-42}$, T-Tau, and pTau181, as determined by IMR. Meanwhile, glymphatic system dysfunction promotes the accumulation of $A\beta_{1-42}$ and tau protein in the brain, resulting in accelerated development of cognitive impairment after ischemic stroke. Therefore, poststroke patients with lower levels of detected plasma $A\beta_{1-42}$, T-Tau, and pTau181 have a greater risk of developing cognitive impairment.

5.6.4 Astrocytic Marker for the Glymphatic System

Advances in neuroimaging permit noninvasive assessment of the glymphatic system in stroke patients in clinical practice. Diffusion tensor imaging analysis along the perivascular space (DTI-ALPS) is the most popular method for measuring perivascular glymphatic activity in clinical practice. DTI-ALPS, initially proposed by Taoka et al., quantifies the diffusion of water within the PVS along deep medullary veins and shows robust correlations with glymphatic clearance, as determined by dynamic contrast-enhanced imaging, an established method [81].

Although DTI-ALPS is useful for evaluating perivascular glymphatic activity, the availability and convenience of neuroimaging impact the frequency of DTI-ALPS. Therefore, in clinical practice, DTI-ALPS is performed every 3–6 months after stroke. Ideally, alternative examinations could be performed more frequently, e.g., once per month, to monitor glymphatic activity in poststroke patients. Clinicians could thereby obtain instant information and provide or modify suitable treatments for the patients.

Wei et al. enrolled 71 participants and performed DTI-ALPS and blood tests [82]. They reported that the levels of glial fibrillary acidic protein (GFAP) in plasma negatively correlated with the DTI-ALPS indices ($r = -0.201$, $p < 0.05$). GFAP is an astrocyte-specific cellular marker whose release into blood indicates astrocytic injury. Astrocytes use the active water transporter AQP4 on their end feet to facilitate brain fluid flow in the glymphatic system. Injury to the glymphatic system causes the death of astrocytes, resulting in the release of GFAP. Thus, the astrocytic biomarker plasma GFAP is a promising marker of the glymphatic system.

Wei et al. further reported that the plasma GFAP level was negatively correlated with the MoCA score ($r = -0.26$; $p < 0.05$) and positively correlated with the ADAS-cog score ($r = 0.3$; $p < 0.05$). These results indicate that cognitive impairment can be contributed from astrocyte damage. Therefore, in patients with ischemic stroke, the glymphatic system is injured. As the activity of the glymphatic system deteriorates, astrocytes are damaged and release GFAP, leading to increased

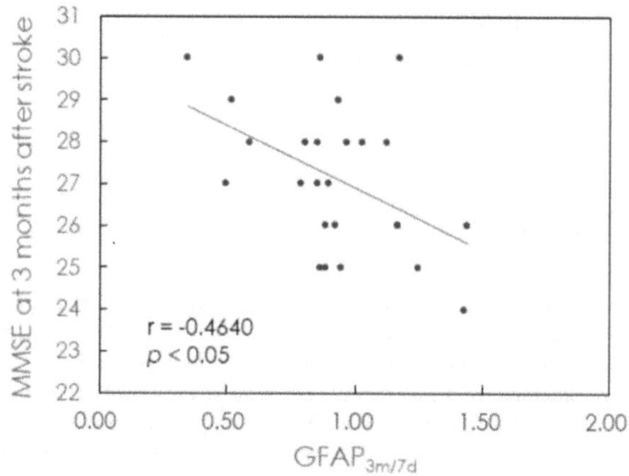

Fig. 5.20 MMSE scores at 3 months after ischemic stroke versus the ratio of plasma GFAP levels of 3 months to 7 days after ischemic stroke

plasma GFAP levels. Moreover, the risk of developing cognitive impairment in patients increases.

On the other hand, the longitudinal decreases in plasma GFAP levels might imply the recovery of the glymphatic system, possibly leading to the better cognition. Huang et al. enrolled 26 poststroke patients and performed MMSE tests 3 months after ischemic stroke [83]. Assay of plasma GFAP using IMR was done for every participant at 7 days and 3 weeks after ischemic stroke.

The MMSE scores at 3 months after stroke range from 24 to 30. The measured plasma GFAP levels at 7 days after stroke range from 14.65 to 45.80 pg/ml. The measured plasma GFAP levels at 3 weeks after stroke range from 13.47 to 43.22 pg/ml. The measured plasma GFAP levels do not significantly differ at 7 days and 3 weeks after stroke ($p > 0.05$). The GFAP levels do not significantly correlate with MMSE scores at 3 months after stroke. However, the ratio in plasma GFAP levels of 3 months to 7 days, referred to as $GFAP_{3m/7d}$, significantly and negatively correlates with MMSE score at 3 months after stroke ($r = -0.464$, $p < 0.05$), as shown in Fig. 5.20. Thus, the longitudinal decreases in plasma GFAP levels result in higher MMSE scores. The longitudinal decreases in plasma GFAP levels imply the recovery of glymphatic system, in turn, leading to better cognition.

In Fig. 5.20, in the case of $GFAP_{3m/7d}$ greater than 1 (>1), which denotes the increases in plasma GFAP levels at 3 months as compared to those at 7 days after stroke, four of six patients (66.7%) showed MMSE scores lower than 26. These four patients can be referred to as having cognitive impairment. In the case of $GFAP_{3w/7d} \leq 1$, which denotes the decreases in plasma GFAP levels at 3 months as compared to those at 7 days after stroke, 5 of 18 patients (27.8%) showed MMSE scores lower than 26. The odds ratio in developing cognitive impairment between $GFAP_{3w/7d} > 1$ and $GFAP_{3w/7d} \leq 1$ is 2.4. The results provide clear evidence of the decreases in plasma GFAP levels during recovery, which prevents cognitive impairment.

Notably, compared with plasma T-Tau or pTau, the correlation between plasma GFAP and cognitive impairment is not as strong ($r = -0.2$ to -0.4, $p < 0.05$). This suggests that the damage to astrocytes might not be directly attributed to cognitive impairment. Instead, neuronal damage is more directly attributed to the development of cognitive impairment.

References

1. Jack, C. R., Jr., Knopman, D. S., Jagust, W. J., Shaw, L. M., Aisen, P. S., Weiner, M. W., Petersen, R. C., & Trojanowski, J. Q. (2010). Hypothetical model of dynamic biomarkers of the Alzheimer's pathological cascade. *Lancet Neurology, 9*, 119–128.
2. Bateman, R. J., Xiong, C., Benzinger, T. L. S., Fagan, A. M., Goate, A., Fox, N. C., Marcus, D. S., Cairns, N. J., Xie, X., Blazey, T. M., Holtzman, D. M., Santacruz, A., Buckles, V., Oliver, A., Moulder, K., Aisen, P. S., Ghetti, B., Klunk, W. E., McDade, E., Martins, R. N., Masters, C. L., Mayeux, R., Ringman, J. M., Rossor, M. N., Schofield, P. R., Sperling, R. A., Salloway, S., & Morris, J. C. (2012). Clinical and biomarker changes in dominantly inherited Alzheimer's disease. *The New England Journal of Medicine, 367*, 795–804.
3. Jack, C. R., Jr., Bennett, D. A., Blennow, K., Carrillo, M. C., Dunn, B., Haeberlein, S. B., Holtzman, D. M., Jagust, W., Jessen, F., Karlawish, J., Liu, E., Molinuevo, J. L., Montine, T., Phelps, C., Rankin, K. P., Rowe, C. C., Scheltens, P., Siemers, E., Snyder, H. M., & Sperling, R. (2018). NIA-AA research framework: Toward a biological definition of Alzheimer's disease. *Alzheimer's & Dementia, 14*, 535–562.
4. Polonsky, K. S. (2012). The past 200 years in diabetes. *The New England Journal of Medicine, 367*, 1332–1340.
5. Jack, C. R., Jr., Andrews, J. S., Beach, T. G., Buracchio, T., Dunn, B., Graf, A., Hansson, O., Ho, C., Jagust, W., McDade, E., Molinuevo, J. L., Okonkwo, O. C., Pani, L., Rafii, M. S., Scheltens, P., Siemers, E., Snyder, H. M., Sperling, R., Teunissen, C. E., & Carrillo, M. C. (2024). Revised criteria for diagnosis and staging of Alzheimer's disease: Alzheimer's association workgroup. *Alzheimer's & Dementia, 20*, 5143–5169.
6. Palmqvist, S., Whitson, H. E., Allen, L. A., Suarez-Calvet, M., Galasko, D., Karikari, T. K., Okrahvi, H. R., Paczynski, M., Schindler, S. E., Teunissen, C. E., Zetterberg, H., Carrillo, M. C., Edelmayer, R. M., Mahinrad, S., McAteer, M. B., Kahale, L. A., Pahlke, S., & Tampi, M. P. (2025). Alzheimer's association clinical practice guideline on the use of blood-based biomarkers in the diagnostic workup of suspected Alzheimer's disease within specialized care settings. *Alzheimer's & Dementia, 21*, e70535-1-17.
7. Sperling, R. A., Aisen, P. S., Beckett, L. A., Bennett, D. A., Craft, S., Fagan, A. M., Iwatsubo, T., Jack, C. R., Jr., Kaye, J., Montine, T. J., Park, D. C., Reiman, E. M., Rowe, C. C., Siemers, E., Stern, Y., Yaffe, K., Carrillo, M. C., Thies, B., Morrison-Bogorad, M., Wagster, M. V., & Phelps, C. H. (2011). Toward defining the preclinical stages of Alzheimer's disease: Recommendations from the National Institute on Aging-Alzheimer's Association workgroups on diagnostic guidelines for Alzheimer's disease. *Alzheimer's & Dementia, 7*, 280–292.
8. McKhann, G. M., Knopman, D. S., Chertkow, H., Hyman, B. T., Jack, C. R., Jr., Kawas, C. H., Klunk, W. E., Koroshetz, W. J., Manly, J. J., Mayeux, R., Mohs, R. C., Morris, J. C., Rossor, M. N., Scheltens, P., Carrillo, M. C., Thies, B., Weintraub, S., & Phelps, C. H. (2011). The diagnosis of dementia due to Alzheimer's disease: Recommendations from the National Institute on Aging-Alzheimer's Association workgroups on diagnostic guidelines for Alzheimer's disease. *Alzheimer's & Dementia, 7*, 263–269.
9. Jack, C. R., Jr., Albert, M. S., Knopman, D. S., McKhann, G. M., Sperling, R. A., Carrillo, M. C., Thies, B., & Phelps, C. H. (2011). Introduction to the recommendations from the National Institute on Aging-Alzheimer's Association workgroups on diagnostic guidelines for Alzheimer's disease. *Alzheimer's & Dementia, 7*, 257–262.

10. Dubois, B., Feldman, H. H., Jacova, C., Hampel, H., Molinuevo, J. L., Blennow, K., DeKosky, S. T., Gauthier, S., Selkoe, D., Bateman, R., Cappa, S., Crutch, S., Engelborghs, S., Frisoni, G. B., Fox, N. C., Galasko, D., Habert, M. O., Jicha, G. A., Nordberg, A., Pasquier, F., Rabinovici, G., Robert, P., Rowe, C., Salloway, S., Sarazin, M., Epelbaum, S., de Souza, L. C., Vellas, B., Visser, P. J., Schneider, L., Stern, Y., Scheltens, P., & Cummings, J. L. (2014). Advancing research diagnostic criteria for Alzheimer's disease: The IWG-2 criteria. *Lancet Neurology, 13*, 614–629.
11. Chiu, M. J., Yang, S. Y., Horng, H. E., Yang, C. C., Chen, T. F., Chieh, J. J., Chen, H. H., Chen, T. C., Ho, C. S., Chang, S. F., Liu, H. C., Hong, C. Y., & Yang, H. C. (2013). Combined plasma biomarkers for diagnosing mild cognition impairment and Alzheimer's disease. *ACS Chemical Neuroscience, 4*, 1530–1536.
12. Lue, L. F., Sabbagh, M. N., Chiu, M. J., Jing, N., Snyder, N. L., Schmitz, C., Guerra, A., Belden, C. M., Chen, T. F., Yang, C. C., Yang, S. Y., Walker, D. G., Chen, K., & Reiman, E. M. (2017). Plasma levels of Aβ42 and tau identified probable Alzheimer's dementia: Findings in two cohorts. *Frontiers in Aging Neuroscience, 9*, 2261-1-9.
13. Jiao, F., Yi, F., Wang, Y., Zhang, S., Guo, Y., Du, W., Gao, Y., Ren, J., Zhang, H., Liu, L., Song, H., & Wang, L. (2020). The validation of multifactor model of plasma Aβ42 and total-tau in combination with MoCA for diagnosing probable Alzheimer disease. *Frontiers in Aging Neuroscience, 12*, 212-1-9.
14. Chiu, M. J., Chen, T. F., Hu, C. J., Yan, S. H., Sun, Y., Liu, B. H., Chang, Y. T., Yang, C. C., & Yang, S. Y. (2020). Nanoparticle-based immunomagnetic assay of plasma biomarkers for differentiating dementia and prodromal states of Alzheimer's disease – A cross-validation study. *Nanomedicine: Nanotechnology, Biology, and Medicine, 26*, 102182-1–102182-9.
15. Tsai, C. L., Liang, C. S., Yang, C. P., Lee, J. T., Ho, T. H., Su, M. W., Lin, G. Y., Lin, Y. K., Chu, H. T., Hsu, Y. W., & Yang, F. C. (2020). Indicators of rapid cognitive decline in amnestic mild cognitive impairment: The role of plasma biomarkers using magnetically labeled immunoassays. *Journal of Psychiatric Research, 129*, 66–72.
16. Lee, Y. J., Lin, S. Y., Peng, S. W., Lin, Y. C., Chen, T. B., Wang, P. N., & Cheng, I. H. (2023). Predictive utility of plasma amyloid and tau for cognitive decline in cognitively normal adults. *The Journal of Prevention of Alzheimer's Disease, 10*, 178–185.
17. Yang, S. Y., & Liu, H. C. (2025). Plasma-to-autopsy study of neurodegenerative diseases. *Journal of Clinical Neuroscience, 9*, 1–8.
18. Salvadó, G., Janelidze, S., Bali, D., Dolado, A. O., Therriault, J., Brum, W. S., Pichet Binette, A., Stomrud, E., Mattsson-Carlgren, N., Palmqvist, S., Coomans, E. M., Teunissen, C. E., van der Flier, W. M., Rahmouni, N., Benzinger, T. L. S., Gispert, J. D., Blennow, K., Doré, V., Feizpour, A., Rowe, C. C., Alcolea, D., Fortea, J., Villeneuve, S., Johnson, S. C., Rosa-Neto, P., Petersen, R. C., Jack, C. R., Jr., Schindler, S. E., Suárez-Calvet, M., Ossenkoppele, R., & Hansson, O. (2025). Plasma phosphorylated tau 217 to identify preclinical Alzheimer disease. *JAMA Neurology, 82*, 1122–1134.
19. Kaup, A. R., Nettiksimmons, J., LeBlanc, E. S., & Yaffe, K. (2015). Memory complaints and risk of cognitive impairment after nearly 2 decades among older women. *The Neurologist, 85*, 1852–1858.
20. Gifford, K. A., Liu, D., Lu, Z., Tripodis, Y., Cantwell, N. G., Palmisano, J., Kowall, N., & Jefferson, A. L. (2014). The source of cognitive complaints predicts diagnostic conversion differentially among nondemented older adults. *Alzheimer's & Dementia, 10*, 319–327.
21. Buckley, R. F., Ellis, K. A., Ames, D., Rowe, C. C., Lautenschlager, N. T., Maruff, P., Villemagne, V. L., Macaulay, S. L., Szoeke, C., Martins, R. N., Masters, C. L., Savage, G., Rainey-Smith, S. R., Rembach, A., & Saling, M. M. (2015). Phenomenological characterization of memory complaints in preclinical and prodromal Alzheimer's disease. *Neuropsychology, 29*, 571–581.
22. Samieri, C., Proust-Lima, C., Glymour, M., et al. (2014). Subjective cognitive concerns, episodic memory, and the APOE ε4 allele. *Alzheimer's & Dementia, 10*, 752–759.
23. Mitchell, A. J., Beaumont, H., Ferguson, D., Yadegarfar, M., & Stubbs, B. (2014). Risk of dementia and mild cognitive impairment in older people with subjective memory complaints: Meta-analysis. *Acta Psychiatrica Scandinavica, 130*, 439–451.

24. Bernard, M. A., Boutajangout, A., Debure, L., Briggs, A. Q., Vedvyas, A., Marsh, K., Osorio, R., Shao, Y., Betensky, R. A., Wisniewski, T., & Masurkar, A. V. (2024). *Alzheimer's & Dementia, 20*(Suppl. 2), e093492-1-2.
25. Hsieh, C. H., Ko, C. A., Liang, C. S., Yeh, P. K., Tsai, C. K., Tsai, C. L., Lin, G. Y., Lin, Y. K., Tsai, M. C., & Yang, F. C. (2024). Longitudinal assessment of plasma biomarkers for early detection of cognitive changes in subjective cognitive decline. *Frontiers in Aging Neuroscience, 16*, 1389595-1-18.
26. Kivipelto, M., Ngandu, T., Laatikainen, T., Winblad, B., Soininen, H., & Tuomilehto, J. (2006). Risk score for the prediction of dementia risk in 20 years among middle aged people: A longitudinal, population-based study. *Lancet Neurology, 5*, 735–741.
27. Weaver, D. F. (2024). Thirty risk factors for Alzheimer's disease unified by a common neuroimmune-neuroinflammationmechanism. Brain Sci. 14, 41-1-25
28. Durães, J., Tábuas-Pereira, M., Araújo, R., Duro, D., Baldeiras, I., Santiago, B., & Santana, I. (2018). The head turning sign in dementia and mild cognitive impairment: Its relationship to cognition, behavior, and cerebrospinal fluid biomarkers. *Dementia and Geriatric Cognitive Disorders, 46*, 42–49.
29. Lue, L. F., Pai, M. C., Chen, T. F., Hu, C. J., Huang, L. K., Lin, W. C., Wu, C. C., Jeng, J. S., Blennow, K., Sabbagh, M. N., Yan, S. H., Wang, P. N., Yang, S. Y., Hatsuta, H., Morimoto, S., Takeda, A., Itoh, Y., Liu, J., Xie, H., & Chiu, M. J. (2019). Age-dependent relationship between plasma Aβ40 and Aβ42 and total tau levels in cognitively normal subjects. *Frontiers in Aging Neuroscience, 11*, 222-1-8.
30. Hu, C. J., Chiu, M. J., Pai, M. C., Yan, S. H., Wang, P. M., Chiu, P. Y., Lin, C. H., Chen, T. F., Yang, F. C., Huang, K. L., Hsu, Y. T., Hou, Y. C., Lin, W. C., Lu, C. H., Huang, L. K., & Yang, S. Y. (2021). Assessment of high risk for Alzheimer's disease using plasma biomarkers in subjects with normal cognition in Taiwan: A preliminary study. *Journal of Alzheimer's Disease Reports, 5*, 761–770.
31. Pai, M. C., Wu, C. C., Hou, Y. C., Jeng, J. S., Tang, S. C., Lin, W. C., Lu, C. H., Chiu, M. J., Chen, T. F., Yan, S. H., Hu, C. J., & Yang, S. Y. (2022). Evidence of plasma biomarkers indicating high risk of dementia in cognitively normal subjects. *Scientific Reports, 12*, 1192-1-7.
32. Jack, C. R., Jr., Wiste, H. J., Weigand, S. D., Knopman, D. S., Vemuri, P., Mielke, M. M., Lowe, V., Senjem, M. L., Gunter, J. L., Machulda, M. M., Gregg, B. E., Pankratz, V. S., Rocca, W. A., & Petersen, R. C. (2015). Age, sex, and APOE ε4 effects on memory, brain structure, and β-amyloid across the adult life span. *JAMA Neurology, 72*, 511–519.
33. Lin, S. Y., Lin, K. J., Lin, P. C., Huang, C. C., Chang, C. C., Lee, Y. C., Hsiao, I. T., Yen, T. C., Huang, W. S., Yang, B. H., & Wang, P. N. (2019). Plasma amyloid assay as a pre-screening tool for amyloid positron emission tomography imaging in early stage Alzheimer's disease. *Alzheimer's Research & Therapy, 11*, 111.
34. Jiang, M., Liu, J. (2025). In Preparation.
35. Chen, T. B., Lee, Y. J., Lin, S. Y., Chen, J. P., Hu, C. J., Wang, P. N., & Cheng, I. H. (2019). Plasma Aβ42 and total tau predict cognitive decline in amnestic mild cognitive impairment. *Science Reports, 9*, 13984-1-10.
36. Chu, C. S., Lin, Y. K., Tsai, C. L., Sung, Y. F., Tsai, C. K., Lin, G. Y., Ko, C. A., Liu, Y., Liang, C. S., & Yang, F. C. (2025). Association of rapidly elevated plasma tau protein with cognitive decline in patients with amnestic mild cognitive impairment and Alzheimer's disease. *Psychiatric Investigation, 22*, 130–139.
37. Kivipelto, M., Solomon, A., Ahtiluoto, S., Ngandu, T., Lehtisalo, J., Antikainen, R., Bäckman, L., Hänninen, T., Jula, A., Laatikainen, T., Lindström, J., Mangialasche, F., Nissinen, A., Paajanen, T., Pajala, S., Peltonen, M., Rauramaa, R., Stigsdotter-Neely, A., Strandberg, T., Tuomilehto, J., & Soininen, H. (2013). The Finnish geriatric intervention study to prevent cognitive impairment and disability (FINGER): Study design and progress. *Alzheimer's & Dementia, 9*, 657–665.
38. Ngandu, T., Lehtisalo, J., Solomon, A., Levälahti, E., Ahtiluoto, S., Antikainen, R., Bäckman, L., Hänninen, T., Jula, A., Laatikainen, T., Lindström, J., Mangialasche, F., Paajanen, T.,

Pajala, S., Peltonen, M., Rauramaa, R., Stigsdotter-Neely, A., Strandberg, T., Tuomilehto, J., Soininen, H., & Kivipelto, M. (2015). A 2 year multidomain intervention of diet, exercise, cognitive training, and vascular risk monitoring versus control to prevent cognitive decline in at-risk elderly people (FINGER): A randomised controlled trial. *Lancet, 385*, 2255–2263.

39. Solomon, A., Turunen, H., Ngandu, T., Peltonen, M., Levälahti, E., Helisalmi, S., Antikainen, R., Bäckman, L., Hänninen, T., Jula, A., Laatikainen, T., Lehtisalo, J., Lindström, J., Paajanen, T., Pajala, S., Stigsdotter-Neely, A., Strandberg, T., Tuomilehto, J., Soininen, H., & Kivipelto, M. (2018). Effect of the apolipoprotein E genotype on cognitive change during a multidomain lifestyle intervention: A subgroup analysis of a randomized clinical trial. *JAMA Neurology, 75*, 462–470.
40. Baker, L. D., Espeland, M. A., Whitmer, R. A., Snyder, H. M., Leng, X., Lovato, L., Papp, K. V., Yu, M., Kivipelto, M., Alexander, A. S., Antkowiak, S., Cleveland, M., Day, C., Elbein, R., Tomaszewski Farias, S., Felton, D., Garcia, K. R., Gitelman, D. R., Graef, S., Howard, M., Katula, J., Lambert, K., Matongo, O., McDonald, A. M., Pavlik, V., Raman, R., Salloway, S., Tangney, C., Ventrelle, J., Wilmoth, S., Willliams, B. J., Wing, R., Woolard, N., & Carrillo, M. C. (2025). Structured vs self-guided multidomain lifestyle interventions for global cognitive function: The US POINTER randomized clinical trial. *Journal of the American Medical Association, 334*, 681–691.
41. van Dyck, C. H., Swanson, C. J., Aisen, P., Bateman, R. J., Chen, C., Gee, M., Kanekiyo, M., Li, D., Reyderman, L., Cohen, S., Froelich, L., Katayama, S., Sabbagh, M. N., Vellas, B., Watson, D., Dhadda, S., Irizarry, M., Kramer, L. D., & Iwatsubo, T. (2023). Lecanemab in early Alzheimer's disease. *The New England Journal of Medicine, 388*, 9–21.
42. Bregman, N., Nathan, T., Shir, D., Omer, N., Levy, M. H., David, A. B., Aizenstien, O., Lotan, E., Alcalay, Y., Awad, A. A., Gadoth, A., Ash, E., & Shiner, T. (2025). Lecanemab in clinical practice: Real-world outcomes in early Alzheimer's disease. *Alzheimer's Research & Therapy, 17*, 119-1-10.
43. Mintun, M. A., Lo, A. C., Evans, C. D., Wessels, A. M., Ardayfio, P. A., Andersen, S. W., Shcherbinin, S., Sparks, J. D., Sims, J. R., Brys, M., Apostolova, L. G., Salloway, S. P., & Skovronsky, D. M. (2023). Donanemab in early Alzheimer's disease. *The New England Journal of Medicine, 384*, 1691–1704.
44. Rabinovici, G. D., Selkoe, D. J., Schindler, S. E., Aisen, P., Apostolova, L. G., Atri, A., Greenberg, S. M., Hendrix, S. B., Petersen, R. C., Weiner, M., Salloway, S., & Cummings, J. (2025). Donanemab: Appropriate use recommendations. *Journal of Prevention of Alzheimer's Disease, 12*, 100150-1-18.
45. Hu, Y., Cho, M., Sachdev, P., Dage, J., Hendrix, S., Hansson, O., Bateman, R. J., & Hampel, H. (2024). Fluid biomarkers in the context of amyloid-targeting disease-modifying treatments in Alzheimer's disease. *The Medicus, 5*, 1206–1226.
46. Pontecorvo, M. J., Lu, M., Burnham, S. C., Schade, A. E., Dage, J. L., Shcherbinin, S., Collins, E. C., Sims, J. R., & Mintun, M. A. (2022). Association of donanemab treatment with exploratory plasma biomarkers in early symptomatic Alzheimer disease: A secondary analysis of the TRAILBLAZER-ALZ randomized clinical trial. *JAMA Neurology, 79*, 1250–1259.
47. Liu, W. T., Huang, H. T., Hung, H. Y., Lin, S. Y., Hsu, W. H., Lee, F. Y., Kuan, Y. C., Lin, Y. T., Hsu, C. R., Stettler, M., Yang, C. M., Wang, J., Duh, P. J., Lee, K. Y., Wu, D., Lee, H. C., Kang, J. H., Lee, S. S., Wong, H. J., Tsai, C. Y., & Majumdar, A. (2023). Continuous positive airway pressure reduces plasma neurochemical levels in patients with OSA: A pilot study. *Life., 13*, 613-1-13.
48. Lin, L. J., & Li, K. Y. (2022). Comparing the effects of olfactory-based sensory stimulation and board game training on cognition, emotion, and blood biomarkers among individuals with dementia: A pilot randomized controlled trial. *Frontiers in Psychology, 13*, 1003325-1-11.
49. Smith, A. D., Smith, S. M., de Jager, C. A., Whitbread, P., Johnston, C., Agacinski, G., Oulhaj, A., Bradley, K. M., Jacoby, R., & Refsum, H. (2010). Homocysteine-lowering by B vitamins slows the rate of accelerated brain atrophy in mild cognitive impairment: A randomized controlled trial. *PLoS One, 5*, e12244-1-10.

50. Kehoe, P. G., Turner, N., Howden, B., Jarutyte, L., Clegg, S. L., Malone, I. B., Barnes, J., Nielsen, C., Sudre, C. H., Wilson, A., Thai, N. J., Blair, P. S., Coulthard, E., Lane, J. A., Passmore, P., Taylor, J., Mutsaerts, H. J., Thomas, D. L., Fox, M. C., Wilkinson, I., & Ben-Shlomo, Y. (2021). Safety and efficacy of losartan for the reduction of brain atrophy in clinically diagnosed Alzheimer's disease (the RADAR trial): A double-blind, randomised, placebo-controlled, phase 2 trial. *Lancet Neurology, 20*, 895–906.
51. Soininen, H., Solomon, A., Visser, P. J., Hendrix, S. B., Blennow, K., Kivipelto, M., & Hartmann, T. (2021). LipiDiDiet clinical study group. 36-month LipiDiDiet multinutrient clinical trial in prodromal Alzheimer's disease. *Alzheimer's & Dementia, 17*, 29–40.
52. Abryanto, F., Wigata, A., & Gurning, S. S. (2025). What are the most effective pharmacological and non-pharmacological interventions for managing progression of cerebral atrophy in patients with neurodegenerative disorders? A systematic review. *Indonesian Journal of General Medicine, 12*, 49–89.
53. Neiman, J., Lang, A. E., Fornazzari, L., & Carlen, P. L. (1990). Movement disorders in alcoholism. *Neurology, 40*, 741–741.
54. Gu, S. C., Shi, R., Gao, C., Yuan, X. L., Wụ, Y., Liu, Z. G., Wang, C. D., Zhao, S. R., Chen, X., Yuan, C. X., & Ye, Q. (2023). Autonomic function and motor subtypes in Parkinson's disease: A multicentre cross-sectional study. *Science Reports, 13*, 14548-1-11.
55. Qin, Y., Meng, D. T., Jin, Z. H., Du, W. J., & Fang, B. Y. (2024). Association between autonomic dysfunction with motor and non-motor symptoms in patients with Parkinson's disease. *Journal of Neural Transmission (Vienna), 131*, 323–334.
56. Palomar, F. J., Suárez, A., Franco, E., Carrillo, F., Gil-Néciga, E., & Mir, P. (2013). Abnormal sensorimotor plasticity in CADASIL correlates with neuropsychological impairment. *Journal of Neurology, Neurosurgery, and Psychiatry, 84*, 329–336.
57. Reddy, H., de Stefano, N., Mortilla, M., Federico, A., & Matthews, P. M. (2002). Functional reorganization of motor cortex increases with greater axonal injury from CADASIL. *Stroke, 33*, 502–508.
58. Fu, M. H., & Chang, Y. Y. (2023). Adult-onset genetic leukoencephalopathies with movement disorders. *Journal of Movement Disorders, 16*, 115–132.
59. Muehlschlegel, S., Okun, M. S., Foote, K. D., Coco, D., Yachnis, A. T., & Fernandez, H. H. (2005). Paraneoplastic chorea with leukoencephalopathy presenting with obsessive–compulsive and behavioral disorder. *Movement Disorders, 20*, 1523–1527.
60. Lin, C. H., Liu, H. C., Yang, S. Y., Yang, K. C., Wu, C. C., & Chiu, M. J. (2019). Plasma pS129-α-Synuclein is a surrogate biofluid marker of motor severity and progression in Parkinson's disease. *Journal of Clinical Medicine, 8*, 1601-1-11.
61. Hong, C. T., Chung, C. C., Hsieh, Y. C., & Chan, L. (2025). Plasma extracellular vesicle neurofilament light chain as the biomarkers of the progression of Parkinson's disease. *Biomolecules and Biomedicine, 25*, 588–594.
62. Chen, N. C., Chen, H. L., Li, S. H., Chang, Y. H., Chen, M. H., Tsai, N. W., Yu, C. C., Yang, S. Y., Lu, C. H., & Lin, W. C. (2020). Plasma levels of α-synuclein, Aβ-40 and T-tau as biomarkers to predict cognitive impairment in Parkinson's disease. *Frontiers in Aging Neuroscience, 12*, 112-1-8.
63. Lin, C. H., Yang, S. Y., Horng, H. E., Yang, C. C., Chieh, J. J., Chen, H. H., Liu, B. H., & Chiu, M. J. (2017). Plasma α-synuclein predicts cognitive decline in Parkinson's disease. *Journal of Neurology, Neurosurgery, and Psychiatry, 88*, 818–824.
64. Lin, W. T., Shaw, J. S., Cheng, F. U., & Chen, P. H. (2021). Plasma total tau predicts executive dysfunction in Parkinson's disease. *Acta Neurologica Scandinavica, 00*, 1–8.
65. Lin, W. C., Chiu, P. Y., Chiu, M. J., Hu, C. J., Wang, P. N., Chen, T. F., Yang, F. C., Huang, L. K., Lu, C. H., Liu, H. C., & Yang, S. Y. (2025). Assay of neurodegenerative biomarkers in human plasma of patients with Parkinson's disease. *Journal of Biotechnology and Biomedicine, 8*, 159–170.
66. Guttuso, T., Jr., Shepherd, R., Frick, L., Feltri, L., Frerichs, V., Ramanathan, M., Zivadinov, R., & Bergsland, N. (2023). Lithium's effects on therapeutic targets and MRI biomarkers in Parkinson's disease: A pilot clinical trial. *IBRO Neuroscience Reports, 14*, 429–434.

67. Eijsvogel, P., Misra, P., Concha-Marambio, L., Boyd, J. D., Ding, S., Fedor, L., Hsieh, Y. T., Sun, Y. S., Vroom, M. M., Farris, C. M., Ma, Y., de Kam, M. L., Radanovic, I., Vissers, M. F. J. M., Mirski, D., Shareghi, G., Shahnawaz, M., Singer, W., Kremer, P., Groeneveld, G. N., Yu, H. J., & Dodart, J. C. (2024). Target engagement and immunogenicity of an active immunotherapeutic targeting pathological α-synuclein: A phase 1 placebo-controlled trial. *Nature Medicine, 30*, 2631–2640.
68. Malaty, G. R., Decourt, B., Shill, H. A., & Sabbagh, M. N. (2024). Biomarker assessment in Parkinson's disease dementia and dementia with Lewy bodies by the immunomagnetic reduction assay and clinical measures. *Journal of Alzheimer's Disease Reports, 8*, 1361–1371.
69. Kannarkat, G. T., Zack, R., Skrinak, R. T., Morley, J. F., Davila-Rivera, R., Arezoumandan, S., Dorfman, K., Luk, K., Wolk, D. S., Weintraub, D., Tropea, T. F., Lee, E. B., Xie, S. X., Chandrasekaran, G., Lee, V. M. Y., Irwin, D., Akhtar, R. S., & Chen-Plotkin, A. S. (2025). Blood α-synuclein separates Parkinson's disease from dementia with Lewy bodies. *Annals of Neurology, 00*, 1–17.
70. Chouliaras, L., Thomas, A., Malpetti, M., Donaghy, P., Kane, J., Mak, E., Savulich, G., Prats-Sedano, M. A., Heslegrave, A. J., Zetterberg, H., Su, L., Rowe, J. B., & O'Brien, O. T. (2022). Differential levels of plasma biomarkers of neurodegeneration in Lewy body dementia, Alzheimer's disease, frontotemporal dementia and progressive supranuclear palsy. *Journal of Neurology, Neurosurgery, and Psychiatry, 93*, 651–658.
71. Chen, Y. T., Orimo, S., Wei, C. Y., Hung, G. U., Yang, S. Y., & Chiu, P. Y. (2022). Synuclein motor dysfunction composite scale for the discrimination of dementia with Lewy bodies from Alzheimer's disease. *Frontiers in Aging Neuroscience, 14*, 920591-1–920591-8.
72. Foulds, P., McAuley, E., Gibbons, L., Davidson, Y., Pickering-Brown, S. M., Neary, D., Snowden, J. S., Allsop, D., & Mann, D. M. A. (2008). TDP-43 protein in plasma may index TDP-43 brain pathology in Alzheimer's disease and frontotemporal lobar degeneration. *Acta Neuropathologica, 116*, 141–146.
73. Nelson, P. T., Dickson, D. W., Trojanowski, J. Q., Jack, C. R., Jr., Boyle, P. A., Arfanakis, K., Rademakers, R., Alafuzoff, I., Attems, J., Brayne, C., Coyle-Gilchrist, I. T. S., Chui, H. C., Fardo, D. W., Flanagan, M. E., Halliday, G., Hokkanen, S. R. K., Hunter, S., Jicha, G. A., Katsumata, Y., Kawas, C. H., Keene, C. D., Kovacs, G. G., Kukull, W. A., Levey, A. I., Makkinejad, N., Montine, T. J., Murayama, S., Murray, M. E., Nag, S., Rissman, R. A., Seeley, W. W., Sperling, R. A., White, C. L., 3rd, Yu, L., & Schneider, J. A. (2019). Limbic-predominant age-related TDP-43 encephalopathy (LATE): Consensus working group report. *Brain, 142*, 1503–1527.
74. Gomez-Tortosa, E., Aguero-Rabes, P., Ruiz-Gonzalez, A., Wagner-Reguero, S., Tellez, R., Mahillo, I., Ruiz-Calvo, A., Sainz, M. J., Nystrom, A. L., del Ser, T., & Sanchez-Juan, P. (2025). Plasma biomarkers in the distinction of Alzheimer's disease and frontotemporal dementia. *Journal of Molecular Sciences, 26*, 1231-1-13.
75. Thijssen, E. H., Verberk, I. M. W., Kindermans, J., Abramian, A., Vanbrabant, J., Ball, A. J., Pijnenburg, Y., Lemstra, A. W., van der Flier, W. M., Stoops, E., Hirtz, C., & Teunissen, C. E. (2022). Differential diagnostic performance of a panel of plasma biomarkers for different types of dementia. *Alzheimer's & Dementia, 14*, e12285-1-11.
76. Yang, S. Y., Lu, H. C., Lin, C. Y., Chiu, M. J., Chen, T. F., Lin, C. H., & Chen, H. H. (2020). Development of assaying plasma TDP-43 utilizing immunomagnetic reduction. *Journal of Neurological Disorders, 8*, 7-1-8.
77. Chi, N. F., Chao, S. P., Huang, L. K., Chan, L., Chen, Y. R., Chiou, H. Y., & Hu, C. J. (2019). Plasma amyloid beta and tau levels are predictors of post-stroke cognitive impairment: A longitudinal study. *Frontiers in Neurology, 10*, 715-1-10.
78. Huang, L. K., Chao, S. P., Hu, C. J., Chien, L. N., Chiou, H. Y., Lo, Y. C., & Hsieh, Y. C. (2022). Plasma phosphorylated-tau181 is a predictor of post-stroke cognitive impairment: A longitudinal study. *Frontiers in Aging Neuroscience, 14*, 889101-1-10.
79. Xia, Y., Lyu, C., Chen, P. Y., Jiang, Y., Qu, C., Lyu, X., Cheng, Z., & Gao, B. (2025). The glymphatic system was impaired in spontaneously hypertensive rats. *Science Reports, 15*, 18321-1-12.

80. Zhang, S., Yu, W., Zhang, X., Cheng, H., Zhang, Z., Gao, S., Zheng, J., Yu, L., Wang, P., Geng, Y., Zhang, J., & Gao, F. (2025). Glymphatic dysfunction as a biomarker for post-stroke cognitive impairment. *Science Reports, 15*, 19382-1-12.
81. Taoka, T., Masutani, Y., Kawai, H., Nakane, T., Matsuoka, K., Yasuno, F., Kishimoto, T., & Naganawa, S. (2017). Evaluation of glymphatic system activity with the diffusion MR technique: Diffusion tensor image analysis along the perivascular space (DTI-ALPS) in Alzheimer's disease cases. *Japanese Journal of Radiology, 35*, 172–178.
82. Wei, Y. C., Hsu, C. C. H., Huang, W. Y., Lin, C., Chen, C. K., Chen, Y. L., Chen, P. Y., Shyu, Y. C., & Lin, C. P. (2023). Vascular risk factors and astrocytic marker for the glymphatic system activity. *La Radiologia Medica, 128*, 1148–1161.
83. Huang, G. L. (n.d.). Private Communication.

Index

C. S.-Y. Yang, M. N. Sabbagh, *Assessing Neurodegenerative Diseases: High-Sensitivity Technologies That Assay Pathological Biomarkers in Blood*,
https://doi.org/10.1007/978-3-032-23863-4

D

E

F

Q

R

S

T

MIX
Papier aus verantwortungsvollen Quellen
Paper from responsible sources
FSC® C105338

If you have any concerns about our products,
you can contact us on
ProductSafety@springernature.com

In case Publisher is established outside the EU,
the EU authorized representative is:
Springer Nature Customer Service Center GmbH
Europaplatz 3, 69115 Heidelberg, Germany

Printed by Libri Plureos GmbH
in Hamburg, Germany